THE PHENOMENOLOGY OF MAN AND OF THE HUMAN CONDITION
PART II

ANALECTA HUSSERLIANA

THE YEARBOOK OF PHENOMENOLOGICAL RESEARCH

VOLUME XXI

Editor-in-Chief

ANNA-TERESA TYMIENIECKA
The World Institute for Advanced Phenomenological Research and Learning
Belmont, Massachusetts

THE PHENOMENOLOGY OF MAN AND OF THE HUMAN CONDITION

II. The Meeting Point Between Occidental and Oriental Philosophies

Edited by

ANNA-TERESA TYMIENIECKA

The World Phenomenology Institute

Published under the auspices of
The World Institute for Advanced Phenomenological Research and Learning
A-T. Tymieniecka, President

D. REIDEL PUBLISHING COMPANY

A MEMBER OF THE KLUWER ACADEMIC PUBLISHERS GROUP

DORDRECHT / BOSTON / LANCASTER / TOKYO

Library of Congress Cataloging-in-Publication Data
(Revised for vol. 2)
The Phenomenology of man and of the human condition.
(Analecta Husserliana ; v. 14, 21)
"The jubilee volume 1968-1978 presenting the research program of the Jubilee Year 1977/1978"-T.p., v. 1.
Includes index.
Contents: 1. Plotting the territory for interdisciplinary communication -- 2. The meeting point between occidental and oriental philosophies.
1. Phenomenology. 2. Philosophical anthropology. 3. Philosophy of nature. I. Tymieniecka, Anna-Teresa. II. Series: Analecta Husserliana ; v. 14, etc.
B3279.H94A129 vol. 14 [B829.5] 142'.7 82-9817
ISBN 90-277-1447-9 (v. 1)
ISBN 90-277-2185-8 (v. 2)

Published by D. Reidel Publishing Company
P.O. Box 17, 3300 AA Dordrecht, Holland

Sold and distributed in the U.S.A. and Canada
by Kluwer Academic Publishers,
101 Philip Drive, Assinippi Park, Norwell, MA 02061, U.S.A.

In all other countries, sold and distributed
by Kluwer Academic Publishers Group,
P.O. Box 322, 3300 AH Dordrecht, Holland

Printed in The Netherlands

TABLE OF CONTENTS

PART III
THE SPATIO-TEMPORAL ARTERIES OF INDIVIDUAL EXISTENCE

PART IV
SOME FURTHER PERSPECTIVES UPON THE HUMAN CONDITION

CLOSURE

THE THEME

PHENOMENOLOGY OF MAN AND OF THE HUMAN CONDITION

Philosophy in a New Key

The work of the *World Phenomenology Institute*, which is being carried out on five continents, aims at *the phenomenological reconstruction of philosophy.* Much of modern philosophy has remained preoccupied with borderline issues and marginal approaches and has lost its bearings within the ever-expanding horizon of human cognition. Beginning in 1960 with a critical assessment of the classic inheritance of phenomenology, we immediately found a need for a new focus of attention.

Classical phenomenology brought forth the network of philosophical questions out of which Husserl's thinking evolved: man within his life-world. Husserl, together with his collaborators and followers, approached philosophy from several different perspectives (e.g., the ideal principles of objectivity; the laws of subjective constitution; the bodily extension of intentionality — explored by Husserl; action — investigated by Max Scheler; being — examined by Heidegger; psycho-corporeal motility — discussed by Merleau-Ponty; etc.). Each of these perspectives, which offered a new point of view, terminated in an impass: "unity" or "harmony" remained out of sight. The quest for a "radical beginning" and "originary grounding" of givenness was aborted; its deep-seeded reasons, however, indicated an urgent need for a change of direction and the postulation of an Archimedean point. The theme "The Phenomenology of Man and of the Human Condition" emerged from within the wealth of fragmented probings and draws upon the wealth of the phenomenological inheritance; yet it overturns in its formulation of the issues the directions marked out by the classic beginnings.

Without assuming a particular approach or perspective to the questions which the universe and man's place in it pose, "the third phase of phenomenology", as I call it,* proposes to elucidate them all at their very source: the *Human Condition within the system of the*

* Cf. A-T. Tymieniecka, "Beyond Ingarden's Idealism/Realism Controversy with Husserl — The Third Phase of Phenomenology," *Analecta Husserliana*, Vol. IV, 1976.

constructive unfolding of life. A new, all-encompassing field of philosophical inquiry, which undertakes a dialogue with the human sciences, is hereby opened up. This field stretches perpendicularly into the "phenomenology of life," *understood in its evolutive genesis of evermore complex forms.* One of its phases is marked by the surging of the *Human Condition.* The assessment of its *creative virtualities* opens horizontally upon the vast realm projected by the human being, unfolding them *as uniquely that being's own creative powers.*

To bring together and then to differentiate is the great and enigmatic work of life. Man, who enters into this working, is but one of the constructive links of the *unity of everything there is alive.* Thus, the classic phenomenological anthropocentrism is overturned. And yet, entering life's constructive progress as a *creator, the human being assumes a crucial position with respect to its course*: he introduces the *human significance of life*: culture, history, his own destiny, the quest after the Ultimate Telos of the All.

In *Analecta Husserliana*, volume 14, Part 1 of "The Phenomenology of Man and of the Human Condition," we established the groundwork for this two-directional research by approaching in a many-sided way the relationship between Nature and the human being in a dialogue with the human sciences. Each of our subsequent volumes brought an essential building stone to the edifice of reconstruction. Let me mention here in particular *Phenomenology of Life and of the Human Condition, in a Dialogue between Occidental and Chinese Philosophies* (*Analecta*, volume 17), in which the entire spectrum of issues concerning the phenomenology of life — of human life in particular — was surveyed.

With the present collection of studies we mark an essential step in probing the Human Condition through discussions dealing with the creative act of man and origin of rationality, individualization of life, temporality and spatiality, historicity, axiology and the human condition, etc. Here, as in our preceding research programs, philosophers from different cultures representing Occidental and Oriental thought — Western and Eastern Europe, North America, Korea, and China — discuss a variety of common problems.

Husserl was right to suggest that phenomenological sophistication is the fruit of the Occidental spirit. His later historicism, with its emphasis upon cultural differentiations of experience, may seem restrictive. Yet in our focus upon *the virtualities of the Human Condition as the pristine point of reference for the investigation of the origin, destiny, role, and place in the schema of life that the human being occupies,* all the cultural biases are lifted. We reach together to the fountain of life and of the community of human minds.

A-T. T.

ACKNOWLEDGMENTS

The studies contained in this volume were presented at the XVIIth World Congress of Philosophy, held by the International Federation of Philosophical Societies (FISP) in August 1983 in Montreal, Canada, as part of the program offered by the *International Husserl and Phenomenological Research Society.*

Our program included two roundtables: "The Moral Sense, Existence and the Life-World" (with the participation of Dallas Laskey, Emmanuel Levinas, Yvon Belaval, Hiroshi Kojima, Saen Yang Kha, and A-T. Tymieniecka) and "Métaphysique et Littérature" (with the participation of Eugene Kaelin, Maria Teresa Bertelloni, Marlies Kronnegger, and Hiroshi Kojima). These roundtables functioned as official events of the Congress and will appear in its proceedings.

Warmest thanks and appreciation are extended to our collaborators, who came from five continents to participate in the discussions and share with the world community of philosophers the particular orientation, scope, and concerns of the phenomenological investigations carried on by the *World Institute for Advanced Phenomenological Research and Learning.*

A-T. T.

TRACTATUS BREVIS

ANNA-TERESA TYMIENIECKA

FIRST PRINCIPLES OF THE METAPHYSICS OF LIFE CHARTING THE HUMAN CONDITION

Man's Creative Act and the Origin of Rationalities

INTRODUCTION: THE CALL FOR METAPHYSICS

1. *Postulates for "Every Future Metaphysics"*. The issue which we will treat concerns the urgency to investigate the nature and origin of the multiple rationalities with which the enormous growth of human knowledge about nature, the world, and man confronts us today. To approach it, the basic mode of access to such a query is of primary importance. It means no more and no less than finding the key to unlock the door which would open the realm of metaphysics. Indeed, my master, Roman Ingarden, in his less than successful attempt of discovery sought the "specific type of knowledge" that would reveal the key to metaphysics understood by him as the direct investigation of reality and of real existence. Does Ingarden's failure to find such a type of cognition mean that for phenomenology with its radical criticism of philosophical procedures there is no access to the first principles? Seeing the unsuccessful attempts of Husserl himself and Heidegger, as well as Merleau-Ponty after them, to bring together the eidos with the concrete fact would confirm such a doubt. However, these great thinkers do not seem to have delved deeply enough into the genesis of the constitution of objectivity; this objectivity meant for them the essential "givenness." In contrast, I propose its *inner workings* as the locus whence *eidos* and fact simultaneously spring. This amounts to saying that not constitutive intentionality but the *constructive advance of life which carries it may alone reveal to us the first principles of all things*; not the givenness of the world but the "poetics of life."

In view of the advanced clarification of cognitive as well as methodological requirements of the philosophical thinking in our time, the question emerges whether metaphysics is still necessary and at all possible. This question comes up again after we have been disillusioned with traditional forms of speculation and the systematic model of thinking, on the one hand, and philosophical imagery concerning being, on the other hand. Metaphysics has been lost from view in all the

A-T. Tymieniecka (ed.), Analecta Husserliana, Vol. XXI, 3–9.

lively predominating currents of today: in existentialist thought, in the analytical school, and in the empirical approaches (for example, in Marxism). As stated before, classic phenomenology did not reach it. As for the other aforesaid trends, they may be accused of a tendency toward the hasty reductionism of themes and evidence that they judge irrelevant or bothersome. Phenomenology, in contrast, is open to all forms of human experience, tendencies, nostalgias, etc., with a serious and rigorous intent of honestly doing them justice. Consequently, is not the field of phenomenology the appropriate one where the question of the necessity of resuscitating the topic of metaphysics may be proposed and duly debated? But it is precisely in this very field of phenomenology, the most ample and widest of all which in principle circumscribes all philosophical questions, that the necessity or possibility of some special enquiry which would not be absorbed by phenomenology itself, and would remain clear and distinct in itself from the total body of phenomenology differentiated into special fields such as regional ontologies, constitutive genesis, phenomenological hermeneutics, for example, appears problematic. And all this owing to good reasons which are too well known to be worth dwelling upon.

Nevertheless, with its radicalism and discernment phenomenology allows us to examine critically the procedures of classic endeavors in their establishing of the first principles of all things as well as laying down in a rigorous fashion the main postulates which an investigation, undertaken in keeping with the enormous progress of knowledge accomplished in our century to reclaim the field, would have to satisfy. Ultimately it offers us royal access to all metaphysical queries.

Below I list the basic postulates for every metaphysics that may, in keeping with the great classical tradition as well as with the progress of our times, legitimately enter the philosophical scene. Metaphysics is to be characterized by: (1) the central position of its content (subject matter) in relation to all the other fields of philosophy and of knowledge in general; (2) the profound relevance of its affirmations to all fields of knowledge, whether theoretical, contemplative, or practical; (3) the relevance of its direct and indirect considerations to the ultimate questions that human reason (theoretical, practical, and contemplative) is capable of conceiving; (4) the absolute evidence with which its "discovery" ought to be affirmed; a discovery that announces itself as a key to the labyrinth of entities: world, life, cosmos, and human destiny; (5) finally, the idea of linkages or of concatenations of the formal expression of any future metaphysics requires the mode of necessity with which the articulations of the elements of its construct, as an organic whole, may be projected.

These essential traits for all future metaphysics which may surge from the ground of philosophic thought — traits rigorously classified in keeping with the intransigent rigor of phenomenology including the concrete evidence of the given — clearly indicate that metaphysics could justifiably distinguish itself from the rest of the concerns of phenomenology only insofar as it would begin from, be rooted in, and would proceed in accordance with radically concrete and absolute evidence, if such an evidence is accessible to mankind. To show that this is the point that makes questionable the accessibility to such a center of universal gravitation it is sufficient to mention the philosophic efforts of Husserl, Ingarden, Merleau-Ponty, and others who in their gigantic programs judged it a culminating point and a master key for the unity of their fragmented edifices which, nevertheless, they were not able to find. What, however, would be the issue which could not be appropriately tested but by metaphysics?

2. *The Critique of Reason and Metaphysics of Life.* Emphasizing in his *Prolegomena* that every future metaphysics, a body of thought claiming to offer the first principles for the human understanding, would have to repose upon a critique of reason, Immanuel Kant introduced a model for metaphysical reflection valid as long as human reason itself. The way, however, to formulate appropriately the basic questions to revive and focus this critique as well as to outline the context as proposed by Kant appears in the present day outdated, to fall short of the great task of metaphysics. The reason for Kant's inquiry into the means by which transcendental consciousness orders the formless manifold of the originally given so as to project a world structure has been for him, as for Leibniz, to find a common ground for the two apparently different types of rationale dividing the empirical from the mathematical sciences. Edmund Husserl's phenomenology had been inspired by the same concern. His conception of intentionality was to bring the two types together at a deeper level. Yet the rationale of intentionality had been ultimately denounced by Husserl as responsible for the overgrown rationalisation of what he came to differentiate as the basic "life-world" supposedly underlying the intentional structure itself. This was held responsible for what, in the footsteps of Spengler, Husserl called the "crisis of Occidental culture!"

It was Ortega y Gasset who had broken this classic schema according to which rationality had to be identified with ideal mathematical forms or with transcendental/intentional a priori order. Shifting the Husserlian emphasis on the crisis of culture to that of reason, he proclaimed reason as the "theme of our times." However, the radical crisis of reason into which humanity had run took on forms which the great Spanish thinker did not foresee. Ortega's view had proven prophetic. It is no more the question of finding a theoretical foundation upon which "universal characters" (to use Leibniz's words) of the ideal order of intellectual constructs projected by the human being would meet the pre-structured, pre-thematised data of man's sensing and feeling of the empirical natural order, an antithesis that he has himself overcome. Yet his formulation of the question of reason and rationality also fails to respond to the needs of the present-day situation of knowledge.

As I have been saying for some time, we may consider the mid-sixties of our century as the period in which the innumerable forces which underly our civilization, and which since the beginning of modernity have remained in fermentation, violently protested the issue of reason or rationality.

First, the ever-progressing differentiation of knowledge into scientific fields not only obliterates the classic antithesis of the empirical/mathematical within which the classic problem of reason has been framed but brings forth into focus an ever-increasing variety of types of rationale on the basis of which this differentiation is operated.

Second, with the constantly advancing progress of everyday's technologies, new rational approaches to concrete questions of everyday existence emerge. The rationality of values being thereby infinitely differentiated, the human being remains disoriented by the criteria of choice in solving the problems of his concrete life-course.

Lastly, the coming into prominence upon the world scene of various cultures which have remained hitherto in the shadow of the Occident, together with the receding significance of the latter, brought to light the ingenuity, originality and richness of cultural forms of life. This prominence has revealed the multiple ways in which reason enters into the spontaneities of life — differentiating races, cultures and societies, each according to its intrinsic projects.

In our analysis it appears that the great progress accomplished by man in this century, in his knowledge about the world, cosmos, life, society and in his powers to use their inner workings, has the dramatic effect of opening before him innumerable perspectives reposed upon its own rationale and corresponding language. As there is no communication among them, so there is no opportunity for today's man to make a synthesis: to reach a holistic view of his own existence within the world and life. Overwhelmed by his powers, the human being may consider himself a center of the universe or the "master of nature," to use Descartes' words. Yet in fact he has severed the existential links to *everything that is alive* and he thus hangs in the void, having lost his bearings upon earth.

Therefore, at the present turn of our civilization we no longer face a theoretical issue seeking the explanation of the common intelligibility of the ideal and empirical order. Instead, prompted by the vital necessity of the human being to redefine his place within the life-world, we are urged to seek for the Archimedean point from which spring all the rationalities carrying on life itself. A renewed attempt at the critique of reason is called for. However, this attempt could not now focus on the level which the Leibnizean, Kantian and Husserlian attempts favored, namely at the structure and source of objectivity as it is given in human experience. In a radical contrast, in order to reach the unifying source of the innumerable rationalities that life itself prompts through its constructive advance, we have to delve into the *conditions of life itself.* Not the givenness of the already man-projected life-world but the *Human Condition* which holds the virtualities that enable the human being to unfold himself together with his life-world lies at the source of this proliferation.

3. *Neither philosophical anthropologies*, focusing upon the human being as if he were the center of the universe, nor ontologies, stressing the objective structures, which are the fruit of the most advanced forms of life, can do justice to this urgent call, issuing from the disarray in which human knowledge and man find themselves, for the definitive clarification of the sources of rationality. The metaphysical investigation of the Human Condition, in which no type of rationale enjoys a privileged position and each finds its appropriate role in life's progress, on the one hand, and no arbitrarily chosen platform — be it givenness or nature — sets the standards of life's accomplishment, on the other, offers the final critique of reason we are now ready to undertake.

Presenting a new critique of reason I propose the following points elaborated in my previous writings to establish its context:

a. Life consists in a self-individualizing advance through which all types of rationalities emerge in their respective functional roles;
b. Man's creative act allows a unique and absolutely evident access to the existential constructivity of life;
c. From among the sequences of life's constructive progress, the phase at which the Human Condition emerges stands out. The unfolding of its unique set of virtualities with the Imaginatio Creatrix at the center marks the turning point in life's constructivism: from the life-subservient vital rationalities we move to the invention of specifically human avenues of life;
d. The creative orchestration of the individual's functioning is instrumental in the surging of the *source-experience*, from which phase the specifically human rationalities originate;
e. The source-experience is the watershed between the vitally relevant rationalities of life and the inventive virtualities of the Human Condition. It is also the locus where all the rationalities which account for the diversification of the cultural forms of humanity from within their circumambient conditions of the living system on earth acquire their specific sense.

Let us note at the very outset of our inquiry that the above-delineated principles of the metaphysics of life satisfy all the postulates inherited from the history of Occidental philosophy and of classic phenomenology for "every metaphysics which would ever plausibly assume that name" that we have stated to begin with. Indeed, the creative act as the quintessence of the "creative tremor" or "tremor"/"thrill" of life opens up for us the Human Condition at the threshold of life's differentiation with a unique type of certainty, necessity and absolute evidence. Not stopping at the elucidation of the nature of meanings or of linguistic formation but approaching directly the very "sense of sense" which differentiates through their structures, this critique of reason goes to the very end of the road: the origin of the individualizing life.

This first philosophy of which foundational logic, ontology, epistemology, philosophical anthropology and transcendental phenomenology are but offsprings can be properly identified only by the name of "metaphysics of life." Substantiating the above statement in what follows, I will begin by presenting in the first chapter the philosophical vision which inspires this reflection and lays out the germinal intuitions and insights concerning life's order and the unity of *everything there is alive*. In the second chapter the origin of the multiple rationalities of life will be exfoliated.

CHAPTER I

IN PRAISE OF LIFE*

PART I: THE METAPHYSICAL VISION OF LIFE

INTRODUCTION: "VISION" INSTEAD OF "METHOD"

In philosophy we stress the laws of discourse by clarifying the way in which it *departs*: some or other type of "intuition." We try to justify what in our view is its unique validity by elaborating the nature of and the way in which this intuition yields us the subject matter for investigation; that is, we "adopt a method" as our starting point. In contrast, I propose that the truly decisive point for the philosophical enterprise is not the manner of intellectual presentation, but the perspective which the initiating intuition opens of the universe of man, and of all that is of concern to him: a VISION.

Classic phenomenology has worked with two visions. First, it took "essential intuition" as its lead: it opened a STRUCTURAL vision of the universe of things and beings. Second, by enlarging the objectifying intuition with the intuition of experience Husserl has redirected his query. Constitutive consciousness with the I as its center spread itself, like a peacock his feathers, in the TRANSCENDENTAL vision of the life-world supposedly comprising all. They met in the intentionality of experience; intentionality, a network connecting the processes of life's progress, reaches, however, only as far as this connectedness is man's doing. The transcendental/structural vision, which assumed the human consciousness as its root-source, placed the human being as the beginning, the initiator, a demiurgos.

However, the same intuitive departure of phenomenology may lead to a different perspective and, consequently, yield a contrasting philosophical vision. The vision which I present here shows the human subject not as the root-source of the life-world and of the universal order, but merely cooperating in the genesis of life, Nature, and the life-world: the *life-order*. In this new vision man is not demiurgos by whose workings his entire universe emerges; instead he appears caught up in the turmoil of a generating progress, and thus as one of innumerable

A-T. Tymieniecka (ed.), Analecta Husserliana, Vol. XXI, 10—26.

moments of the immensurable stream of life. And yet, he does not appear as just one of the segments within some or other subordinate sequence; he surges from the entanglements of all the networks as the VORTEX in which all the networks of life's orderings meet; as their "knot," man, appears in the "position" of *vortex of the universal sense.*

In spite of the radical transposition of man's role from that of a "source" to this of a "vortex," the classic and new vision coincide. Nothing of what has been discovered with the old one needs to be discarded in the new one; it will be merely "resituated" within a larger field of inquiry.

Philosophical vision bears at its junctures its own points of query. Its cohesion is suspended upon their peculiar formulation and their solution. Hence, we must first expose the vision; we will then reflect upon the basic questions which subtend its body.

THE GLORY OF LIFE AND THE HUMAN CONDITION

I. *Individualization and Precariousness*

The human being seeks the unity, interconnectedness, and harmony of everything that surrounds him. He orders everything upon which his vital process depends and everything accessible to him. For his vital progress he needs the reassurance of certainties. His vision, which encompasses the possible and its menacing limits, forces his imagination to strain to expand the narrow hold he has upon realities; he thus invents infinitely varied ideal realms "below" and "above" his narrow compass. For his life and the sense of his existence he takes his chance at the throw of the dice to choose a unifying net in which he would find a secure place. In reaching through the workings of his mind, the Archimedean point of the "unity of all" as the "primeval unity" below, and by extending the net up to the "heavenly unity" above, he aims at achieving an ultimate self-satisfaction and equipoise.

Yet in doing so he forfeits the unique GLORY of life. For like a silkworm he spins around himself an incomparably complex cocoon, within which he is caught immobilized, enclosed, a butterfly in its crysallis, a pearl in its shell. Not enlightened by exposure to the vast span of rays of the sun, but blinded by his own anguish, incapacities, fear of life, nostalgia — unfitness to delve into the gushing stream of existence.

No imaginative horizons of idle and fruitless repose but the trepidating progress of life is man's true element! Man, you who condemn yourself to a rest, lull yourself into drowsiness, dull your experience into slow motion, you see merely a faint reflection of life, whereas at each and every fraction of an instant myriads of forces, streams of energies, fragments of projects, surge at odds, struggle to break through, support, or defeat each other mercilessly. Life surges, man emerges, all whirls. No certitude, no foothold, no break, no repose. Yet a design takes from through which the PRIMEVAL SPONTANEITIES pulsate. Lawless, merciless, restless. It prompts all: pushes, mixes, stirs, turns over, melts, freezes, binds, and dissolves in order to make its own way. It is the way of life.

How precarious man's every step, at the mercy of every and each circumstance, at the turn of each event. What a miracle that anything may come to be and persist in becoming.

2. *Existence Is a Thrill/Tremor*

Man, the weakling, will lull himself to sleep; man, the idler, will forever slumber; man, the coward, will forever seek to camouflage every perilousness and smooth a step. He needs fictitious universes to drown the gist of life. But, the fact, the essence, the marrow of our bones — existence — is a THRILL.

No matter how intricate the security measures we install through our incessant speculative effort, the ACT OF LIFE occurs each time as an incalculable risk; hence its thrill.

There is the thrill of working with all our organs in their coordinated might to break the indifferent Element in its heart and convert it into the human element; the thrill to catch the escaping wind into wings, water into urns; to forge the resisting iron and steel, to break the marble and stone; tame the growth of plants and watch them blossom and yield fruit. All our force is being gained and spent in this brief, precarious, unlawfully breakable pact with "mother Nature." Contentment, pain, joy, and ever-recurring indomitable suffering emanating from within the human frame as this transient beingness make us one — what a thrill it is. It holds us by our bowels and keeps us breathless: a thrill to exist.

3. *The Heroic Fervor*

The human being does not merely float with the tides of the gushing waters. Nor is he subdued to the rules of the recurrent games of force. Nor is he caught unaware in the strategies of the spontaneities. On the contrary, he builds dams to stop up rivers; he diverts the waters; he braves the elements. He challenges the entire world for the sake of his dreams. With a HEROIC FERVOR he forges ever-new channels for his life, till his force is spent. In his striving no effort appears to him too great, no project impossible, no aim unattainable, and no sacrifice too painful. He throws his love, his hopes, his desires, and his life into the noble task of CREATION; creation of his own "nest" on the battlefield of life.

4. *The Ardent Joy of Existence*

Man: leave aside for a moment the petty calculations of security, restfulness, peace, and survival! For the great game of existence, look into thyself. However, do not seek to find yourself as the triumphant architect of life-strategies — life itself outwits them all. Do not expect to find yourself in the ideals which you nurture — life discredits them as transient self-illusions. Abandon the propensity to picture yourself in dreams of elevated tasks, noble actions, worthy accomplishments — what you accomplish will not be "meritorious" but the outcome of a game of chance. Ideals, dreams, desires of accomplishment are but strategies of life that you yourself create. To find yourself look into the game of existence in which you are caught. The toss of a dice, the turn of a current, the flowing together and separating of myriads of occasional instants make the turmoil in which the game is played. However, within the turmoil, your hand follows your eyes, your leg follows your desire to move: all your vital vibrations fall into an operational pattern and canvass a stupendously rich and cogent fabric: you open your eyelids and a universe surrounds you. As insignificant as your forces might prove in the great challenges of the life-course do not ever give up to impotence, turn sour in deception, despair; feel in each of your pulsations, the entire universe pulsate. With each breath you suck the sap of the spheres; every turn of your neck turns together with suns, planets, galaxies. You taste with each mouthful the germinating substance of plants, animals, birds, and fishes which lived, live, and will be born; their fragrance melts in your mouth each time you salivate. You loose your vision in the infinite horizon of beingness and you inhale waves, the rays, and the dust of the spheres. And within all this you still feel the power of bringing it all unto yourself: the power to manipulate — and command — the infinitely small till it becomes infinitely extended in your imagination on your own battlefield of existence.

Man's beingness is itself an exultation in his power of experience: rejoicing in his forces, sensibilities, strengths, weaknesses, joy and suffering, pain and delight, losses and gains as they come, *in being his own self.*

5. *Communion of Everything-there-is-alive*

In the midst of the growing and declining streamlets of life all seems at odds; in competition beyond solicitude and justice one vanishes, the other takes its place yet all stays in the communion of a common effort. Generosity and ruthlessness, love and hatred, forgiveness and revenge succeed each other seemingly disconnected yet they tie the knots of growth, decay, birth, extinction, rebirth for the sake of the rise of a new life-segment. Discord, wear and tear, use and abuse, thrust and deceit, ambush and welcome are all the means of conjunction in the chain operations of the ruthless push in which the living beingness self-individualizes itself. They consolidate its scheming to promote ultimately the aims of life. Their oppositions melt in one spontaneous gush of life that passes on the unquenchable thirst; *in this thirst and push toward life communes all.*

To be, to become, to unfold is no one being's task alone; it is the incalculable effort of everything that strives to emerge and to assume the status of being alive.

Each instant that we are granted to breathe as "ourselves" is a miracle of life: I, a distinct being within the infinitely varying whirls of life, in my very own, uniquely own existence!

This breath, this sight, this thought seemingly freed from all constraints while filling this singular frame, this feeling of my self as a unique other, different from everything else, is nevertheless the fruit of, as well as it remains sustained by, the infinitely extended array of life's meshes; of the innumerable circuits of life's pulsations within all other selves. In the core of its most intimate self-uniqueness it partakes of an entire schema chaining, cross-breeding and generatively interlacing life-individualizations of an entire life-circuit which, itself, partakes of other circuits generating life. It remains so interwoven, intertwined, interdependent on others to the very core of its own individualized existence and simultaneously sustains them. Tear one of the loops of the intergenerative life-chain away and the entire chain will loosen its ties, wither away, and vanish.

The great rush of life is chained by operating segments, organs, faculties. More complex faculties spread further the thirst for growth, life-challenge, expansion, and complexity. The human being is not a mere life-segment in which some lines of the life-articulations meet. He emerges and unfolds at the cross-currents of all the vital growth articulations: lines of ordering. However, in the unity of life-purposes and schemes in *everything-there-is-alive,* he still occupies a unique position: *he invents further ways and schema to propagate life.*

6. *The Human Beingness within the Unity of Everything-there-is-alive*

Within the beingness of *man-the-inventor* the universal life-generating system finds its center of articulations: *center of sense.* However, beyond gathering all the vital lines of articulation within its complexity, the human beingness unfolds to its own form by transforming them into an additional, peculiar system of faculties: faculties which *invent* new avenues for life to develop. Thus human beingness becomes the vortex of all ordering. *Man the inventor* is not doomed to be caught within his fictitious world he might invent; he has the unique way open to him to plunge into the game of generating life in all the dimensions of its deployment and to comprise the thrill of them all. He draws upon forces and their life-articulations first; he draws them in and reworks them. Second, by reworking them, he generates new forces and makes them flow into *new channels of universal life-significance,* which he invents.

In fact, at the instant my umbilical cord by which I have been directly connected with an other beingness, my genitrix, and have communicated through her infinitely complex system of life-arteries with the life-system of everything else, is detached, I am ready to project my own arteries into the system of life directly. I am an individual segment of life; I am ready to sustain my own beingness, my own life. My tentacles into Nature are introduced into its schema before I acquired the *individual status.* Now, I have to *participate in the progress of life on my own.*

Indeed, even before I unfold my functions (of knowing, desiring, enjoying, and disliking), all my acts which stream from my self have a significance: the significance of participating in the generative pulsations of being alive — *together* — *with everything there is* (e.g., sucking, I vibrate with the energies which operate in the living substance of milk — whether in the form of aromatic freshness or of that of dried powder dissolved in living water; through the mild sweetness of the oatmeal dissolving in my mouth, I partake of the growth of oats and all fruits of the earth).

Throughout his existence, the human being as a living individual is carried by the cyclic life of *everything alive.*

The gardener times his life by his patience and wakefulness in following the advance of sprout to a leaf, then to a blossom, and finally, a fruit. The pulsations of his veins fall in rhythm with those of juices of the fruit while it turns slowly from the green to the red, ripens, and weighs heavily from the branch. He feels the thirst of the roots drinking with them every drop of the rain. The forester lives day by day with his trees maturing with their getting of age to be lumbered; he starts a 'new' life-season planting a new forest.

The farmer lives with the seasons of the crops and renderings of the harvest. Beginning over and over again the sowing and germinating and irrigating and harvesting, he rhythms his existence by the cycles of vegetation. A sailor lives by the whims of the weather and the majestic unpredictability of the sea-Element.

A dweller of the cities could even forget the timing of the year and cycles of life, yet in his vital functioning he partakes of the rhythm of *everything-there-is alive.* He is suspended with each of his fibers upon the *total orchestration of life's rushing onwards.*

The human individual is entangled with everything there is living; he remains at the mercy of the *universal equipoise of the life-system,* and yet he also maneuvers it as its master.

This limited submission together with this limited mastery belong also to the equipoise of life.

In fact, the human being not only follows the spontaneously pre-traced schemas of universal life; he *invents and creates new ones.* First, he invents his 'own' life-strategies: ideas, ideals, principles, dichotomies, triads, diversification of "realms," transcendences. From within his vital framework he establishes a vast fictitious universe within which his life is immersed; his faculty of invention expands the elementary vital framework of life. His beingness culminates in transforming the elements according to his inventive devices: in CREATION. The human being incessantly creates life-devices, artifacts, works of art; the apex of this thrill of existence lies in the THRILL OF CREATION.

Nevertheless, the core-significance of human creation does not consist in producing artifacts but in transforming the vital struggle into the *significance of human existence.* Human creation ties the final knot of the universal communion of *everything-there-is alive*; the axis around which the life and social world evolve is incarnated in the maxim: *everyone his due*!

7. *The Ek-stasis of the Ultimate Surrender*

It appears then that *individualization-in-becoming* is in essence ambivalent. It proceeds in a struggle in simultaneously two seemingly contrary directions; the one of the ever more refined differentiation; the other in a concurrent working out of ever tighter links within ever larger circles of interactions with *everything-there-is alive.* In fact, the stronger "self-centeredness" and quasi "autonomy" the human beingness acquires, the stronger primeval longings surge from within it to retrieve the final principle of unity: scope, telos, meaningfulness of "it all." Their ever growing urgency within tears everything apart. Thus man forsakes all he fought for and dissolves the victoriously conquered selfhood in the EKSTASIS OF THE ULTIMATE SURRENDER; yet, not in ascending imaginary worlds but in descending into the forge of the Human Condition.

8. *Marvelling about the Human Condition*

How extraordinary that I exist! I, a distinctive, self-enclosed beingness with the bastion of assumed stability within myself: the fulcrum in which all rays pass through and catch each other, where all substances mix. Yet I am not submerged by the influx of substances, ideas, values. I use them to my own advantage. They are neither consumed or rejected instantly; they "work for me" and they remain "unconsumed" for "a while" spurning my reactivity. I am the cross current of elements which, springing forth, tie and dissolve in turns the already established segments in an infinite process, yet remain insoluble for "a while." This truce for "a while" lends to my processlike beingness a *simile* of repose, stability, appearance.

In fact, my beingness resides in the indomitable spontaneities which, flowing into multiple centers of intergeneration, absorb everything that generates, but at its own time and at its own carefully measured impulse. How extraordinary, then, the Human Condition: *individual* and yet *in the unity with everything there is alive*; carrying by its own progress that of the *all* and yet not absorbed by it, but surpassing its rules by expanding its own; an integral segment of the vital life-order, and yet not enslaved by it but creatively originating from within a fictitious universe which, however, assumes the status of "reality."

While man, the repose-seeker, the dreamer, the idler spins the ideal of a distant happiness, which lasts an instant and turns into frustration, of a golden age for humanity, the possibility of which is belied already by the present, the philosopher seeks enlightenment.

He has cleared the ground from fiction and illusion of his own doing. He delves into the great game of existence played within himself; he keeps alert not to miss any of its magnificent instants: the thrill to participate in the momentous struggle of man together with all living beingness. This struggle to live, to create, to be is the consummation of the glory of life. This is the HUMAN CONDITION.

But that which is evident for experience, might be a paradox for reason; hence the philosophical inquisitiveness sets in.

PART II: THE THEORETICAL QUEST AFTER ORDER IN A SEQUENCE OF ISSUES

The metaphysical vision of the HUMAN CONDITION WITHIN THE GAME OF LIFE presented above is in the perspective of the human intellect suspended upon a set of questions which have emerged together with the human reflection and in some or other form or disguise reappear with every type of query into the origins and conditions of life. Although, as it seems, most intimately interwoven either by the very nature of the subject into which we inquire, or by the links among its elements stemming from the investigating human mind, or from both, the issues themselves partake in their formulation from the vision which makes them emerge. In the first place, these are the perennial metaphysical issues which are at stake: "individualization," "the one and the many," "being and becoming," "the possible and the actual," "differentiation versus unity," "continuity versus chaos," "life and the origin of order," "life and modes of beingness," "harmony versus discord," etc.

To respond to the state of affairs on the one hand, and to the limits of the human intellect on the other hand, our discourse has to be fragmentary and furthermore we have to restrict its means and merely pinpoint the *crossing of intuitions* emerging from the metaphysical vision giving a novel shape to the perennial arguments.

It appears, however, not only that the subterranean query of this metaphysical upturn has its vortex in the issues concerning the Human Condition, but that consequently it is in the explication of the human condition that it finds a framework for its discrete fragments, the unity of which escapes the human horizon.

Hence in the following discourse we will, first, outline the set of the issues; second, we will situate them within the COORDINATES OF THE HUMAN CONDITION.

I. INDIVIDUALIZATION VERSUS UNITY

1. *Individualization; Individualization in Becoming; Becoming as Deriving from Life or as a Mode of Life; Life as Becoming*

(a) In the foregoing description we assume the becoming of life. Yet can we not believe with equal right that "everything is and always was," as well as "nothing is and never was"? I answer: we cannot, because "always" and "never" take their meaning from the "Is," and "is" and "becoming" are "one."

(b) Becoming is neither a singular process nor an event. Although it is always the becoming of "something," yet there is no singular "something" alone that it would be the becoming of. Becoming involves "everything"; it draws in, expands out, reverberates in all: it is the *inward and outward of Nature-forming life.*

(c) The unity of the living individual, specifically of the human being, and conversely the "ONE," which is to be found within him distributing the unity of the all, is however neither "planted" there, as one of his principles, as Confucius thought; nor is it a ready-made pattern of a "monad" to unfold, as intended by Leibniz. The turmoil of "pre-life" — to use this expression vaguely — becomes "one" while myriads of elements, forces, and energies criss-cross, tie links, and appropriately dissolve; while forces pull, energies pass, but nothing "is" until they meet in the "one." Man the inventor is the vortex of this meeting.

(d) Here is emphasized the current conception that forming is "creation." Yet creation "is" not, "becomes" not. It is the wringing out of what is *not yet* — but only "virtually" — of *that all* of Nature that becomes.

2. *Principle of Unity; Differentiation-correlate of Oneness; the Primacy of Something "There is"*

(a) But how can forces, energies, germinal pulsations which run wild bring themselves (or be inclined or drawn) into any constructive schema? Should we assume that they free themselves from a "primeval" unity of a substance, a "matter," a spirit, or a reservoir of possibles which only needs an impulse to differentiate from within into the innumerable ways of forming life? All these concepts are misleading; they stem from a confusion about the status of "what is" and that of the postulates which "being there" implies. No singular instance of beingness may preceed in the absolute sense the entire schema of becoming as "being there." "There is" sets in only in becoming; what is not *beingness*, that is already a segment of the becoming of life, "is" not.

3. *Impossibility of a Universal Harmony, and Yet, Life Postulates a Factor of Cohesion for the Life-strivings*

(a) Man seeks everywhere a universal balance, equipoise, harmony. But, in fact, there *is* none. There is only the unending discord of tendencies, competition between tasks and ends, contrary strivings of forces, waste of energies. In this constant turmoil as forceful as life itself, there is no truce or peace. At the mercy of hazard with which the concurrent life-forces encounter the issue, each outcome at the end of a seemingly logical channel is haphazard: the "just" judgment is built upon unfair reasons; the merited reward is attributed for petty calculation; the innocent is condemned for the sake of a noble cause. No balance between effort and result, between merit and reward, between crime and punishment.

(b) It is the "unity of *everything-there-is-alive*" that appears as the crucial principle of the cohesion and articulation of becoming as well as of any beingness. But how can there ever come to be a cohesive unity between the life-arteries of *everything-there-is-alive*, such that it centers within oneness? The interdependence of both is evident. Unless we can discern measures of balance for the centrifugal and centripetal tensions, random differentiations, ground for reconciliation for spectra differentiating the extreme opposites, for growth and deterioration, construction and decomposition, etc., "nothing" would become.

It amounts to presuming that *individualization-in-becoming is suspended upon a SET OF COORDINATES.*

II. THE THREEFOLD SET OF THE CONSTRUCTIVE COORDINATES OF THE HUMAN CONDITION

1. *Source- or Root-Conditions*

(a) Original "elements" with which generation may "work" are the initial "presupposition" of life. By this I mean the generative factors to initiate — and carry on — the constructive course of beingness, its concrete dynamic progress, which is, however, already from the initial moment accomplishment oriented through a telic principle of Nature/life. Yet we discover this telic mode of beingness only when we situate it within the life/Nature schema. Since we are bound to the artifacts of a substantival linguistic expression we have to talk here about "conditions," giving this expression, however, within this context a special meaning. By "source-conditions" we usually mean what is to be found as germinal elements, the initiating moment, the "ground" from which the initiating impulse springs. It may also mean a depository of elements. What we traditionally seek to grasp by the notion of "source" reveals itself within this framework in a peculiar sense; this sense itself comports differentiations with respect to the system of the Human condition.

(b) The Human condition appears in fact, in its full-fledged outlay of what we ambiguously call "sources." Yet its *root-conditions* spread as of themselves into (a) "operation-conditions" or (b) "process-conditions," or, to put it specifically (c) "individualization conditions." In fact, what is at stake is the *possibility of individualization of beingness* within the whirl of innumerable instantaneous and segmental operations, bursts of energy, pull of forces, etc. Is the outburst of the Initial Spontaneity totally unbound? How could beingness *as the relatively stable foothold of constructivism* emerge? Without it, however, how could the system of life/Nature organize itself? Could life ever come to emerge and progress without some "systematic" articulations? Thus the root-conditions spread into conditions which are simultaneously "prior" and "posterior" to the source-conditions, namely, the *coordinates* of generation/construction: of beingness.

(c) Then what about the "unity" or "harmony" or "measure" of the stream of becoming of *everything-there-is-alive*? Questions of "cohesion," "unity," and "harmony" of this vertiginous whirl of constructive generation carrying man and which is carried by him comes here indeed to an unprecedented significance. Neither that which gives significance to this apparently obvious, but in fact confused amalgam of notions — because loaded with sedimentations of meanings flown into it from classic philosophies of matter, substance, extension, duration, etc. — nor the nature of the cohension should be sought in the *descriptive features of the medium* that is assumed beforehand to "continue" (e.g., motion, connectedness, extension, duration, etc.) Furthermore, "unity" or "harmony" do not stem from the nature of units, particles, complexes, movements, etc., that is, from *what* is brought into unity.

The preoccupation of Jacques Derrida with finding the underlying unity below the dismantled interconnections of essential structures, ideal links, and those of speculative being — the connectives of our own intentionality, imaginative thinking and sensorial directives — belongs still to this residuum of Cartesian philosophical imagery. Indeed, seeking unity while stressing the structuration already being dismantled, he remains bound to the principles of its articulations. The notion of "deferment" being meant to dissolve the antithetic notions of discretion/continuity, disruption/extension, instaneity/duration, etc., which are the descriptive features of the unity of structurizing, draws its sense from the underlying extension-duration. Extension-duration has been for Leibniz the axis of the order in the world of substances; but the substances and the world have been in the perspective of becoming considered by Leibniz with reference to the already structurized form they were to assume. When, in contrast, we attempt to reach the workings of life at its "source" — at its pre-constitutive phase, prior to the forms emergent from the genesis — we are led into the generative play of the life-current itself. The game in which its forms are projected has to be initiated and pregnant with its germinal elements beneath the constituting functions; its cohension, unity, order can be sought only within the current of the game itself, in its constructive function or in the *discrete* constructive *thread which runs through it.*

Indeed, with the first set of meanings of the source-condition, intertwining each other, stay also the conditions of the constructive progress, of the devices to project and follow it, the conditions of expanding into the all-carrying grand design of life.

2. *Constructive Coordinates of the Human Condition*

In fact, from the perspective of the human being, who, not only entangled in unfolding the route of his beingness with the all, but essentially contributing to the universal schema of the life-routes by his inventive role, we distinguish concurrent conditions of constructivity in a set of three: 1. the coordinate of the otherness; 2. coordinates of *Nature as a system of everything-there-is-alive*; 3. coordinates of the life-world. They present circuits of virtual as well as possible articulations. It is, indeed, with respect to these virtual all-embracing circuits that the operations oriented by the life processes and these life-processes organize themselves.

(At this point it becomes obvious that the gigantic body of phenomenological research is by no means forgotten or discarded; to the contrary, the present vision could emerge only upon the groundwork of its results. However, the great tenets of the phenomenological investigation from Husserl till Merleau-Ponty, cannot be assumed as the foundation of the ultimate cohesion and generation we are concerned with but merely as its CONSTRUCTIVE COORDINATES.)

a. *Otherness as Coordinate of Constructivity*

The otherness appears within the life-generation as the crucial moment of differentiation, movement and co-ordination. It is the primordial moment of articulation in all phrases of the generative/regressive progress of life. It is also crucial within a reciprocal transactions-network with an other self. In fact, it acquires a crucial significance with respect to the Human Condition. That is, I absorb-reject, I react-act, I throw a tentacle-withdraw, I solicitate-repulse, I accept-deny — all these elementary and primary impulses of my life-operations, within which I unfold and practice my own self as an individualized agency, work within a network and participate in the processes of this network. This agency is in its modalities relevant in the radius of its operations to its spread in various degrees; the outcome is conditioned by the sum of its circuits. To be myself, an individualized beingness, I can be only over against and enmeshed together with multiple other beingnesses: "the other" is my very own counterpart. Yet his individualization does not prevent my autonomy, frees it instead. With myself/the other axis of incipient regulations, pulling each other asunder and coming together, opposites, similarities, common interests, antagonisms . . . the system of Nature and the life community emerge. I myself and the Other are the dynamic axis of the entire spread of this system that the game of Nature/Life establishes and propagates expanding it, adjusting.

b. *The Life-world as the Second Constructive Coordinate of the Human Condition: The Counterpart to the Vortex of Individuality*

The network of operations within which the pushing and whirling sprouts of life find a cohesion, life-promoting channels through which the life-forces may flow is called the *life-world.* What else is the discrete and yet cohesive web of palpitating energies leading to, pumping them into outlets of a constructive life-design, into life-growth-ramification where a life-rest can be found, toward a life-fulcrum, if not to what we call the "life-world"? The life-world takes on here, however, a specific meaning. It is not seen in the Husserlian sense as a product of man's constitutive power. It is not meant in the Heideggerian sense as the ground within which the human individual "dwells"; neither do we see it with Merleau-Ponty as the originary ground from which the living individual matter and form would stem, as its measure and its circumscribing factor. In this vision the life-world appears as one of the factors upon which the type of beingness of a human individual is suspended, but also inversely, as the outcome of forces, tendencies, aims, and projects of life within which the individual himself plays a crucial existential function. In this vertiginous game of Life/Nature the individual Beingness is the vortex where all the coordinates meet; within their incalculable and imponderable play they carry on the great game while conditioning each other.

c. *The Third Constructive Coordinate of the Human Condition: The Other Self and the Social Circuit of the Life-World*

In fact, the other self is not merely an impassive respondent by catching the empty loop thrown to him. On the contrary, whatever the type of the encounter, the other self stirs within myself the deeply rooted and complex threads of my human condition — specifically, my human condition which unfolds. The other stirs within me all the threads upon which my operational system of vital existence holds and the springs of a pandora box of "gifts" open, stirrings of imagination, inventive ideas, thought, faith, nostalgia, benevolence and of beauty.

The *benevolent sentiment* is the new bridge which I and the other self project toward each other. Upon it I construct with the other a new world: a world of *social transactions* where the "gifts" find application (use).

Indeed, the benevolent exchange with the other self means the beginning of the specifically social communication; vital communication emerges already as the modality of the interlinking within the networks of Nature/Life. But it grows roots of its own within the strictly "human" operational schema between the self and other selves.

Communication as the conductor running through the circuit of the selves means mind, *imaginatio creatrix*, moral sense, will. The Other becomes a unique type of link within the chain of life: link of sharing, understanding, inventing. In fact, he alone is the coordinate-factor of the "meaning of life." It is then up to him to introduce us into the meaningfulness of existence, and allow us to tie the knots of the One, as well as to dissolve them one by one together with all its self-fabricated "reasons," in order to bear witness to our ultimate innocence.

Closure

Man's existence is a work of every infinitely divisible fraction of an instant; each of them precarious and fragile. It is like the flight of a butterfly which opens its wings to life at high noon, flutters them "for a while" by going from blossom to blossom, and is finally spent. Man's instant does not belong to him and is not his doing. But he does not just glide along a thread which ties the innumerable variety of his mute and loud acts together; he invents their significance. The *meaning of life* is up to him. Yet the thread of the infinitely intricate game-span of the

Human Condition — what is it leading to? what is it attached to? does it hang at a loose end? who holds it? so that man may glide along it knotting and unknotting the chains of life without falling into the abyss of non-existence? None of the routes he takes lead to an end. Everyone gets to a point and is lost from sight.

Industrious as a bee, man surpasses her as inventor — and yet he had to invent Prometheus to answer for himself the puzzle of the primeval invention of life.

In his might man surpasses the lion, attempting to master the elements — yet he had to invent indomitable dragons to measure the limits of his force.

Committed by his innermost to truth, man had to invent the Sphinx (and to identify himself with Oedipus) to concede that total truth is for him inaccessible: he is bound to error.

Although the spirit might fill the "empty" spaces, the "idle instants," what holds them together? What brought them in succession? No universal harmony could be preestablished upon loose ends. I propose: *the universal design projects itself in the Great Game of life and of human existence which ties the knots at the top, throwing their hooks down into the bottomless.*

Our mind falls short to go any further, but let us cherish infinitely each of life's thrills!

Having covered the most conspicuous colors of the metaphysical vision of life, which spread themselves like a rainbow before our creative intuition, we have now to come back to the question of order, its origin and its rules. Yet within our approach, in which it is man's creative act playing the role of the vortex of the metaphysical investigation, the question of order does not get lost in the artificial realm of abstract speculations; the question of its origin centers upon the *creative act* itself. With the Husserlian radicalism we will go however further than he went: all types of orders being brought to the origin of sense meet in its source which is the individualising progress of life itself. Thus we pass now to the investigation of the "making" or of "poetics" of life.

The existential coordinates set up within the Human Condition for the distribution of ordering lines being revealed above, let us now discover the arteries through which the ordering segments and links are being processed and the three types of sense which flow through them carrying on the sense of life.

CHAPTER II

MAN'S CREATIVE TREMOR AS THE ACCESS TO THE ORIGIN OF LIFE'S RATIONALITIES

INTRODUCTION

Before entering into an exposition of my metaphysics of life I wish to recall that Ortega y Gasset has already proposed such an unheard-of undertaking as the transposition of the focus of philosophical attention from objectivity to life.

My own elaboration of the metaphysics of life proposed later than that of Ortega was arrived at prior to my acquaintance with his work. Both of us have, albeit independently, unfolded our reflection from within a critique of Husserlian phenomenology absorbing nevertheless its major acquisitions. Both of us, albeit in different ways, entered into our task through the critique of reason. What can be thought more provocative and assuring to a thinker who dares, as Kant would say, to navigate unexplored waters, than the knowledge that prior to him another sailor had bravely committed himself to the exploration of the same expanse where he found the very same inspiration? Consequently I will enter into my own reflection by way of an argument with the thought of Ortega y Gasset. The comparison of our respective discoveries at the main stages of our respective progress can be particularly fruitful in the endeavor to understand fully and deeply the meaning, scope, and accomplishments of such an adventure of philosophical minds.

I reached the heart of the "poetics of life" through an arduous itinerary. First there came a dissatisfaction with the rigid rationality of *eidos* as separate from concrete existence that I discovered in the thought of my mentor, Ingarden. Then, during my study of the first transcription of Husserl's *Crisis* in 1952—53 in the Husserl Archives at Louvain, I discovered a work that not only inspired me toward a critique of science and of all objectification but also indicated the pathway for a pre-predicative and *unthematizable* "actual existence" which was in unison with my considerations in the field of artistic and literary criticism. From such a starting point, in one of my early works, *Phenomenology and Science in Contemporary European Thought*,[1] I interpreted Husserl's phenomenological method in the footsteps of the

A-T. Tymieniecka (ed.), Analecta Husserliana, Vol. XXI, 27—73.

celebrated psychiatrist, Ludwig Binswanger, as the "categorical intuition" and I distinguished, with the inspiration of Roman Ingarden, various levels of phenomenological analysis. With this approach, I isolated three principal themes with which phenomenology has renovated the sciences of man, breaking the rational-objectifying field: the themes of the "other" in K. Jaspers and G. Marcel and of existential communication; the theme of human being-in-the-world in Heidegger; and, finally, the Husserlian notion of being-in-the-world as a base and foundation for all "pre-thematic" knowledge. This itinerary I do not offer as a philosophical autobiography, but rather as the philosophical meandering of intuitive elucidation of insights. The rupture with objectivism led me to open a cosmological path with my book *Why Is There Something Rather Than Nothing*.[2] It also led me toward the decisive leap into the domain of the pre-predicative, pre-intellective, and pre-intentional, there where human faculties have their origin, where those life capacities specifically human confirm their linkages, there where man's constructive aspirations surge jointly with the origin of the "Imaginatio Creatrix." I was ready to approach directly the creative function of man.

Thence, after a thorough examination of man's creative function in its ramified directions along the current of *Initial Spontaneity*,[3] I breached the last barrier of the intellectual veil of the goddess of Parmenides and I established, with firm conviction, the significance of the *"primogenital" factor of life*, the *human individual's act of living*, and, finally, the *tremor*, the *vibration* of life itself, as the key opening the field of metaphysics.

Sharing with Ortega the initial question after the elucidation of the origin and nature of reason, it will be illuminating for my own position which differs from his essentially to explain it in an argument with this great thinker.

I begin with my conception of the life-context.

PART I: THE LIVING HUMAN INDIVIDUAL WITHIN THE LIFE-WORLD OF VIRTUAL EXISTENCE

1. SOURCE-EXPERIENCE

In his anthropological conception of the human being, Husserl has established the origin and genetic progress of the human being around the axis of (1) transcendental consciousness, with its widespread constitutive system and (2) its correlate, the life-world, as simultaneously the effect and the ground for the constructive unfolding of consciousness. Since, in Husserl's conception, it is this genetic unfolding of the transcendental consciousness that establishes the life-world as a meaningful system — the major function of consciousness being meaning-bestowing within one homogeneous network of intentionality — these two, the consciousness and the world, are seen as one undissociable network carrying the route of the human individual, but also one within which is caught man's entire *virtual* spread of experience. Having thus singled out the cognitive-rational intentionality function of man as representative in its nature of the human existence, Husserl, as is well known, struggled in vain to account for the preconstitutive, antepredicative, that is, preconscious, "subliminal" — as I call it — vital and organic functions which, as I have consistently emphasized, originating directly from man's differentiating himself from the forces and energies of Nature on the one hand, and following, on the other hand, the universal life-system, escape the rational grasp by not entering into the intentional system; this latter represents, in my view, only an organization of life at a higher level. Yet these forces and energies of Nature and the life-system are the source and the origin of both the transcendental consciousness and the life-world.

In contrast to Husserl, I have, in fact, proposed to approach the human by the route which he unfolds as a *self-directed agent* from within the natural, organic, and vital conditions by INDIVIDUALIZING HIMSELF from them through his existential progress in all its functional lines. The organic, vital, sentient, affective, emotive, rational, and spiritual processes put this entire functional orchestration to work in delineating step-by-step man's genesis from within alien conditions and according to an inward built-in code to follow; not the higher form of life exemplified in the phenomenological notion of the "life-world," but its very *condition*, appears on my analysis as a *proto-phenomenological, existential meaning-bestowing* articulation of the *circumambient world* of man.

In fact, at the organic and vital levels of his progress toward developing his complete functional pattern of operations, man's existential individualizing seems to comply chiefly with two referential systems: the otherwise inchoate forces and energies of elementary nature which "are there" present as an indefinite well from which to draw, on the one hand, and the inner code of self-directing program of operations to unfold, with which the individual appears to originate, on the other hand. The *circumambient world of life-conditions*, which allow him to develop his virtualities and to install new ones, emerges from the interplay of the two.

This interplay itself lies within a universal "life-project." With the development of the ecological and geo-ethnic investigations, we become more and more aware of the existentially operational interlacing in a complementary fashion of various vital factors indispensable to some or other configurations of life; we are also becoming progressively more and more aware of intrinsic life-interlinkings between the coexistence of some beings, plants, and organic conditions. It is obvious that no individual living being "exists" entirely on its own: for its vital existence an entire array of other *living beings in existence* is indispensable; with the disappearance of one of them, the entire set of life-conditions is altered and other living beings cannot exist.

In considering, then, the self-oriented and self-projecting individual existential route, we must at a certain point of its progress assume it to be a special project of the *universal life-system*. The life-system comprising the *virtual* individual life-projects with respect to their *virtual coexistence* thereby constructing together the milieu of actual existence, I call — with reference to Husserl's expression, but to which I give quite a different connotation and denotation — "*the life-world of virtual existence*."

It is through all the vital processes and acts that, from the incipient moment of their entering into operation, the individual living being articulates into a meaningful script his own progress, as well as the radius of his operations, into a circumambient world. Yet it is only at a level of growth at which his operations have been articulated into specific organs that the *subliminal virtualities* of the living individual as a specifically human being may emerge. They surface through *source-experiential* patterns. At this phase of his unfolding we may pursue retroactively the conditions and factors of this meaningful route, which I have elsewhere called — emphasizing the self-centered agency — his "self-individualization" or "self-interpretation in existence".[4] With reference to Husserl's conception of "originary experience," I call this stage "source experience": it is the stage at which, due to the specific type of articulation of the entire spectrum of the functioning, we reach the unique type of configuration that marks the crucial phase in the existential route by the intrusion of *self-awareness* into man's *existential meaningfulness*.

2. THE SIGNIFICANCE OF "SOURCE-EXPERIENCE" FOR THE SPECIFICALLY HUMAN SELF-INTERPRETATION IN EXISTENCE

From our above indications it follows that the source-experience as the crucial phase of man's articulating his functional system into a consciously directed — and not merely blind unfolding according to a prescribed code — existential route, marks a crucial stage in the shaping of the circumambiant world.

From the action/reaction system of the living agent caught in the attempt to use the available energies of nature for his advantage according to a project of growth and development, there emerges an experiential self-awareness in which all the functions of the living being participate. We could, in fact, consider the self-awareness as triggered by the fusion of the organic, vital, sentient, affective, emotive, aesthetic, spiritual, religious yearnings, pulsations and strivings, as a simultaneous surging of intentional rationality. This latter appears as a constructively decisive and final organizing function, and yet it does not play other than a formal role. The actual quality of experience which decides about the life-attitudes, propensities, inclinations, and appreciation of the qualitative aspects of ourselves and of *givenness* is molded in the "crucibles," of source experience.

Last, it must be emphasized that the will and inventiveness does not come from the formal rational function either. On the contrary, it stems from the specific configurations which the above subliminal factors enter into in order to form the specific patterns of source-experience.

3. THE LIFE-WORLD OF VIRTUAL EXISTENCE

This configuration expresses the preponderant role which one of them confers upon the mold which the others share. On the one extreme is the vital and *subliminal source* of the living human being, its originary spontaneities shaped by working their way through the experiential patterns. On the other extreme, however, this shaping is oriented by the *postulates of the conditions* entailed by the life-system, in which it emerges and within which it is to play a *constructive* role. That is, the source-experience differentiates in various patterns, not only with respect to the preponderant role which one of the specific forms of the *subliminal spontaneities* plays within it,[5] but it will be still further differentiated and specified with respect to the postulates which the *life-world of virtual existence* presents. In other words, not only is the source-experience expressing what we could call "congenital conditions" of the present life-world which is its groundwork, with its present ecological, ethnic, cultural, and social situation, but it also regulates itself spontaneously with reference to the postulates which this specific life-world includes as the horizon of *existential virtualities*; it regulates itself in such a way as to prepare the human being toward the fulfilling of an appropriating role within the various sections and phases of organic development: plants, animals, things, and organic elements intertwined within one existential system.

The nature of source-experience is thus coupled with the specific *life-world of virtual existence.* With the human mind limited in its reconstructing schemas, ever expanding in its analytic spread by focusing upon new fields of interest while leaving behind the focus of the previously envisaged, as well as always singling some of the seemingly infinite number of horizons of the future, we will never be able to grasp either the total expanse of the human conditions in Nature or its virtualities. And yet, the negative sketching of our mind's limitations already gives us a perspective within which the possibility of assessing some of the crucial features of life, beingness, and existence of the real individual being, that is, of man, is opened to philosophy.

With the perpetual progress and conflicting observations of science, philosophy might be perplexed to conjecture the universal order of things and beings as we know them; in doing so, it may fall into the partiality for either rationalism or subjectivism or poetics of speculation. It may also fall into the seductive trap of a realistic naiveté. Nevertheless, with a more restricted scope in view, our present point of departure, which takes the genesis of the individualizing life as the key, and the source-experience as the decisive phase of the real individual unfolding of a human form from within, appears to offer us a proper *conjectural* starting point for assessing that *the human individual according to his analytically immediately inspectable experiential crucibles postulates a life-system for his virtual existence.* As I have just pointed out, the existential route of the living being differentiates according to the natural situation, "natural" both with respect to the particular life-conditions (on a singular part of the earth) and with respect to the coexistential situation/ecology. With reference to both of them, the source-experience, as the constructive entrance, not only into a circumambiant natural milieu, but into the life-world in Husserl's sense, of the fully developed socio-cultural world of the human being, differentiates into various patterns.

This referential axis according to which the human existence is shaping itself into existentially complete, organic, ethnic, societal, and cultural forms of the life-world of virtual existence, far from shaping itself into existentially complete organic, ethnic, societal, self-enclosed universe, as the ground and horizon of this existential shaping, constitutes the network of existential forces differentiated into specific constitutive molds, such that expressing specific natural conditions, a coexistential individual life may emerge, unfold, and propagate itself.

In this framework opened on both ends and yet mysteriously self-constraining at its crucibles the human being may ever advance in *ciphering* inventively his *life script.* Its peak is reached by the ciphers of philosophy, which has been working at it from the very beginning.[6]

PART II: THE QUESTION OF THE ACCESS TO THE RADICAL FACT OF LIFE AND TO ITS LOGOS; A DIALOGUE WITH ORTEGA Y GASSET

1. HUMAN LIFE AS THE PROPER GROUND OF METAPHYSICS; ITS THEMATIC KNOT

In one of his early works Ortega y Gasset had stated simultaneously the three basic convictions toward the establishment of his metaphysics, which was to be centered upon the "radical fact of life." With one stroke of imagination Ortega enters into the heart of the issue. In the first place, he affirms that only with Adam — the first man — did the significance of that which we call "life" emerge. Adam is in paradise the only living creature which takes note of what it means "to live" — who assumes the fact of his being alive, which means that he identifies himself with the fact of his being alive. In the second place, it is precisely due to this discovery made by Adam that he is a living creature and has assumed perfectly within one synthesized glance the fact of being alive — a fact which reposes upon experience being extended in all directions of the living universe — that this universe emerged as an organised living system vibrating with sounds, colors, fragrances, etc.: the universe of life. Within this experience of Adam the life of the universe proper began. Finally, this assuming of life in the self-experience of Adam is operated through the simultaneously emerging conscious act (or, act of consciousness). As we will see later on, the crucial question concerning the validity of the "fact of life" as the gate to metaphysics will be about the understanding of this "assuming".

This triple vision of Ortega necessitates an explication. In order to satisfy the demands of metaphysics, which he understands as a "first philosophy," it is necessary, in the first place to explain how consciousness, a conscious being, could have undertaken the task of the ordering of life. Second, what relation does consciousness maintain with other types of human functioning? Third, we have to ask — and Ortega also raises this question — In virtue of what could consciousness, thought, mind, the cogito emerge solely within the human being?

Considering the question of the ordering of life, Ortega asks, Could intellectual or rational logic of consciousness introduce order into the palpitating flux of life, taken as a concrete, radical fact? Obviously in order to answer this question we have to interpret reason in a different way than identifying it with the intellectual and logical order. We have to take into consideration simultaneously the order of life and the order of thought by means of which man assuming his own life tries to understand, interpret and explain the events and happenings which otherwise appear fortuitous and arbitrary. To this effect, Ortega dethrones the supremacy of the intellect, giving priority to the "vital reason" (*razón vital*), and refuses supremacy to the strict and rigid structures and rules of logic as well. By the same stroke the cogito is in its hitherto unquestionably assumed role of the royal access to reality replaced by the mode of "narration" which is within the temporal flux regulated by a specifically human rationality: "historical reason" (*razón historica*).

In this way bringing to light the knot of theses to serve as the basis of his project, Ortega situates it at the very center of the crisis of our times.

Questions which underly his distinction between the vital and historical reason are the crucial questions which our culture has to confront and to solve.

In the first chapter of our study we have already hit at the central point of the metaphysical vision: the question of order. This question, intertwined with the most fundamental philosophical issues which occupied the center of traditional philosophical controversies, cannot be directly confronted; we cannot even approach it adequately without finding the Archimedean point whence comes the thread with which the entire fabric of the human universe reconstructing philosophically the workings of life is woven and through which it maintains itself. As we have already stated in the introduction to our study, it is the nature of reason and of rationality which is at stake. I am on this point in perfect agreement with Ortega y Gasset. However, having followed a different philosophical itinerary from him, we diverge essentially in the key in which we respectively propose to exfoliate life in its logos as well as in the results which we respectively obtain.

Ortega confronts and attempts to explain the radical fact of life by means of his conception of vital and historical reason. I propose, in contrast to Ortega, the *creative act of man* as the 'radical fact of life', which contains the quintessence of the logos of life and allows us to exfoliate it.

We will begin by following Ortega's line of thought which is initiated by the question, How did the first man reach the crucial point of the order of life? The entire metaphysics of Ortega dealing with the rationality of life unfolds in the answer to this question.

2. THE RADICAL BEGINNING AND THE NECESSITY TO LEGITIMATE THE POINT OF DEPARTURE

Although we denounce the misunderstandings created in classical philosophy by rationalism in its various rigorously logical or Platonic forms and reject these forms as outdated in order to replace them by others as the task of our times, we are bound, if we are engaged in a serious philosophical pursuit, to espouse some of the major postulates which have proved to be the strong-holds of foundational philosophies throughout history. Ortega also is acutely concerned with maintaining the principal tenets or exigencies of traditional philosophy. In fact, in spite of his having, in his revolutionizing aim, appropriately substituted the pedantry of the scholarly exposition of thought by an intuitive literary style, he proposes to give to his metaphysics the backbone of "absolutely certain" evidences and the "radicalism" of the beginning point. He seeks the "indubitability" and the "self-givenness" of data for his novel metaphysical edifice in "absolutely certain evidences." Maintaining the criteria of "absolute certainty" and of the "radical beginning" — although giving them a different twist — he simultaneously meets the postulates of the classic phenomenology of Edmund Husserl, on the one hand, and, on the other hand, assumes the philosophical standards of every *philosophia prima.* There are, then, essential points of affinity as well as simultaneous points of divergence between the Ortegan and the phenomenological project.

For the above-mentioned reasons, not only does Ortega abandon the artifacts of interconnectedness of the rational speculation that in traditional metaphysics has been providing links among disparate insights toward a cogent whole — that is, he renounces, like Husserl, anything which would not be immediately inspectable in intuition — but he dispenses with the traditional kinds of rational linkage altogether. Yet, does he provide an adequate novel linkage of intuitive evidence to establish his views? I have here in mind a specific and, for my argument, most significant point. To begin with, we will investigate Ortega's effort to separate his own point of departure from the Cartesian as well as from the Husserlian "cogito" and to legitimate it simultaneously.

In the first place, Ortega, like Husserl, discusses the question of the Cartesian doubt, rightly seeing in it not only the illuminating emergence of the essential role of consciousness but also the point of convergence of the innumerable perspectives upon the universe with their respective types of *rationale.*

In his search after "datas of the Universe" ("datos de universo"), which could serve as the basis for a take-off, and which would satisfy the exigency of absolute (apodictic-) certainty — which means a doubly founded or proto-founded evidence, establishing their, as he calls it, "archi-certain certainty" — Ortega investigates the types of "what there is" within the universe of knowledge to the degree of validity they present.[7] Is it not with respect to the cognitive validity that we differentiate various levels of reals of "reality"? Yet, diverging in some respects, are they to be seen as separated?

In his *Meditaciones de Don Quijote,* Ortega raises the question of reality and certainty simultaneously. He argues that, although the existence of a chimera (*quimera*) lacks the same mode of reality which the garden where the poet (Jimēnez) places her possesses, and leads us to conclude that it exists only in our mind, this deficiency does not dispose of existence. Instead, the fact that Don Quixote ". . . should not have seen a giant where he saw a windmill!" raises the question: "But where does man get his giants?" The conclusion which he draws is that the existential status of a chimera is, like the "forces, atoms, electrons . . . which the physicist assures us of," such that they are "ultimate foundations" of reality, merely "the chimera of physical theory"; they, like the chimera of a poet, simply cannot count for the "radical data of the universe." In the first place, there is a differentiation to be made in various types of reality. The fact that we cannot invest our full confidence in them led Descartes to his famous doubt. Yet Ortega dismisses Descartes' conclusion that the cogito — the doubt — itself may give us any definitive proof about the radicality of a type of existence. "What is there undoubtedly in the Universe? When one doubts about the entire Universe, what remains? There remains . . . the doubt — the fact of doubting: even if I doubt whether the world exists, I cannot doubt that I doubt —: behold the limits of all possible doubting." In the first place, the act of thought, the act of doubt, does not hold, according to Ortega, any privilege that would elude the doubt itself. The act of cognition, in which any object whose existence we doubt "appears to me," expands into the infinite "Universe." This appearance "in consciousness" — consciousness, which belongs to the network of realities and is nothing other than a special mode of existence in the Universe — places the object within this network of existence. It might well be ". . . that among the things that exist or purport to exist in the Universe, there is one whose mode of being is radically different from the rest: thought . . .".[8] Nevertheless, when we ask what "exists" means, for example in the phrase "this amphitheater 'exists,'" we mean by exists "éstà ahí:" *Ahí* means: "there in the world, there in the Universe, in the general realm of realities . . ." By this stroke Ortega believes he has placed this special mode of existence of cognition, of subjectivity, of consciousness together with real things and beings, within the same grand schema of realities. Thus Ortega believes he has overcome the Cartesian division between mind and body as well as the Husserlian division between empirical and pure consciousness and body and spirit, a division which led Husserl to the "suspension" of the validity of real, actually existing empirical consciousness in its correlated world. Recognized by Ortega as a *center for the theory*, subjectivity is seen as only one mode of existence among others within the totality of the Universe.

In the second place, he interprets the Cartesian *cogito* toward its specific connection with affirmation of reality as such. "El parecer" of things in the cogito means their existing in various modalities not only "para mí" but also "para sí," that is, in their own right; ". . . the mode of being of all things — in the Universe — far from consisting in their being for themselves or in knowing themselves — consists rather in the very opposite: in their being for another."[9] Indeed, this "other" (*otro*) means not only the "parecer a mí" but also means the existential interconnectedness with the rest of things and beings (we find here the direct correspondence to Heidegger's conception of the "Dasein" in *Being and Time*).

These two crucial points of discussion lead Ortega to legitimate theoretically his great intuitions. He has attempted also to differentiate them philosophically from rationalism, on the one hand, and idealism on the other. It is this legitimacy which is at stake here. In the conclusion of his analysis, he summarizes: ". . . To doubt means *it appears to me* that something is doubtful and problematic. To have something *appear to me* and to think it, are one and the same thing. Doubt is nothing but a thought . . . the entire Universe may *appear to me* as doubtful — except the fact of its *appearing to me*."[10] So far he agrees with Descartes. But with this conversion of the doubting into appearing to myself, a link is formed beyond consciousness toward the external reality of things and being by the way in which Ortega envisions cognition. To "appear to me" implies the Cartesian discovery of consciousness, not with absolute but only relative significance, *because it implies the existence of external reality.* Says Ortega: ". . . In order for me to come into cognizant possession of something, it is indispensable that this something should manifest itself or present itself to me in its entirety, such as it is, as it purports to be, with no part of its substance remaining concealed."[11]

Although consciousness is conceived as an interiorization, as reflexivity and intimacy without which there would be no cognition of the external world and ourselves within it, yet reality of thinking is radical *merely with respect to cognition; it is theoretical* ("It is understandable that that which consists precisely in being known should be a radical datum for knowledge or cognition . . .") because the mind is not "the center and support of all reality . . .". On the other hand, Ortega insists that "the mode of being of all things, is far from consisting in their being for another . . .".[12] The discovery of consciousness is in the final analysis interpreted by Ortega in a direction opposite to Descartes: In his quest after the proof of reality, Descartes emphasizes the universality of the changeable features of real objects. Ortega, in contrast, insists upon the *radical concreteness of the hic et nunc* in which the question of reality should be raised. Consciousness may be the center for cognitive-theoretical appearance of the data of the Universe, but for the case of *reality in its own right,* these data have to be considered in each *hic et nunc.*

Ortega operates then the crucial movement: He concludes that not only the "parecer a mi" means existence of things in their own right, but, most importantly, that the act of the *cogito* appears itself as an *act of my concrete life.*[13] Ortega opposes to the Cartesian formula "cogito, therefore I am" — I am in a conscious, theoretical mode — the formula: "I think, therefore I live"; this latter formula means a concrete actual existence of living in its own right.

It seems that in this discussion Ortega offers the legitimating of his foundational statements of metaphysics: *the direct intuition of concrete existence as the radical "archi-certain" fact of life.* It would also appear that making the significance of the cogito deviate from the direction toward the quest after absolute certainty of cognition and consciousness and redirecting it, instead, toward that of the *immersion of subjectivity in the intimately enmeshed existential connectedness of everything in the Universe,* he is laying the ground for the explanation of this same interlinkage. Here lie the origin and significance of his conception of the "razón vital-razón histórica." In fact, it is the cogito as such, as an act of existence, which is supposed to open for us Pandora's box of the radical reality, of the concrete life in its unique individuality. As an act of existence, the cogito is seen by Ortega as an element of this network of concrete reality.

I would in no way contest that the Cartesian cogito yields the intuition of the subject as cognizing; neither can anyone deny Husserl's view of cogito as opening the entire universe of things, beings, processes in their intentional — that is, as far as constituted, cognizable — form. However, the act of cogito does not yield any evidence that the cognizing subject is performing its act (an act of imagination, of a dream, or a brute fact of life) as a real living entity.[14] The strongest intuition that it yields *by itself alone* is the one which Husserl has asserted: that the cogito constitutes for us human beings *all* that we may know (taking the term "know" in a very strict, intellectual, and complete sense). There is however, *no* continuous evidence present leading from the *act* of cogito to the affirmation of the nature of the intuition, nor is there evidence available that the act of cognition is a concrete and real act of living of the subject who performs it.

Following Ortega's argument, I fail to see how from the theoretical evidence of the existence of consciousness/subjectivity/mind, we may conclude that its continuing immersion is an instance of the real, actual existence within *reality in its own right.* I also find that the link is missing in the passage from the purely conscious, mental level of the "I think" — the *relative* mode of reality — to the *radical* reality of life: "therefore I live." Within his tight intuitive chain of the argument, this intuitive link is missing. And yet should not the strength of the argument consist in placing consciousness in the network of life's interconnectedness? This interlinkage of the elements of the fact of life is shown by Ortega himself in the argument discussed underlying even the work of thinking, or philosophy, in which we do not essentially "live" but "theorize."

The weakness of the legitimacy as discussed above appears even more obvious when we consider that this reality, of which consciousness is just a special mode of existence, is posed forcefully by Ortega as a set of archi-certain evidences, as themselves indubitable and rooted in the *radical fact of life.* Until we provide this relationship, the origin of these evidences and of the system of references with which the legitimacy of the archi-certainty and radicality of the fact of life would be established, remains an open question.

The discontinuity in the sequence of evidences, this lack of the link indispensable to conducting the evidential progress — step by step legitimately, without putting in doubt the crucial thesis of Ortega's metaphysics — makes its foundation deficient and forces us to seek further after the definitive access to the concrete reality. In the progress of my argument I will show how only a turn in a different direction may allow us this crucial insight into life in a self-legitimating fashion.

I do not mean by this analysis to undermine the legitimacy either of Ortega's criticism of the Cartesian interpretation of the cogito (with its far-reaching rationalistic consequences) or the striking truth of his affirmation of the radical intuition of the fact of life. On the contrary, following to the end his own rejection of the primacy of intellectualism, I affirm that *there is* a radical access by means of which we may, as living human beings in our unique concreteness, make this radical plunge into the fountain of life a point of access that would yield its very own absolute intuition, indubitable evidence . . . and open for our discovery Pandora's box of all its riches of existence.

However, before drawing any general conclusion, we have to follow Ortega's quest to its end. The cogito and the thinking subject might not have provided the link to the radical assertion of the fact of life; nevertheless it has led us to a significant development of thought. In fact, Ortega has in the above argument conceived of the concrete reality – of the concreteness of life's entanglements – in terms of the totality of the various types of realities linked into one network. The thinking subject is meant by Ortega to stay within the network of life's interconnectedness. Our attention is then attracted to the relationship that in his thought seems to exist between this interlinkage among elements of life as its intrinsic mode of existential connectedness and the way in which he has answered the "question of our times" in proposing a new type of rationality. In fact, in his approach to this crucial point of modern controversies, Ortega (as shown above) has displaced consciousness, mind, and intellect as the center or source of his insights. Going to the extreme opposite, he has proposed to abolish the sovereignty of intellectual/conscious rationality and to replace it by recourse to the radical reality of life. In point of fact, he proposes to substitute for the abstract, universal, artificial rationality instituted by the intellect, the rationale of the concrete life itself: the "vital" and "historical" reason. The question which emerges is whether we should not understand the vital and historical reason as the vehicle of this existential interlinkage seen as the articulations of life itself.

Maybe the new conception of reason will give us the access to the radical fact of life.

3. THE DIALECTIC BETWEEN THE VITAL AND HISTORICAL REASON IN CARRYING ON THE PROGRESS OF LIFE: ARE THEY AN ACCESS TO THE RADICAL "REALITY" OF LIFE'S SELF-INDIVIDUALISM?

A. *"The Vital Reason" (Razón vital)*

We do not approach, in Ortega's conception, the human being by means of his nature or essence. The human being and his *circumstancia* as the concrete individual, a "cada uno," each and everyone at any point of time, is approached through his life (*su vida*). That means: I am identical with my life, and: "Living," Ortega writes, ". . . is the unceasing discovery that we make of ourselves and of the surrounding world."[15] In this way man is the point of departure of his metaphysics as well as its center; it is his *life* alone that offers the way of exfoliating what the human individual consists of. This is understandable when we consider that Ortega has turned his back upon the isolating of the particular nature of man from the rest of things and beings, and, on the contrary, inserts man into his circumambient sphere of life. Through the analysis of such particular modes of insertions Ortega proposes to describe what man is. He tells us: "our point of departure was what is habitually called — what each of us calls — in times of sorrow and of joy, in the fullness of hope and in that of anguish: *his life.* That (Ortega goes on) is the radical reality. That is what we find as already existing — not as more or less a theory, a hypothesis, a supposition — but as what is always there, before all theory; that is, one's *being,* the *real* thing."[16] The nature of the human being is in its "essence" immersed in the commerce with his surrounding forces and elements of the life-world. But immediately an implication follows. A human being identified with his life is in a perpetual flux. Although man is the identical point in the universal flux, yet it is identical because his being, that is his life ". . . is not, but he becomes and *un-becomes,* he is passing and happening, he is a continuous flux. Man is what he is today, precisely because yesterday he was something else." These mutations of the human being in which he maintains his identity occur in all aspects of life. In his vital/natural development from the phases of infancy, childhood, youth, maturation, as well as in his cultural transformation, as we see in the history of the Occident, he traverses all possible stages. "Man experiences and goes through all . . . forms of being; a pilgrim in search of being, he keeps becoming and un-becoming all these stages, that is, he lives through them."[17] Consequently, in order to account for the direction of this life discovery to be made at every step which the living being must take to survive, Ortega repudiates on the one hand the validity of "mathematical reason," and, on the other hand, he repudiates the empirical causality of biology, which being mechanical lacks the adjustability necessary for the fluctuation of life. Emphasizing the processlike progress of life, which has to find its way within the innumerable factors of its individual "circumstancia," Ortega postulates the necessity to accept a special type of reason that could account for these features of life.[18] That is he distinguishes within this progress the essential factor of discernment and selection; it is of primordial importance, since "Living is having no choice but to reason before the inexorable *circumstancia.*" And, as the illustrious interpreter of Ortega, Julian Marias, writes: "Man, therefore, because he does not have a previously formed being, cannot live without orienting himself, that is, without thinking, without reasoning."[19] This type of reason, which man follows, is in its nature submitted to its *service* to life. When we ask further what it consists of to weigh the alternatives, to seek solutions to our problems, to make selections and project further steps in our life-course — what this new type of reason consists of — the answer is simple: this type of reason is life itself. Hence Ortega proposes to us a new type of rationality: the "vital reason" of the progress of life. Assuming the radicality of the fact of life, from which all other types of reality would process, Ortega naturally considers the vital reason as the basic and all-encompassing rationality; all the other types are at its service. "The vital reason is at the same time the means of access to this radical reality of the life of each one and of each moment."

B. *Historical Reason*

As we have, however, hinted at above, the specificity of the human being, namely his synthesizing self-awareness of himself and his circumstances, which prompts *his specific urge to understand and to know the truth*, leads Ortega to attribute to the human being a special form of the vital reason: the "historical reason."

And here lies a radical distinction, which Ortega is introducing between the human being and all other living individuals: man does not consist in his natural being alone, but in *his historicity*. . . "because history is the mode of being of an entity which is constitutively" — but, first of all — "*radically* (my emphasis) *mobility and change*."[20] For this reason, maintains Ortega, man cannot be understood either by rationalistic metaphysical speculations or in biological laboratories. There is no logical chain of interconnectedness that could account for this unpredictable changeability of man that occurs through his life as immersed in the life of others and the whole range of vital circumstances. There is neither an ideal set of reasons nor an empirical set of causal connectedness that could disentangle the innumerable knots that are tied in man's vital course so that he may progress and in which his life and his identity would reside.[21] To establish this progress, man is in possession of a specific function by means of which he attempts to disentangle the chains of mutual interpenetrating and intergenerating links of his progress and to understand himself: *It is narration.* Ortega tells us, "Man is what he is today, precisely because yesterday he was something else. Ah, then, to understand what he is today it is enough for us to be told what he was yesterday. It is enough, and there appears, there transpires what we are being now."[22]

He assumes, then, that there is a specific type of interconnectedness explicatory of the singular facts, that gives the continuity to the life progress of understanding — continuity — which means that there is a specific type of rationality or *razón*. Since it works by bringing dialectically the past, which is already vanishing, and the projection of future designs, which give a special constructive horizon to the life-course in the present, thus establishing the continuity of what we call "history," Ortega proposes that this specific reason, through which man's narrative function works, is to be called "historical reason." As Julian Marias emphasizes, the vital reason is the "omnipotent" reason of which "pure reason," "empirical reason," "physico-mathematical reason," "geometrical reason," etc. would be derivative; historical reason is supposed to be at the same time vital reason. In fact, it is difficult, as José Ferrater Mora pointed out, to differentiate sharply in Ortega's thought historical reason from the vital one. Nevertheless, following Julian Marias, who emphasizes that the vital reason is, as Ferrater puts it, "an operation of life itself," such a differentiation seems obvious. Although, historical reason and vital reason work for the sake of life — here vital deliberation, there narration — in the very mode of their proceeding lies a sharp differentiation of the historicity of man's life — historicity which makes him specifically "human" — from his "vitality," which makes him alive and survive by just following the rules of the cyclic course of nature. What is the key to this differentiation? What is the factor of man's self-conscious historical quest for understanding himself? If conscious self-assessment is for Ortega the vehicle of the historical reason, what are the various factors of the vital reason which allow the living being to progress in a constructive unfolding toward the seminal self-awareness of Adam? Since I have proposed that the vital and historical types of reason which are partly synonymous, partly divergent, be investigated in Ortega's thoughts as possibly giving access to the radical fact of life, these questions must be answered or the link of differentiation between the vital and historical rationality and the existential interlinkage is lacking.[23]

In my own approach, I agree completely, in principle, with the intention of this differentiation. To explain it, however, we need a means of access to it at the point at which this differentiation springs forth from the progress of life itself. I propose such a means of access through "man's self-interpretation-in existence," which presents the *self-individualization progress of the living being from the rest of life's schema as endowing each life's operation by an appropriate life-significance.* Thus is established the *meaningfulness of life, as well as of the circumambient world sphere in which the individualization occurs.*[24] Within the network of life — knitting itself through several phases of the individual's meaning-bestowing upon his progress and its factors, phases which follow each other in the discrete continuity of the constructive life's progress — we detect a phase of the passage from the "vital significance" of life in general and the "cultural" — or "socio-cultural" — life's significance which the individual "invents".[25] The first type of significance is merely projected by the individual into his life-conditions through his functional discrimination and selectiveness; the second, the specifically human significance of life, *emerges only when the inventive function surges within the vital functioning as a unique factor of meaning-bestowing.* This function not only establishes the avenues of natural life but converts the life-course of the living individual into a "historical sequence." This preliminary statement of my own position allows us to formulate the questions that the differentiation of the types of life's rationality has to answer.

Ortega's general philosophical views are too well known to be in need of exposition, had it not been for the sake of the metaphysical argument I am concerned with. I agree fully with Ortega's general insight, namely, that in contrast to the animal nature, the human being is "historical" in the sense that he not only follows in the development of his life the cycle order of *naturaleza* but *above* and *within* it introduces — and in this every man is the "first man," Adam — multilinear and unpredictably varied networks of avenues of life itself. Man forges his own being accordingly, and this institutes his very specific type of being as expressing its progress: history. Nevertheless, the radical explanation of the fact of life as the ultimate reality — the ultimate reality of this specific human being — is not given by the *razón vital. Razón vital* merely interprets by comparison, association, references among innumerable incidents of the past and projects forward the future that which has already taken place. And to the view that the instituting of history through narration stems, according to Ortega, from man's crucial need for truth, we must object that truth is never reached through the historical reason as seen by Ortega. More importantly, the work of narration does not touch the rationale of life in its "radical," primogenital form. If there is such a rationale of life — and its specific truth — it precedes all specific types of rationality, of meaningfulness established by man, whereas the work of narration consists merely in a further interpretation of what has already acquired a meaning.

Ortega has attributed to metaphor the role of separating the passage from one type of life expressions to others, from the one-to-one subordination of meaning to elements of life, to the transposition of such a universalized expression to another level at which it applies to an entire configuration of meanings. In short, he sees in metaphor the vehicle of different ways of interpreting facts so that within the vital meaningfulness (of the living individual) we may bring in the "biographical" — auto-biographical — that is "historicized" meaningfulness of the individual's life-course. I cannot but agree with Ortega, first, that metaphor — that is, the metaphorical process — is the vehicle of flexibility in the self-interpretative process of life. Second, it is also certainly the instrument operative in endowing the individual biographies with a specific "style", a style which is shared by an entire "generation." However, the crucial questions concerning the rationale intrinsic to life is why does the metaphoric process come about? and What are the factors of its specific operation?

In recent times metaphor became the focus of numerous thinkers who have devoted considerable attention to it. Paul Ricoeur, for instance, couples the metaphor with imagination. However, unless we explain the origin and role of imagination within the *entire schema of life's self-individualization*, we have not taken into consideration essential insights to formulate adequately the questions pointed out above. We are here after the very essence of rationality.

Not only does the metaphor as such merely touch the ratio as such but Ortega's own theory of perspectivism in operation points to the innumerable points of view with which the narrative explanation — whether in real life or in fiction, in literature or in art — is working. The work of narration certainly establishes the knots of relatedness among meaningful — that is *already established as such* — segments of life's progress, yet none of which reaches the point of the radical fact, or the arch-certainty of the primogenital reason. And yet it is through the vital/historical reason that we might have expected to accede to the radical fact of life. Let us push our succinct probing further. Ortega's positing of "historical reason" as the guidepost of man's "humanity" requires us to search after the "primogenital" fact of its surging within the first man, and the human being as such, as distinguished from the rest of the living beings. In our quest after the radical status of the basic truth of metaphysics we have to raise and emphasize the question: In virtue of what is man a historical being? Since the narrative function with its elementary indispensability in shaping everyday life's assessment, recording, ordering in a communicative way with others — as it has been so carefully studied and discussed recently, in particular by Paul Ricoeur — appears not sufficient to account for man's quest after "understanding," we ask whether there are not to be differentiated types of "understanding," one, oriented toward *the establishing of an interpretative interconnectedness of life's events*, and another, specifically human, *ultimately oriented by the need to discover the possible way toward "ultimate telos."* In other terms, does not the specifically human responding to this need to "understand" mean to "invent" man's uniquely own destiny?

With these questions we have here reached the end of one line of Ortega's investigation and see that in order to treat these issues any further, it is necessary to change the perspective. From the assessment of the vital and historical rationality which appears neither self-explanatory nor conclusive, we must dig deeper and reach their very roots. Only by switching radically from the cognitive understanding in all its modes — to which historical reason also belongs, even if Ortega sees it rooted in natural consciousness — to the *creative making of man*, do we discover the *seminal sources of the multiple rationalities, pervading the human life-world and instituting man's discourse with himself and others, in the virtualities of the Human Condition.*[26]

Before we enter into the exposition of my own approach, there remains still one more avenue in Ortega's thought along which we may continue our search after the archi-certain evidence of the radical fact of life.

4. "EXPERIENCE OF LIFE." CAN THERE BE A "RADICAL EVIDENCE" IN THE "EXPERIENCE OF LIFE"? IS THERE AN "EXPERIENCE OF LIFE"?

In our ongoing dialogue in depth with the great Spanish thinker, a prompt objection to my above reflection rises immediately. If the rationality factor of life seen in the form of the vital/historical reason does not suffice to reach it, is there a special type of "experience" that would reveal this *radical datum which is life itself*, as well as other primogenital "data of the Universe" in their absolute evidence and archi-certainty? In fact, we may seek once again to answer this question by the way in which Ortega envisages experience and consciousness. He devotes, indeed, serious attention to the problem of consciousness as the source of experiences. To avoid the pitfalls of idealism he accepts as the only form of consciousness the "primary consciousness," which he sees as an ingredient of reality — in opposition to Husserl's merely contemplative or "pure" consciousness which stays aloof from actual existence. However, could the so-conceived consciousness possibly warrant the desirable radicality or certainty of experience?[27] The data of the primary consciousness, consciousness which feels, wants, desires are — as Descartes has already emphasized — as changeable, fluctuating, nonrepetitive, etc., as the stream of life itself.[28] "Primacy consciousness" in Ortega's understanding appears, on the one hand, to be no other than "empirical consciousness" which Husserl has "suspended" in its empirical features in order to isolate from it and bring to its proper significance the universe of cognition in the unchangeable intentional skeleton and its rational permanence and stability. The primary consciousness of Ortega seems, on the other hand, to point to the originary pre-intentional realm, the concern of the later Husserl. In neither of these forms, however, can it, as we know it from phenomenological investigation, yield the guarantee of certainty or radicality of its givenness.

Nevertheless, there is an original and most interesting turn in the fashion in which Ortega seeks to approach life through experience. He attempts to elucidate its idea in reference to the ancient notion of the "*emperia.*" "Emperia" indicates by its very root (*raíz*) "transition," "peligro," "riesgo" as well as "passage," "ascent." In short, the original significance of "experience" is seen by him as a "traveling without roads (*caminos*)."[29] With such essential characteristics, we are leaving behind the objectifying function of "experience." Experience in general consists in establishing "roads" for our orientation in life, or in recognizing them when established. That is, experience as the primary vehicle of life strictly refers to the establishing and "recognizing" of the "objects," of the objective world and its roads for the path of life's progress. But "experience of life" cannot refer to the routes established by things and objects; Ortega assumes it to be prior to them. If we conceive of "experience of life" as characterized by these features, we obviously mean a generalization of life's ways of advancing. This view offers a penetrating insight into the nature of the hazardous and ever-tentative carrying on of the life-course by a living being taken as a whole. Nevertheless, it seems that these features characterize a universalized view of life as such, not the palpitating, singular instance of living which is acutely bare of features and yet bears upon the total outline of the past and present experience and the sum total of the actual state of the living man who performs it.

Corroborating his idea of the "radical fact of life" in various ways, Ortega himself expresses this intuition in one of his writings: "As in a tree all resides in the expansion of a seed which is hidden in the soil, so everything in man is nothing else but the ramification of a sensation or of a radical sentiment in front of life." What is this "radical sentiment" of life? It is defined by the author as the synthetic sentiment of the life-expansion, as the expansion of the living being through his internal/external functioning. The emphasis falls upon the continuity or inter-linkage of the functional elements of this expansion itself, which, for the understanding of its life-character, brings us back to the sentiment. The conception of the "radicality" of "experience" is thus extrapolated/transferred from the *singular* instance of what I call the "inner workings of life" to its *network*. Nevertheless, as much as this expansion is otherwise significant for the grasping of the universal life spread in its great swing, it does not allow us to penetrate and rejoin intuitively the concrete "act" of living itself. The sentiment does not reach deeper than the cogito or the vital reason.

In our pursuit of the royal access to the "actual existence" as I call it, or to the "radical reality of life," as it is termed in the Ortegan philosophical schema, this above-mentioned scrutiny of Ortega's main perspective leaves us at a crossroads, although we seem to have followed the proper guidelines. That it must also have been experienced this way by his followers indicates the collective effort of his disciples in interrogating themselves directly about what is the "experience of life." It may be stated that all of their penetrating probings share one characteristic feature. In describing the "experience of life" they all emphasize the *historical unfolding* of the individual: life-experience is assumed to reside in the sedimentations of the individual's history in all its possible autobiographical perspectives. Experience in its singular instances of the concrete life proper, that of the individual, is out of sight; life-experience is assumed from the start as a sum total of life's sedimented route. It is the fruit of "maturity," it is the "wisdom of life." We must object that this is a "secondary" type of experience of life. This type of life's experience is possible because we have developed and carry on, to begin with, a "primary" type of experience and a certain business of life; we encompass it and our experience of it assumes an elevated tonality which lifts it above the acute, actual act. In this respect, we find a most illuminating treatment of the question by Julian Marias, which allows us to take an important step ahead in our pursuit.

In a masterly, subtle inquiry Marias investigates what possibly would be "experience of life." To begin with, he introduces a distinction between "experience *about* life," and "life's own experience" ("the experience that life itself has or acquires").[30]

To the "experiencia *acerca* de la vida," Marias concedes the historical maturation comporting an "elevation," "una altura" of experience, which confers upon it a special status with respect to the "direct" type of our experiences. Whereas, if there be an "experiencia propia de la vida," Marias postulates some of its crucial features: (1) it should show that "the actual encounter with reality — not merely mental or theoretical — occurs *in living*." Consequently (2) it should be ". . . the non-theoretical form of the vital reason, when it is applied to the totality of what is real . . . the alveolum inside which the latter functions."[31]

Distinguishing these two features, Marias has brought our inquiry to the last step on the road before crossing the Rubicon. Yet so far Marias does not cross it; he offers us merely the postulates which such an experience should fulfill, but he does not indicate the type of experience actually available to man which would satisfy them. The reason is clear: if existing, such an instance of life would not be "experience," as this term is understood in any current philosophical way.

With the above-discussed attempt of Marias to interpret Ortega's intuitions on the crucially important point of the access to the radical fact of life, we have completed our probing and we have found three decisive indications for future inquiry. It is clear that (1) we still have to find direct access to the concrete, instantaneous, irreducible, actual — that is "radical" — fact of life, which with Ortega's investigations remains extant; (2) this access has to bear within itself its absolute certainty and, consequently, its own legitimization; (3) being simultaneously concrete and yet universally valid in expressing this one uniquely particular instance but also every act of life as an act of life, it has to be situated at the crossing of all the arteries of the individualizing life; moreover, *it has to bear within itself the virtual joints of all the arteries which pass through it and of their existential articulations necessary to carry on life.* That is, it has to be the *virtual knot of all actual, past and future rationalities of life's progress.*

PART III: PHENOMENOLOGY OF THE CREATIVE TREMOR AS THE KEY TO THE LOGOS OF LIFE AND THE ACCESS TO THE METAPHYSICS OF LIFE

Before entering directly into the presentation of my own position concerning the radical insight into the "fact of life" and the status of life's *rationale* that it allows one to exfoliate, I have, first, to state some features of the "philosophical overturn" of priorities of my own approach and of the radical reformulation of questions which follow it. As emphasized before, it is the *creative act of man which constitutes the Archimedean point of this overturn.* Second, I have to state briefly the main steps of the tortuous itinerary which the *creative act delineates along the ramifications of the creative function of man until we reach in its poignant concreteness its very core within the creative forge of the Human Condition: the tremor/the thrill of creation = the tremor/the thrill of life.*[32]

Whereas the main branches of the philosophical/phenomenological questioning focus principally upon the elucidation of givenness and upon the interpretation of its modalities, that is upon cognition, or knowledge, of *what there is* in and about us, the metaphysics of life, as I have proposed it, with its roots plunged into their discovery, consists in distilling from it:

(a) the inner workings of *man's self-individualization* from within the *unity of everything-there-is alive*;
(b) the re-cognition of man's specific status within the universal life-schema;
(c) the culmination in and conversion of all the other queries into the metaphysics of life's most specific task: delving into the *Human Condition* which reveals the ultimate status of man, by conjecturing its *virtualities.*

In this query it is indeed a "conversion" which is accomplished: from the mere elucidation of the states of affairs and facts (cognition of the given), we are moving to the uncovering of life's network: *the unity of operation—cognition—appreciation.*

1. THE TREMOR/THRILL OF THE CREATIVE ACT, OR THE RADICAL ACCESS TO THE "FACT" OF LIFE

A. As Julian Marias rightly pointed out, Occidental philosophy needed two thousand years to reach an approach to the philosophy of life, although it is life that constitutes the ground and sense of philosophizing. This delay has, however, very good reasons. As we have already seen, there cannot be an "experience" of life. Experience, as we know it, is, in fact, in its progress an operational sequence of "objectifying" and a result of an "objectivation" of elements present within an operational segment of our life-process which impinge themselves upon our actual present and arrest our attention. This objectivation aims at presenting a complete, finished construct which posits an instance of our circumambient world, or of our commerce with it, without which, this positing would elude our attention. Whereas, when we ask ourselves, What is life? we see, first, that, in contrast, life is nothing "objective"; it is the most intimate, obvious factuality of our existence, ever present by captivating our innermost attentiveness. As such, it eludes every objectivation; it advances incessantly with every one of our vital acts; it cannot be arrested even for an instant so that we could focus upon the ray of our intellectual attention. It eludes any attempt to be followed along its flux and grasped at some instant as a whole, because it never reaches any point of completion or totality. Furthermore, it spreads in all directions of which none represents it on its own, and it appears infinitely expansive; seeking a way to express the supreme and endless differentiation, richness, variability of its elements and a dynamic not to be encompassed by the human mind, we say "large as life."

Second, we see that it is only our vital, operational attentiveness to the task of living that keeps us breathless by life's work.

The reflexive attentiveness which awakens whenever our actual consciousness is presented in its actual field with elements "of interest" remains aloof, at rest, concerning the innumerable variety of the elements which incessantly surge in, play a role and vanish from our vital life-progress: "to live," to "be alive" is in its natural advance so smooth that it is consumed as "natural" in its supreme simplicity which attracts no attention until something in its regulative process is disrupted. Our life is to us so unavoidably present in every experience, pulsation, thought, etc. and facing the "danger of life" or of "losing life," of "imminent death", is in its radical poignancy so bare of any intrinsic feature that we cannot begin philosophizing by raising directly the question, What is life? We have to come to this question, on the contrary, by a roundabout way. But every philosophical thought pushed far enough has to come to it in some or other form. Inheritors of classical phenomenology, which has left us the largest possible legacy of the inquiry into human cognition and the human mind, we may now, after we have found our way out of the classical phenomenological supremacy of the intellect and cognition by discovering the primacy of the *creative act*, formulate the question of life. We are equipped to break through all the rational, objective barriers that separate us from the pulp of our life and penetrate it not in any objectified form but *in its workings.*

I propose to present succinctly the *essential proficiency of the creative act* and to show how it enables us to retrieve the gist of life: the *life's tremor/the thrill of life.*

B. The smoothness that brings all the disparate acts into an ever-overflowing — although not contiguous — spread which makes the "substance" of our "living", as well as of our "being ourselves", resides in the "enjoyment" of our acts. As I have elaborated elsewhere, besides assuming the function of the intellect and that of the "moral sense" with the constructive role, on the one hand, and the "communicative" role, on the other hand, our acts consist in *enjoyment.* The enjoyment of our acts makes them reverberate, stretch, and vibrate; in this vibration our "living substance of feeling ourselves" in a spread resides along their reach. If the enjoyment of our acts attains, at certain periods of our life, a great intensity, we experience our entire being as "dynamic," "activated," bursting with "vitality."[33]

While each act vibrates with its specific enjoyment, at the crossroads of life's arteries surges a unique type of act-enjoyment in which the entire individual life's enactment *is synthetically present at the peak of its operational attentiveness.* The vital enjoyment of a singular act is transformed not only into an acute, all-encompassing alertness of our entire being but it may *master* this alertness because it wells up from a *synthesizing urgency, synergizing our entire functional system under one, specific "signal"*, e.g., a signal of imminent danger, joy, delight, etc.

In this synthetic enjoyment — in contrast to each enjoyment of a singular act with its restricted register — we are *all* mobilized in one acute instance of attentiveness to the imminent urgency facing us as living individuals; there are innumerable modalities of our reaction to it (e.g., sheerest delight with "everything" which we put — all of ourselves with it — into a melody we hum, with the "stroke of genius," which animates us and confronts us with an unforeseen solution to a problem, with a project which lures "all of us" into the vast dimension of the unforeseen, with the stroke of a panic or moral dread when we are caught in a fire or an earthquake, etc., etc.). Most importantly, it springs from this vital ground before the objectifying functions of intentionality with its conscious form which molds it in a certain way; thus it "precedes" the formation work of experience and surges intact *through the functional schema of the latter.*[34]

This most authentic "act of life" makes us at once partake acutely of all our operational and experiencing functions as they themselves partake of the forces which run through the common arteries of "everything-there-is-alive." As such this act does not carry any qualitative content; it is just an actual maze of living; only upon objectifying reflection upon the life-world situation from which it stems, do we endow it with its proper name: dread, fear, delight, etc. In itself it is barren of any objective structure or qualitative feature; it is a virginal outcry of life. In itself it means our total life-concentration, our total life self-immersion conjured upon in an instant: the tremor/thrill of life.

And yet, it does not spring forth as a neutral action/reaction; it emerges as *my very own act of life* interchangeable with any other act of the same intensity, acuteness, synthetic nature. To spring forth "from within" myself, gathering into one, most acute act of existence *all which makes me vibrate and persist in living,* it must also express *my part, my place, my share,* and *my most concretely determined situation* within the *bubbling stream* of "everything-there-is alive."[35]

Although it does pass through the personifying form of source-experience, in which it subsequently plays a role, it does not bear the specific form of our rational, conscious or pre-conscious, intentional functioning; it does "personify" my share in the great schema of life.

Although it is an outcry of our lifehood, it has no "content" of its own; it does not communicate "anything."

Being devoid of any *rationale*, the tremor of life presents the authentic enactment, free of any interpretative distortion, of our selfhood at the very junction of the dumb and mute life-individualization and the beginning of the specifically human significance of this individualization.

It is the outcry of the *workings of life at this crucial borderline between the vital self-interpretation-in existence by the living being and the commencement of the specifically human significance of life.*[36] As our very own, it is both the expression of the one and the carrier of the other. Its authenticity consists in this nakedness; the tremor of life springs forth independent of all cultural artifacts and comprises the specifically human world. Our vital thrill of mortal fear, of meeting the challenge of imminent death, of reaching a heartbreaking decision, of the aesthetic delight in which we feel ourselves "boundless," is never distorted by the infinitely varied sedimentations of experiences, experiential webs of the past, and the experiential whirl of the instant in which it rises. And yet, breaking through all the constructs which the workings of life have established for their advance, it is not disjoined from the virtualities of the Human Condition wherefrom it springs. On the contrary, it synergizes to the supreme point the entire spread of our life enactment. It reveals this spread within its highest manifestation, namely, within the *creative act of man.*

C. The creative tremor/thrill of life, or, as we may as well call it, *the tremor/thrill of the creative act*, expresses the quintessence of the uniquely concrete life progress at its very peak. The creative act synergizes all the forces, aspirations, operations, functions of the human individual, who within it enters into a total control of his faculties, aims, and strivings. Like a demiurge, the artist, writer, scientist feels that he holds in his creative act all the strings of his beingness in progress in order to project/invent a new counterpart to his longings, a new perspective for *his self-interpretation-in-existence. Imaginatio Creatrix* first emerged from the source experience that circumscribed the simultaneously released set of virtualities. These virtualities that establish the new platform from which the progress of life is going to proceed bring in all possible forms proposing themselves for concretization in an objectified fashion.[37] What I call the "creative act" is, in fact, no singular instance of my life's performance. There is no isolated instance of the human consciousness which with a "click" would open a new dimension of inventiveness. "Creativity" is the specific mode of the human functioning that, emerging with the Imaginatio Creatrix from the recesses of vitality, reorganizes the sum total of the individual's constructive endeavors into a network of a specifically renovating proficiency: the "*creative orchestration.*"

Nevertheless, we become aware of it only when it reaches within us, as an artist, writer, poet, scientist, etc., a uniquely specialized form. That is, when the *creative orchestration crystallizes into a properly creative process aiming at the inventive formation of an original and absolutely new, objectively significant form of life.* It is within the creative process, spread over several phases ranging from the "creative vision" of the objective form through the progressive stages of its crystallization into the finished product detached from the creative womb of consciousness and established in intersubjective reality — that the creative orchestration makes itself visible. It may be traced as it reveals itself in the imaginative/inventive ways that our otherwise life-oriented functions conceal. In the creative process, they are re-oriented; they re-fashion themselves; they are re-modeled while the progress of deliberating, selecting, planning, and forging new molds in which new significance of life can surge goes on. What I call the "creative act" is this complete concentration of all the human powers in the changeable modes which their creative orchestration introduces under the promptings of the Imaginatio Creatrix. It is in this "act" of utmost density and simultaneous transparency that the creator holds the strings together and continues to be from instant to instant in full possession of its powers and directing its progress.

So understood, the creative act personifies each step of the creative process in its *instantaneous intensity*, as well as in the *entire outline of its operational articulations* already established and/or pointed out for completion, crystallized in the actual, instantaneous grasp by the creative orchestration.

Thus the creative act, representing the entire constructive schema (a) exposes all the ins and outs of the creative functioning in its *operational immediacy* in the *act of operation itself*; (b) *reveals itself as the Archimedean point of the radical transformability of our functioning*; (c) *reveals itself as operating at the borderline phase between the vital — life subservient — functions and their flowing into a novel — specifically human — creative functioning, oriented by the innermost strivings of man's specifically Human Condition*; (d) *exposes not only the workings but also the conditions of the transformability itself.* In brief, *it ties into one knot the unfathomable, pre-individual workings of life itself with its threads which cause the surging of the Human Condition and its original workings.* The creative thrill of the creative act is the thrill of existence at its fullness. *It contains it all in its inventive itinerary for us to explore.*

2. THE MULTIPLE RATIONALITIES OF THE VITAL LIFE PROGRESS IN ITS ORGANIC HISTORICAL PROCESS WITH THEIR RADICAL LEGITIMIZATION FOR A PHILOSOPHIA PRIMA

A. Once we have untied the knot of the vital constructivism of life in its concreteness — as it reveals itself in the ramified network of the *creative intuition* in the *tremor of life* — we discover that life, our uniquely concrete life, consists in a *self-enactment of a constructive self-individualizing.*[38] I "stay" not only in a genetic progress of every instant and in every instance of my functioning movement, but this progress is a *constructive advance.* Its inner workings consist in operations, functions, acts that my "beingness in progress: incessantly performs. In this progress simultaneously I live — exist — I maintain myself, and I establish my difference from my *circumambient conditions.* Seen from the inner workings of its constructivity, it appears that I may do so only in a constant action/interaction with the forces, elements, and substances of the "life-system."[39] To live is the struggle, as Ortega so rightly observes. The vital progress which I forge in this struggle differentiates, however, in as many ways of procedure as there are types of choices, selections, and decisions that must be met in the various operations that I perform. In fact, and here enters the crucial significance of the creative approach, each of these types involves different kinds of my responses according to the "relevance" that the circumambient elements present to my needs, that is, each type of my response carried within itself a *specific type of rationale* which resides in type of response to its "reason" for it. Now, agreeing in principle *with Ortega* that the vital progress carries its own kind of rationality, yet along a different path from Ortega, namely, in following the threads of the creative act, I am in a position to introduce a new and most significant insight: we may differentiate this "rationality" in *as many specific kinds as there are constructive phases in its inner workings.* The term "vital reason" means then globally *their common task in carrying on the individualization of life itself.* I distinguish several kinds of relevances with their specific types of rationale as characteristic principle for three "vital" phases of individualization in existence: the *organic*, the *psycho/sensing* and the *psycho/gregarious* phase.[40] They establish, first, numerous types of *rationale* in respect to their specific constructive functions. Second, doing so, they introduce and carry on the vital sense of life: the *logos* of life's interconnectedness, its continuity, its cyclic order as punctuated by origin, growth, decay, extinction.

But what about man? Our investigation of the self-individualizing arteries of life shows that in order to establish specifically *human* relevances of life-elements at a special point of life's constructive progress a novel vehicle of life emerges: the *Human Condition.* Exfoliating the ways of this constructive mechanism present in the full grasp of the life-tremor in its creative intuition — the "creative tremor" — we do not stop at the "passive synthesis" of the "de-ciphering" of vital significances. On the contrary, I must emphasize that if the tremor of life (thrill of life), in the exfoliation of the threads which the constructive genesis of life spins, brings us to this knot of constructive arteries through which life itself springs forth at each instant and works itself step by step onwards, it pinpoints in a crucially significant way the peak of life's accomplishment in the form of a decisive step toward humanity: the source-experience of the human being.[41] In point of fact, from the central knot of life's inner workings which ties itself at a certain stage of its complexity inaugurating this phase of its constructive unfolding emerges a new, versatile, multidirectional apparatus. This apparatus functions like a watershed distilling the gush of elementary spontaneities of life into new molds, from which experience, as a *specifically human life-instrument,* springs forth. Here it is where all the vital functions at a certain stage of their constructive accomplishment come and release *this new, radically new — complex but irreducible to a mere conjunction of vital functions — faculty of "Imaginatio Creatrix."* We may call it "primogenital," since only with its emergence within life's working apparatus do the two other human faculties, *intellect* and *will*, emerge.

Imaginatio Creatrix and the other two faculties emerge as conjoined instruments of the *new-life-apparatus* and circumscribe the system of virtual unfolding, which I call the "Human Condition," the life's platform from which a novel style of life's expansion unfolds.

B. The creative constructiveness thus inaugurated introduces three new virtualities in order to endow selections, appreciation, and invention of the life's progress *with new types of sense.* Within the universal vital sense — which, however, is then dethroned from its sovereign validity — there spring forth factors of a new valuation of life's demands, of new relevances, and new responses: the MORAL, THE AESTHETIC, and the INTELLECTUAL SENSE. Here it is that, with the advent of the source-experience into naturalness, the human being — the first Adam — awakens to the awareness of life. Not only does the universe resound with responses relevant to his new faculties, as he takes possession of himself as such but, as we will see later on, beyond raising the QUESTION OF LIFE, he also TAKES THAT QUESTION INTO HIS OWN HANDS.

If it is with the first man that all began to sound, pulsate, move, and the entire whirl of life coming to its own, *it is because man's creative function together with the seminal embryo of self-consciousness, having emerged, projected a new way of synergizing his functioning in taking possession of himself in his self-awareness.* This meant to interpret the otherwise "meaningless," because remaining without a focus, events, facts, interactions, and connectedness with respect to the *Imaginatio Creatrix.* With it surged "imaginative possibilities": offering a range of conjectures to project ties between life's operations such that the cogent universe of experiential fabric could have surged, come together, they arouse man's wonder and the inward urge to express their significance for his nostalgias, forebodings, dreams; finally, they prompt his will to "understand" life's occurrences as relevant to his "new" strivings for *his own sake* and not merely for the progress of his life-course. Man's self-individualization in existence, which I have identified with his self-interpretation, reveals then a radically novel phase.

Only with the advent of this inventive/creative apparatus instituting the Human Condition in the individual's specific functioning does "historicity," which Husserl, Heidegger, Ortega, and their followers emphasize as the main feature of man's "humanity," enter into the self-interpretative progress of the living individual. Putting in this way the horse before the cart, I am shifting the priorities in the explanation. That is, historicity appears in this perspective not as a human phenomenon reposing in itself from which further conclusions about man should be drawn but, on the contrary, as a universalized — although comprehensive — expression of an infinitely complex significant system of man's self-constructive "nature" but not this "nature" itself.

We may well concede to Ortega that man's specifically human nature consists in his need for truth: an urge to understand the otherwise opaque entanglements of the innumerable elements which in innumerable perspectives enter into the progress of his self-individualization. We may also agree with the above thinkers that man's historicity, as an expression of this effort at understanding facts proceeds by explaining them within the intergenerative triad of their temporal succession: past/present/future, and through an infinite number of projects which man conjectures as their framework. It is not enough to attribute an important and even decisive role to imagination unless we are able to situate it within the entire system of man's functioning and reveal its origin, its place within it, its various constructive inclinations and trends. Most important of all, we must discover the basic principles which lie at the origin of each of the categories of "reasons", ratio and rationalities which incorporate the very significance of life. There is no "historical sense" of life; in order to take on the "historical style," life devises within the great lines of its basic constructive establishment a most versatile, flexible system of *the origination and transmutation of sense.* Thus we must ask: What are the essential factors of the inner workings of life's emergence and constructive advance?" "How is life's progress capable of accounting for the various types of human — specifically human — urges, aspirations, nostalgias that (from its merely individual self-interpretative expansion of life, cultural expansion) point directly to the role of the creative, inventive apparatus within the Human Condition. I have, in point of fact, attempted in my previous work to answer these questions by outlining the major ways in which the *creative orchestration of human functions* by means of the virtualities of the *moral, intellectual,* and *aesthetic senses nourish the human self-interpretation in existence along an invented and created network of new avenues of life.* These avenues, which follow no more the cyclic routes of the vital course of "naturalezza," but are the expression of man's creative flexibility, ever new, ever original, never predictable, project the constructive progress of man's creative self-interpretation in existence which we usually call "human history."

There remains, however, a most important point to be elucidated. Since invention allows man to free himself from the absorption by his vital needs, what other "needs" does it propose? With reference to what other guidelines does man bestow meaning upon facts, events, experiences? By what standards does the inventive individual measure his specific strivings? What are the principles of relevance which direct the invention/creation of new types of "reason", ratio, significances, rationalities?

Having raised these crucial questions concerning the origin and unfolding of the specifically human "life-world", culture, and human historicity as such, I will point out here succinctly the three main principles which orient it.

C. Man's historicity, as his specific existential feature, resides undoubtedly within and originates from the operational modalities of his *self-interpretation-in-existence* and together with the emergence of the quest to "understand." However, "understanding" has several meanings. By "understanding" I mean neither the grasping of the interconnectedness among events and facts nor establishing cause-effect relationships between the segments of life's operations. Lastly, understanding does not consist in construing and re-construing the "significance" of the events of our existence by varying possible explanations of facts and weaving them into "stories of our life" or of the life of others. We do it incessantly, but it is merely an *expression of a need* to understand and not the *exercise of the understanding function itself.* Although in the vital phases of *man's self-interpretation-in-existence* we seek to follow as closely as possible the existential route as pre-delineated for us by the life-system, and to establish the interconnections of facts according to the vital relevances, so the specifically human — inventive/creative — phase of man's self-interpretative progress is characterized by *his distancing himself from the rudimentary life's conditions. He does not stop at seeking interconnections among facts according to vital relevances. Other promptings than vital needs take precedence and direct his pursuits.* His quest after understanding takes the form of the "search after truth," proposes for itself other aims than the pedestrian "cognition" or "understanding" of facts which is never adequate. For the purposes of the vital significance of life man would be satisfied with such a cognition, but he *has become an inventor who aims at inventing new, other avenues of life.*

With the emergence of the creative function and its specific type of constructivism of the life-routes there enters, indeed, into the inner workings of life *the interrogative quest.*

With its appearance within the functional system of the living individual, Imaginatio Creatrix explodes the straitjacket of vitally relevant regulations to which the selection and decision of the living being had automatically to conform. It prompts his selective initiative to seek a solution to questions beyond the immediate range of answers, and simultaneously proposes various possible options not immediately available. A deliberating mechanism sets in to consider them, evaluate, discriminate, and choose. Already present in germ in the previous phases of the life's unfolding, now it unfolds, fully assuming the role to direct life's progress. Yet, by distancing itself from the strict dependence upon the system of vital relevances according to which life's progress has so far been proceeding, it cannot blindly follow any life's regulations; it has to "interrogate" all the data of the given situation with respect to relevances immediately pertinent as well as "*possible*." This interrogative character, assumed by the deliberating mechanism of man's functioning owing to the opening of the unlimited perspective of options other than those available close at hand, calls for new types of evaluative principles and criteria; they correspond to *new principles or relevance which rise on the wings of human nostalgia, hidden strivings, inclinations, longings, which now come to play their part.*

Our subliminal forces, nostalgias, and longings now released, awakening from a lethargy, drive us toward a search after constructive possibilities, on one side, and imagination, presenting a seemingly unlimited richness of options on the other, together send our hitherto passive mechanism of evaluation and selection into a whirl. We have to interrogate every issue, every option, every tendency . . . at each step of our being confronted with life's problems. We wake up from the passive enactment of life's guidelines; the creative/inventive imagination opens all the gates, breaks all the automatized habits, and further stimulates our subliminal forces, to see realization of their own, that is *to realize man's own unique desires and "dreams" released from the abysmal ground of vital forces.* Within the creative orchestration of all our functions that now encompasses the new turn of life, three new principles of relevance emerge to guide the creative quest to understand: *the relevance of accomplishment, the relevance of conclusiveness,* and *the relevance of self-realization.* With these new guidelines for man's invention of original life avenues, his self-interpretation performs a Copernican turn. *It does so because it inaugurates a phase of life unique-in-nature carried on by an undetermined host of novel rationalities which the human being himself inventively projects.*

1. The first directive of the new line of significance that Man the Creator follows in devising the proper means to satisfy his aspirations is that of *accomplishment*. In contrast to the vital relevances with respect to which the individual being has been appreciating, discriminating, and selecting the proper means to satisfy his needs — each of them being confined to a vitally significant performance or function (even if leading to the construction of an organ) — *the creative constructiveness of life does not limit itself to the functional aspect of a segment; it aims at a result, at an "accomplishment."* By "accomplishment" is meant the *final stroke given to constructive process at its terminal phase so that the result may deliberately assume an autonomous status with respect to the constructive process.* As such it presents a new type of significance *incarnated in an intersubjective, objectified, form.* Of course, the three above distinguished factors of sense, *the moral, the intellectual,* and *the enjoyment/aesthetic, play the leading role.*

The most striking effect of the regulative relevance of accomplishment is the "primogenital" — to use an expression of Ortega — and decisive effect that I find in the establishment of the "social contract" as a type of relatedness among living individual. It breaks the narrow frames of the vital relations and creates a plurivalent pattern of radically new life-significance. From one point of view this pattern offers infinite generative possibilities of the specifically human significance of life's interaction among individuals; from another point of view, it is a platform upon which the specifically human life unfolds. It unfolds by creating an infinitely extensive host of rationalities.

The life-world, which in the psychic/gregarious phase of life's individualization — governed by the vital, life-subservient, relevances — consisted merely in sharing what there is in the circumambient conditions of vital existence — gregariousness — is through the emergence of the social contract elevated to an autonomous, flexible, infinitely transformable, *social world* rooted in the intersubjective relations of human beings among themselves. In point of fact, the type of vital interconnectedness among the particular life-enactments of individuals — their appreciations of vital elements, events, and actions, which within the vital phase of gregariousness is oriented by relevances to their progress — when oriented toward the aim of accomplishment turns into an *intersubjective agreement among independently deliberating partners.* "Agreement" *means an emergence of a novel category of significance; it offers a source for innumerable "reasons" and "rationalities" for contracting further agreements.* From its center, the intersubjective collaboration among individuals, which, taking off from the vital ground, acquires now an autonomous status, unfolds infinite inventive forms of life, with new rationalities to construct ever new avenues of human existence.[41]

2. In everyday life no effort at disentangling the knots of events and facts, no effort at finding an explanation by pointing to a reason, yields a conclusion. When we seek testimonies to events by eye-witnesses there is always a contradicting "evidence" present or at least possible, which nullifies any claim to the conclusiveness of the "statement of facts." Similarly, when we try to grasp our own life in its subjacent reasons, there is an innumerable number of explanations available and none of them are conclusive. And yet, *it is inherent in the essential thirst and concern of the human mind to aim at conclusiveness of explanations.* The search after truth, which characterizes the advent of the creative function within the progress of the individual life, postulates the conclusiveness of the outcome.

In other terms, what I call "understanding" does not mean any type of explanation with reference to the factuality of life or matter-of-fact givenness; the "explanation" of understanding in its full human extent is oriented by the creative relevance toward conclusiveness. There is nothing which stimulates our intellect toward invention of new approaches more than the orientation of its workings by the aim of conclusiveness.

The creative orchestration of all our faculties reaching its peak of intensity and concentration triggers the urge to surpass all the singular, intermediary steps in explaining reality and to grasp it, on the contrary, by its bowels. The urge to "understand" prompts the seeking of the final and lasting reason of the real. Having thus pushed his aspirations to the extreme, *man-the-creator takes the case of reality upon himself.* He disentangles the knots of the given forms and transforms the contingent, evasive, never complete form of any givenness, open to every subjective interpretation, into a new, unique but objectively universalized, pattern, reposing in itself: the object of art, scientific theory, philosophical treatise, etc.

Hence, all the human faculties sharpened in this direction, theory, art and philosophy emerge as the counterparts to the fleeting, unattainable "reality"; they fixate "objectively", in a conclusive form, the aspirations of the human individual. Weaving them with the patterns, shapes, forms, threads which he has inherited from his past into a vast fabric of meanings, man projects their variations into his self-interpretation-in-existence.

The creative function finds, in fact, its definitive, conclusive accomplishment in the creative activity proper: theory, art, literature, technique, philosophy. It is the forge of infinitely varied, new vehicles through which the newly invented forms bring in new types of significance to life's progress. Thus, from the vital life's rationale we pass to the multiple cultural rationalities of human existence. Works of creativity tantalize in their novelty and originality; they attract the attention of the interrogative quest; they become assimilated into the individual weaving of man's self-interpretative schemas. However, assimilated in one generation, this novelty wears out and all the questions which it has meant to answer arise again and again in the next; new strivings to answer them in a "finally" definitive way by a novel type of accomplishment take place. And so human history in which *the personal historicity of man consists in proposing to himself always the same questions — albeit in different guises — and inventing ever new answers to them, is woven and takes ever new forms.*

However, as the myth of Sisyphus illustrates, no absolutely conclusive answer is ever given; nothing is ever accomplished "finally." And yet, man's regulative relevances of conclusiveness and accomplishment continuously stir his strivings, aspirations, forces, faculties to take up the task of the specifically human life again and again.

3. As is clear, the relevances of accomplishment and conclusiveness, if they do not partially overlap in their effectiveness at least complete each other in their regulative effect. Serving as "telic principles" of the significance that we attribute constructively, inventively to the possible, actual and aimed at, constructive goals that we pursue in delineating the course of our specifically human existence, they are intimately linked in their operational regulative roles. Not only do they complement each other in providing guidelines for the criteria to measure the steps by which we advance toward our constructive aims, but they "*imply*" *each other's postulates.* However, for their total effectiveness in circumscribing the complete horizon of man's inventive tending toward the maximal realization of virtualities of the Human Condition there is still indispensable a third essential type of relevance category: *the relevance for self-realization.*

With the deliberating, self-interrogating and inventive lines of the creative orchestration that lift the self-individualizing script of man's progress from the vital to the cultural phase, *it is the centralization and strengthening of the central deliberating and orienting agency that comes to peculiar prominence in the human self.* It is the self which synthesizes life-awareness with its constructive successions of temporality — that is the past, the present and future — but brings its effectiveness to culmination while it *identifies its own unique life with the self-interpretative route accomplished: the historical route.* Either in clear awareness or "through a glass darkly" all of us raise, in our effort to "understand," the question of the relevance of singular facts and events for the "meaning of our life." In other words, we identify our self with our "historical route" already traversed, with its significant script woven into the pattern of the meaningfulness which we have established in our self-interpretative task. Yet, at each instant we are just "on the way"; the pattern is incomplete; the overall meaning is open to further significant threads to be woven in and to various interpretative lines which might complete it in one or other way. At each instant, our quest to interrogate the validity of "it all," as well as our quest to conclude, to accomplish, prompt us to seek the *stage* at which we would finally acknowledge that we have "realized" the nostalgia, the longings, the dreams and, especially, satisfied the search after the ultimate "truth" which runs through them; in brief, we have realized ourselves.

The search after truth running through our entire interrogative system does not stop at the scrutiny of ways to achieve any particular level of our concrete life course and its meaning. On the contrary, *we are carried through an innermost longing to find an ultimate explanation of the very sense of our life.* We are prompted by the urge to seek the "ultimate telos" of our own existence: its very "sense" which we do not any more postulate in the form of any "clarification" or "explanation" of reasons, explanation that would serve to devise some other script; this urge calls to account for our existence with the sense of our very beingness. When we raise the question: "What is all this about?" we — our very selves — long to account for the answer by the *definitive accomplishment of the significance of our existence. We conjecture the possibility of some unique type of self-realization.*

And yet, in whatever avenue we direct this quest, we find that life — now the specifically human life — does not yield any definitively valid answer. Moreover, we discover that no relevant response is available at all and that this urge after a definitive self-realization has no corresponding *ratio, rationale,* or reason; it points out beyond the framework of life, whereas reason, *ratio*, rationality reveal their strict limits: *they are in the service of life*, and in the service of *life's* constructivism: organic, vital, psycho-gregarious, human.

Indeed, with the relevance of ultimate self-realization we circumscribe the origin, unfolding, and reign of rationality as such. Not only our avenues of the vital but also those of the creative understanding find their limit here. Although they do not cease to tend toward it, they lack the means to advance. The "Ultimate" which lurks behind the rational screens eludes *all* understanding. Like mathematical functions tending toward a limit they never reach, the nostalgia, strivings, longings of the human being prompt his mind to tend toward a "limit" or "telos." The human inventive mind may propose an infinite number of ways in which to move. It is in the ever renewed tending that man's forces are spent; in its form and quality he accomplishes a self-realization *relative to it.* But the criteria of the "absolute", "infinite", "definitive", "conclusive", "final" accomplishment float — as its telic principles — on the furthest horizon of man's creative workings, keeping it open for his inventive swings, yet, simultaneously closing it upon the transcendental reach, which his powers circumscribe. With the reach of rationality it closes upon itself.

CONCLUSION

I believe with Ortega y Gasset that the basic insight into the radical reality of the human life is the only authentic ground for metaphysics conceived as the first and last philosophy; I also share his postulate of archi-certainty in elaborating its edifice.

However, having outlined above the main stages of the pursuit in which Ortega attempts to find the direct access to the radical fact of life in a self-legitimate fashion, I pinpointed our divergence in basic insights for setting the principles for the metaphysics of life.

Although Ortega y Gasset has proposed a most refined and far-reaching schema for the philosophy, or metaphysics, of life, this schema lacks the crucial factor indispensable for giving access to the radical "fact of life" and to its rationale. I have proposed as such a unique key to metaphysics: the CREATIVE ACT OF THE HUMAN CONDITION. My phenomenology of creative experience supplies the indispensable access to the "nervous system" of the inner workings of life and uncovers its innermost trajectory of life's self-individualizing, self-interpretative progress. Including it, provides the introduction of the essential principles into the schema of the metaphysics of life.

In summary: I have proposed the creative tremor/thrill of life as the key to its inner workings that lays out the arteries along which the multiple rationalities of its progress originate and unfold. *The creative tremor opens, indeed, the access to the virtualities of the Human Condition.*

Scanning their principal routes, that is, those of the main trends of life's individualizing constructivism, we found, in the first place, the primogenital-vital-rationalities of its differentiation from the circumambient elements. In the second place, following the lead of the creative tremor/thrill upwards through the workings of the creative function, we have seen how by *inventively unfolding the virtualities of the Human Condition the living being transforms the primary avenues of life by bringing in new factors of sense. With these new factors he expands his circumambient conditions into a socio-cultural world, his very own universe within which he seeks his unique self-realization.*

With the creative tremor/thrill uncovering the network of the creative functioning of the inner workings of life, we have discovered an access to the radicality of life which yields the absolute certainty, an access sought in vain by Husserl, Ingarden, Merleau-Ponty, and others. Its evidence, and that of its ramifications, cannot be subjected to intellectual doubt; its evidence is no less than that of the exercise, enactment, of our concrete, individual life, at this very instant and at each of its instants.

We may, at last, proclaim to have arrived at the metaphysics of life and of the Human Condition as the authentic and final philosophical explanation. It takes into account the enormous progress of knowledge accomplished in recent times and the essential postulates posed by the history of philosophy for "any future metaphysics."

Most significantly, however, not only does our *metaphysic of life and of the Human Condition* offer the definitively mature fruit of the phenomenological line of probing, but it meets in an essential way the expectations of Edmund Husserl, who in his last vision saw that "metaphysics" should be the science of the ultimate question of rationality, which was excluded by modern positivism.[42]

World Phenomenology Institute

NOTES

* The first chapter of this monograph has previously been published in the *Phenomenology Information Bulletin: A Review of Philosophical Ideas and Trends*, Issue 6, 1982.

[1] New York: Farrar, Strauss, and Giroux, 1961.

[2] Assen: Royal Van Gorcum, 1962.

[3] By the present writer, *Eros et Logos, introduction à la phénoménologie de l'intériorité créatrice* (Louvain: Nauvelearts, 1967).

[4] For the conception of the "self-interpretation-in-existence" see my "The Creative Self and the Other in Man's Self-interpretation," *Analecta Husserliana*, vol. VI (1977).

[5] For the meaning of the expression 'subliminal spontaneities' see my "The Creative Self and the Other in Man's Self-interpretation," as well as my "The Initial Spontaneity," *Analecta Husserliana*, vol. V (1976).

[6] Man's "self-individualization" from within the circumambient conditions is proposed by me in several of my writings; see *Why Is There Something Rather Than Nothing*? (Assen, 1967) and "Man the Creator and His Triple Telos," in *Analecta Husserliana*, vol. IX (1980).

[7] Postulating as the foundation of philosophy such data of givenness that could — comparably to Husserl's postulate of "apodictic certainty" — carry with them absolute certainty and indubitability, Ortega proposes what he calls "the data of the Universe".

What are these data of the Universe? Initiating this search after absolutely certain givenness, Ortega naturally takes the footsteps of Descartes — the Cartesian doubt; yet — like Husserl — meets Descartes only half way and then goes his own way.

He writes: "Estas cosas cuya existencia es indubitable, que rechazan toda duda posible, que aniquilan y le quitan sentido, estas cosas a prueba de bomba crítica son los datos del Universo . . . los datos no son lo único que hay en el Universo, ni siquiera lo único que seguramente hay, sino que son lo único que indubitablemente hay, cuya existencia se funda en una seguridad especialísima, en una seguridad de cariz indubitable, diríamos en la archiseguridad". "Qué es Filosofía?" p. 150, 151, 152, *Revista de occidente* (Madrid: Obras Ineditas, 1958).

[8] Ibid., p. 156.

[9] Ibid., p. 165.

[10] Ibid., p. 158.

[11] Ibid., p. 162.

[12] Ibid., p. 165.

[13] As we will claim in what follows, for Ortega it is the human life which is this radical reality. "La vida humana es una realidad extraña de la cual lo primero que conviene decir es que es la realidad radical, en el sentido de que a ella tenemos a referir todas las demás, ya que todas las demás realidades, efectivas o presuntas, tienen de uno u otro modo que aparecer en ella". *Ideas y Creencias*, first published in 1934.

[14] "El sustituir a la fórmula de Descartes, 'pienso por tanto soy,' aquella 'pienso porque vivo,' no tiene la misma lógica interna de conexión intuitiva; no tiene la misma evidencia; no tiene la evidencia radical — archicertitud."

[15] *¿Qué es filosofia*? pp. 150—152.

[16] *Unas Lecciones de Metafísica*, p. 195; *Obras Completas*, vol. XII, 1983; *Revista de Occidente.*

[17] Ibid., p. 237.

[18] In his own analysis of life Ortega repudiates, on the one hand, abstract rationalism and, on the other hand, any assumption of "irrationalism" of life; he calls his own conception "ratio-vitalism", emphasising thta life has a reason of its own. Manuel Durán in his introduction to an edition of Ortega's "El Tema de Nuestro Tiempo" summarises perfectly the scope of this work, "El tema de nuestro tiempo exone sus ideas sobre la "razón vital", o su "racio-vitalismo". Ortega afirma en él que el racionalismo ha fracasado en parte porque se ha alejado de su intención inicial, que le dio origen, y que era servir de insrumento al desarrollo de la vida humana. Por otra parte, la pasión, el sentimiento y la intuición no pueden substituirlo, si bien expresan las necessidades vitales de modo más directo. Hay que intentar una síntesis de lo vital y lo racional, de la pasiones instintivas y la inteligencia lógica: ésta es la gran tarea que nos espera, éste es el tema de nuestro tiempo." p. 196, *Ortega y Gasset, sus mejores páginas*, ed. Manuel Durán (Englewood, N.J.: Prentice Hall, 1966).

[19] Julian Marias, *Ortega y la idea de la Razón Vital*, p. 47 (Impreso de Espagna, 1948).

[20] Ibid., p. 237.

[21] In his short essay summarising his thought, "Aurora de la razón historica" posthumously published in the appendix to vol. XII of *Obras Completas*, pp. 326—330, Ortega attributing his debt to Dilthey of the "descubrimiento de una nueva realidad: la vida humana" brings us into the heart of his view of man, writing, "Esa extraña realidad — la vida humana — no es una cosa física ni una cosa psíquica. En absoluto no es una cosa ni un modo, acto o estado de una cosa. Es un puro acontecimiento de carácter dramático . . . no somos sino eso que nos pasa. La más notable peculiaridad de esa realidad que consiste en acontecer está en que posee de suyo, por sí misma, estructura." This structure of life is: ". . . por lo pronto, individual, concreta, de cada vida. Pero hay, al mismo tiempo, una estructura formal de la vida . . .: el hombre al vivir *está* siempre en alguna creencia . . . se vive siempre *desde* ciertas creencias. Este se apoyan unas en otras, porque tienen también estructura y merced a ello puede hablarse de creencias básicas, tierra firme en que descansan todas las demás. Por lo mismo, la modificación más grave que puede experimentar la vida humana es un cambio de sus creencias básicas". (*Ibid.*, p. 327). Since undoubtedly the "creencias básicas" belong to the basic material from which the historical reason seeks to weave explanatory ties, the question emerges, Where do they stem from? Even more important for our understanding of the historical reason is the next question — What causes the human being to undergo such a "grave" modification as to change some of his basic beliefs? Should this question remain left without an answer that would be certain and indubitable, the foundations of the historical reason would be lacking some *crucial* factor.

[22] *Ortega y Gasset*, p. 46 (London: Bowes and Bowes, 1955).

[23] Cf. by the present writer, "Man the Creator and His Triple Telos", in *Analecta Husserliana*, vol. IX, 1979, and, "Poetica Nova, a Treatise in the Phenomenology of the Human Condition," Part I, *Analecta Husserliana*, vol. XII, 1981.

[24] Cf. by the present writer, "Spontaneity and the Translacing Continuity of Beingness", in *Analecta Husserliana*, vol. XIV, 1983.

[25] Julian Marias has formulated the role of the metaphor in a very succinct way which says the essentials, writing: "Las metáforas acumuladas nos llevan a la vivencia que Ortega quiere comunicar. Nos hacen repetir en nuestra propia vida los gestos vitales que lo llevaron originariamente a esa intuición. Una formulación conceptual directa

resultaría probablemente incomprensible, porque las fórmulas no se 'abririan' para irradiar lo que llevan dentro. Es menester que se produzca esa súbita descarga de intelección, que la metáfora desencadena, hasta llegar a articularse en un sistema metafórico." *Ortega, las Trayectorias*, p. 131 (Madrid: Alianza Universidad, 1983).

Stressing the role of the metaphor in cultural communication he pinpoints the fact that the metaphor operates upon "gestos vitales" from which it originated. These "gestos vitales", however, upon my analysis of man's self-individualization in existence as "gestos" are already bearers of significance.

26 Discussing critically the fecundity of Ortega's historical reason in accounting for the specific reality of the human life whose significance it is supposed to establish, Alfonso Lopez Quintas sees its limitation in the incapacity to invest with meaning realms other than empirical: realms which are not in function of the "historical time" (tiempo histórico) as the "religious world" or "poetical world." He writes: "La explicación historica es fecunda y realista por lo que tiene de fiel a las situaciones concretas . . . , pero no puede ser considerada en modo alguno como fácil sustituto del *pensamiento en profundidad.* Justamente, la sensibilidad historica se carga de valor cuando pone en forma la capacidad humana de intuir los campos de possibilidades que la actividas de los hombres — engarzados en comunidades de tradición — van abriendo y entretejiendo a lo largo de la Historia." (p. 403, *El pensamiento filosófico de Ortega y D'Ors*, Colección Universitaria de Bolsillo (Madrid: Punto Omega, Ediciones Guadarama, 1972). Although I do no see any specific difference in the task of the function of interpretation according to the differentiation of facts and events into empirical, aesthetic, spiritual, ethical or religious the way Lopez Quintas does, since in my conception of "man's life-unfolding" as his *self-interpretation-in existence* the very occurrence of any of this type of event means simultaneously a corresponding type of meaningfulness being established by him, I meet at this point the second criticism addressed to Ortega by Quintas, namely: "La condición histórica del hombre implica pues, su inserción de carácter creador en la labor común de fundación de ámbitos de posibilidades, labor sin la cual no hay historicidad verdadera aunque no se detenga el decurso temporal . . ." (*Ibid.*). As I have consistently emphasized in my various writings, without the creative function of man as the *prototype of a human act*, there would be no *inventive flexibility* of choice, hence no *innovation of types* and consequently *no progress in man's cultural development, that is, no history.* (cf. my study: "Imaginatio Creatrix", in *Analecta Husserliana*, vol. III, and previously: *Eros et Logos, introduction à la phénoménologie de l'expérience créatrice* (Louvain: Nauvelearts, 1967). Although I have been presenting my phenomenology of life step by step in various studies, I refer the reader principally to my above cited *Poetica Nova*, where I have outlined the conception of the Human Condition and its "creative forge" of the human "sense" of life.

27 Ortega insists that he did not reach the idea of life as the "fundamental reality" as the "pure event of man's struggle with his circumstances . . . ." "I was directed to it by the very problems that confronted philosophy." Indicating the path of his meditation he demonstrates an aversion to idealism but also his admiring approval of Edmund Husserl's methodological rigor. In spite of this admiration he admits to having found in it (after a serious study in 1912) the same errors, although on a "microscopic scale," which were committed by the previous representatives of idealism. Ortega begins his criticism by stating that the philosopher "sets out in search of an exemplary, primary,

and ultimately solid reality, to which he can refer and on which he can found all other realities." This reality, however, emphasizes Ortega, cannot be something which man "poses" himself; on the contrary, it must *impose itself on him*. Husserl, who claims to have found this reality in pure consciousness remains at the level of positing. In the first place, Ortega states that the role of pure consciousness is merely to be "aware of everything." The "pure I" does not want, desire, see, feel but only notice that it thinks, feels, wants, etc. It is positing objectivities of experience like a mirror reflects things. It is merely contemplative. Furthermore, Ortega denies the "self-positing" role of pure consciousness. According to him it does not posit itself directly, "forcing itself," so to speak upon us. On the contrary, it is the fruit of "manipulation," which we call "phenomenological reduction." Whereas the reality underlying it, says Ortega, is "primary consciousness," "unreflective" and "ingenuous," in which man really believes, what he thinks, wants and feels . . .". The essence of this "primary consciousness," then, is that nothing is for it only an object but rather that everything is reality ". . . an encounter of things with the world." *Prologo para Alemanes*, Obras Completas, vol. VIII. For a succint account of the relation of Ortega to Husserl and Heidegger see Julian Marias in his above-quoted: *Ortega, Las Trayectorias*, pp. 315—325.

[28] *La Idea de principio en Leibniz y la evolución de la teoría deductiva*, pp. 190—194, Buenos Aires, 1958.

[29] *Temas de Viaje*, vol. IV, in *Obras Completas*, II, p. 374.

[30] Julian Marias, "Un excorso de la experiencia de la vida", p. 131, in *Experiencia de la Vida* (Madrid: *Tribuna de la Revista de Occidente*, I, 1969).

[31] Ibid., p. 133.

[32] I have first introduced the conception of the "thrill of life" in my "The Praise of Life, Metaphysics of Life and of the Human Condition" — a treatise in a nutshell — in *Phenomenology Information Bulletin, a Review of Philosophical Ideas and Trends*, issue 6, 1982, Belmont, Mass.: *World Phenomenology Institute.*

[33] I propose the conception of "aesthetic enjoyment" as one of the three basic virtualities of "sense" in my study, "The Aesthetic Enjoyment and the Origin of the Poetic Sense," in *Analecta Husserliana*, vol. XVIII, 1984.

[34] I have presented the conception of the functional system, functional schemas, etc. of man's self-interpretation-in-existence in a sequence of studies. Cf "The Phenomenology of Man and of the Human Condition in Communication with the Human Sciences", in *Analecta Husserliana*, vol. XIV, and "Natural Spontaneity in the Translacing Continuity of Beingness", *Ibid.*

[35] I have introduced the conception of the "unity-of-everything-there-is alive" in the above-quoted "The Praise of Life", and in: "Person and the Human Significance of Life", in *Technological Powers and the Person*, St. Louis, Mo., 1984.

[36] My above-quoted "Poetica Nova".

[37] By the present writer, "Imaginatio Creatrix", in *Analecta Husserliana*, vol. III, 1976.

[38] The above-quoted, "Man the Creator and His Triple Telos".

[39] By the present writer, "What is Philosophy?", in *Phenomenology Information Bulletin*, issue 5, 1981, Belmont, Mass.: *World Phenomenology Institute.*

[40] For my conception of "relevance" and the origin of the social world, cf. my "Moral Sense in the Foundations of the Social World", in *Analecta Husserliana*, vol. XV, 1983. For the conception of the "source-experience", see the above-quoted "The Phenome-

nology of Man and of the Human Condition in Communication with the Human Sciences" and "Natural Spontaneity in the Translacing Continuity of Beingness".

[41] The present writer's "The Moral Sense in the Foundations of the Social World", *Analecta Husserliana*, vol. XV, 1983.

[42] *Crisis*, tr. D. Carr, p. 9; also F. Dastur, *Phenomenology Information Bulletin*, Vol. 7, 1983, pp. 19—26.

PART I

THE NETWORK AND TENTACLES OF INDIVIDUAL EXISTENCE

THADDEUS T'UI-CHIEH HANG (項退結)

MAN AS THE JUNCTION OF YIN-YANG RELATIONSHIPS AND COSMIC HEART

A Phenomenological Interpretation of Some Chinese Ancient Texts about Human Nature

Li-chi, the Book of Rites,[1] is one of the ancient sacred books of China. In Chapter 9, entitled Li-ÿun, we find the following: "Man is the cosmic virtue, the junction of yin-yang relationships and the encounter of spiritual beings" (Jen che T'ien-ti chih teh, yin-yang chih chiao, kui-shen chih hui); "Man is the cosmic heart" (Jen che t'ien-ti chih hsin). Such sayings about human nature may be taken as literary exaggeration, not worthy of serious consideration. However, in the light of the phenomenological (descriptive) concept of life, which I discussed some years ago and restated in a paper delivered during the Second Oriental Phenomenological Conference,[2] I find these statements meaningful, especially in view of the Congress's main topic, "Philosophy and Culture".

My intention in this paper is twofold. First, I want to show the profound meaning of Li-yün's texts, which are at least two thousand years old and which strongly underline the relationships of man with both the material and spiritual worlds. Most of these relationships are essential to every living being, even to every being. Second, I want to point out, that my concept of life is in tune with Chinese tradition.

I. A TENTATIVE CONCEPT OF LIFE

I will begin with a short precis of the paper mentioned. After a critical description of the salient traits of life as described in Chinese, Indian, and Western philosophies, I try to construct a descriptive concept of a living being, a concept which seems comprehensive enough to include the various phenomena of life: "A basic natural unit, which is essentially in close connection with other units, and which has the tendency to produce other units as well as to develop toward more complex and higher levels of unity."

Here, we must begin with an explanation on the three fundamental features of this concept.

A-T. Tymieniecka (ed.), Analecta Husserliana, Vol. XXI, 77–87.

(1) By "basic natural unit" I mean things like atoms, molecules, cells, plants, animals, men, etc. Such units exist in nature; they are neither artificial nor conglomerations like crystal. By stressing the living individual unit I reject such concepts as "universal life," "life as such," etc., as if life could subsist as a real entity outside living individual units. There are many levels of unity and individuality: from the highly developed self-conscious individuality of a human being to the lowest form of an atom. But in all these forms of individuality there exists a relative unity. As Bergson says, every living individual unit consists of different and mutually complementary parts and functions that act upon one another. It is separated and closed off by nature itself. But this separation and consequently its individuality is never perfect in our world. For instance, a plant constitutes an individuality that is separated and closed off; but this individuality can be split up *ad libitum* and at any moment. Such an imperfection does not contradict the fact that a living being has a relatively separated unity and individuality.

(2) Close connection with other individual units belongs also to the essence of every living individual unit. In this view nothing exists in complete isolation; everything is interrelated. This universal interrelation can be explained by two different pairs of terms: that of *yang* and *yin* of ancient Chinese philosophy, on the one hand, and that of "perception" and "appetition" of Leibniz on the other (Whitehead's notions of "prehension," "concrescence," and "nexus" are similar and helpful). As for *yang* and *yin*, the two terms refer not only to the relationship between male and female, but also express that between father and son, mother and daughter, master and pupil, etc. In other words, *yang* and *yin* denote every relationship between active and passive, between giving and receiving under any aspect. In fact, one thing can be actively giving and operating in one respect and passively receiving in some other respect at the same time. Since every individual unit is actively giving and passively receiving under innumerable aspects and on different levels, the universe is actually interwoven through countless *yang* and *yin* connections.

In spite of their shortcomings I would like to use the Leibnizian's terms "perception" and "appetition" to explain universal interrelation. In the Leibnizian view the monads, which are the smallest components of the universe, cannot be distinguished from each other except through their operations of perception and appetition; through these the monads are interrelated with the whole world. This view has the shortcoming that it makes monads "windowless." According to this approach, perception and appetition are nothing but inner modifications of monads without any outside influence. What I mean by perception and appetition is completely different. By "perception" I refer to the original meaning of the Latin word *percipere*, i.e., to take possession of, to seize, to collect, and to absorb. What is meant here is any form of taking in or absorbing outside things or undergoing outside influences at the physical, biological, or noetic levels. By using the Leibnizian term "appetition" I refer to the movement through which an individual unit is attracted by an outside center, or through which two or more individual units attract each other or are simply united in order to constitute a new center of action.

To sum up the two ways of universal interrelationships, the perceiving part plays the passive *yin* role by receiving outside elements or undergoing outside influences; on the other hand, in referring to "appetition," the attracting and the attracted parts play the roles of the active doing of *yang*, and the passive undergoing of *yin* respectively.

By taking or absorbing other things or influences an individual unit maintains and enriches its own individuality at different levels. And by being attracted by others it is pulled toward integration at a higher level of unity. For instance, at the material level an individual unit is attracted by universal gravitation; biologically, it is absorbing or assimilating material needed to maintain its individuality; noetically, it is assimilating knowledge; and socially, man is attracted to other human beings in order to form community.

(3) The tendency to produce other individual units and the tendency to develop toward higher and more complex levels of unity are complementary and condition each other. The living being arrives at a certain level and begins to produce more individual units on the same level; and this process is in turn a precondition for development toward higher levels of unity. For instance, the human embryo at the first moment of conception has only one cell. In this state it produces more cells incessantly by multiplying just like "mirror images." With a sufficient number of cells the embryo is able to develop onto more complex and higher levels of unity, until it finally reaches a high level in the form of human individuality with self-consciousness. At a certain point of development the human being as such begins to grow quantitatively. With a sufficient number of human individuals there is the tendency toward formation of human groups and this group formation

enables human beings to produce ever more complex forms of culture and community.

We must bear in mind, however, that the production of new individual units and their development toward higher forms of unity are necessarily accompanied by the death of already existing individual units or the extinction of forms of life on lower levels. For instance, animals can subsist only through the death of innumerable living plants.

Herewith I conclude my tentative suggestions toward a more comprehensive concept of life.

I am very happy to see that this concept of life gives me a fresh understanding of our ancient philosophical traditions, which professed consistently the fundamental unity of the universe. In asserting such a fundamental unity many philosophers believed that pantheism or an amorphous uniformity of nature were the only possible explanation of this phenomenon. My own approach is to subscribe to the fundamental unity of the universe while at the same time acknowledging essentially different levels of beings. With this position my concept of life is applicable to higher forms of individuality and relationships (even to God himself), while at the same time applicable to some extent also to things which ordinarily are considered lifeless. Evidently the expression "living being" is understood in two different though interrelated meanings, namely, one in a broader sense, and another in the ordinary sense. In a broader sense the above mentioned three features seen in their most primitive state could be applied also to basic natural units, which are ordinarily called lifeless. In an ordinary and stricter sense "living beings" refer to things above the biological level. With the distinction of essentially different levels I can avoid the danger of denying or minimizing differences between biologically living and nonliving things.

II. APPLICATION OF THIS CONCEPT TO HUMAN BEINGS AND ENSUING PHILOSOPHICAL PROBLEMS

I was urged to make such an attempt for two reasons. On the one hand, the most representative contemporary Chinese philosophers (like Hsiung Shih-li, Thomé Fang, and Tang Chün-yi) emphasize life as a fundamental concept of Chinese philosophy. In spite of my great respect for them I cannot fully agree with their positions. In my opinion, they rightly point to cosmic change, productivity, and dynamism as signs of life, but they do not single out with sufficient clarity the role of individual units. On the other hand, while teaching a course on philosophical anthropology I was more than once puzzled by and dissatisfied with the existing Western concepts of life. Obviously, I cannot explain in this paper how my tentative concept could be applied to the manifold phenomena of life. All I can and intend to do is to apply it to the human being. And I shall do it by discussing all three features simultaneously.

Today modern man is suffering from cultural expressions based on philosophical views of man that are often one-dimensional. Man is viewed predominantly on the biological, sentient, sexual, economic, or rational-organizational-technical level. Applied to the human being, our tentative concept of life shows us that a human being is much more complex than is ordinarily understood. In relationships with the whole universe, man is in essentially different levels. At the same time, man as a conscious, thinking power has a special role in the universe. Both realities are pointed out in the old sayings of Li-yün: "Man is the junction of *yin-yang* relationships," and "Man is the cosmic heart." Only such an integrated and truly unified view of a human being will do full justice to the complexity of human reality — otherwise more or less biased forms of culture will dominate the scene and humanity will suffer. One of the worst cultural forms is the one which considers all philosophical discussion as idle talk and which itself is imprisoned in a particular (for example, behavioristic-mechanistic) philosophical view of man. Since such a philosophical view operates unconsciously, it works like a neurotical complex; it is absolute and domineering. On the other hand, we must also discuss the problem of the metaphysical basis of our conception of man. In Chinese philosophy this problem is often related to these cosmological and ethical questions: Where do the universe and man come from? and in what is ethical norm grounded?

1. *The Nature of the Human Being*

Above all, man is an individual natural unit (artificial insemination or implantation are artificial insofar as man's coming into being is concerned, but that does not mean that the human being is therefore artificial), because as a biologically functioning, sentient, and self-conscious thinking being he is totally unified. Note the fact that a sentient and self-conscious man already belongs to two different levels.

On the vegetative levels, however, man is composed of multiple cells. Each of these cells could live its own autonomous life. Man becomes one unified body as innumberable cells are enhanced to a higher unity and so lose their own autonomy. Moreover, the cell is composed of many materials of which every molecule and atom constitutes its own unit. On this level man is constituted by an infinite number of different things. But all these do not retain their autonomy but belong to a higher biological level of unity; and the biological level, in turn, belongs to still higher forms, namely, to the sentient and noetic levels of unity.

At the physical level, i.e., as a full member of the physical world, man is in continual connection with all cosmic realities, at least through universal gravitation. In this respect he is literally attracting everything, though on a very small scale, and, vice versa, is attracted by everything. Man is also absorbing air, water, food, etc. He is really in constant *yin* and *yang* relationships with everything.

But in absorbing foreign substances man shifts toward a higher biological level. At this level the human cells are able to produce new cells. In so doing they are contributing toward ever higher forms of unity; (1) toward the unity of a particular member (for instance, the eye); (2) toward a higher unity of the whole human body; and (3) toward the still higher unity of human self-conscious thought. Yet to understand "human phenomena" properly, we should concentrate on the noetic level: at this level man is self-consciously absorbing every form of knowledge and is attracted to form societal unions with other self-conscious beings.

At all these levels man appears to shift incessantly from one level toward another: from physical to biological, from biological to sentient, from sentient to self-conscious, and vice versa, as well as from the *yin* position to the *yang* position. To sum up all these complex phenomena, we cannot find any more succinct and appropriate formulation than the old saying of Li-yün, that man is the "junction of *yin-yang* relationships." As man appears at the center at all levels, and as he is more properly human at the conscious and noetic level, he may rightly be called "cosmic heart."

Moreover, as is the case with every living being, man is not just an individual unit; he is destined to form still higher levels of unity. Therefore, individualism has no place in this conception of life and of man. On account of his full consciousness and responsibility, man is called a "person." A fully conscious personality belongs to genuine humanhood. But this does not mean that man is an absolute sovereign, as Nietzsche and in his footsteps Sartre seem to say. The fully conscious personality is only a condition *sine qua non* for a genuine "*don de soi*," as a French expression describes it so beautifully. Only with such a "*don de soi*," i.e., with a personal commitment, is a truly societal union possible.

Here we can return to one of the characteristics of a living being: arriving at a certain level the living being has the tendency to produce more individual units of the same level. Even when man produces other human beings at the biological level, he does so also through societal union at the noetic level, i.e., through marriage; and his relationship with the produced human beings is not one between man and things, but a special form that may be termed interhuman, or co-human, or, as Mencius calls it, *jen*. Mencius correctly points out that the consciously cultivated *jen* (*jen* means consciously cultivated affection between two human beings; the character 仁 is composed of two characters: two and man) constitutes what a human being essentially is, and these two things (human being and *jen*) together make up the Tao, i.e. the way a human being should go.[3] With these formulations he indicates that it is man's destiny to practice the *jen*: because only through *jen*-practice he is able to develop as a true human being.

Though this tendency toward societal union is fundamental for every human being, we should avoid going to the extreme, i.e., to collectivist thinking. A genuine societal union is only possible with genuine personal commitment; otherwise it would be a "herd" of animals. Contemporary history offers evidence that, without the expression of full personal commitment, a so-called centralized society does not function efficiently. Properly speaking, such "gregarious" groups are not worthy of the name of a genuine human society.

2. *Some Philosophical Problems*

We must now discuss some problems ensuing from the conception of human being just expounded, namely, the problem concerning the tendency to higher unity and the problem of ethical foundation.

We mentioned already three characteristics of every living being and consequently of every human being: (1) a basic natural unit; (2) a close connection with all other units in the *yin-yang* and perception-appetition relationships on different levels; and (3) a tendency to produce other units as well as to develop toward higher and more complex levels of unity. All of these features are descriptions of empirical facts that in principle can be verified empirically; nothing is postulated a priori. Applied to other levels of living beings, this descriptive approach seems flawless. But we are in trouble if we try to apply it, especially the third feature, to human beings. Man is endowed with self-consciousness and self-determination. Thereupon he is all too inclined to consider himself sovereign and completely independent, a superman in Nietzschean terminology. In fact, small children tend to behave like little emperors, unless you make them think otherwise. It is difficult to have a spoiled child understand that he must behave as a normal member of a family and of an even still larger society. In a certain sense, individualism is innate. It is therefore difficult to understand how we have to develop onto a still higher level of unity, i.e., to that of the interpersonal community.

Of course, everyone finds in himself an irresistible impulse for companionship and societal union with other human beings. But many contemporaries consider this impulse a "group-instinct" with which one is not obliged to comply. They may for instance at one time decide to sacrifice themselves for others, and on another occasion they do the contrary. Today they choose to love someone and tomorrow they decide to destroy him. There is no norm whatsoever and we are not obliged to give an account to anybody. We are completely sovereign and autonomous. That is what Sartre seems to teach us and what many practice in their lives.

There is another opinion that is also diametrically opposed to the human destiny toward ever higher levels of unity, namely, the belief in blind fate. In China Wang Ch'ung (27—100) was the first to develop explicitly this belief. Relying heavily on Taoism, he believed that everything in nature happens spontaneously and he denied categorically the existence of any purpose in the universe. For him blind fate was everything. Combined with the Buddhist belief in transmigration, this fatalistic thinking has influenced the Chinese mind.

Finally, we must turn to two modern doctrines that profess clearly the fundamental unity of the universe while denying the ultimate independency of human individuals. I refer to Marxism and scientism. These doctrines consider the human being just like any other object of nature: they make it completely subject to dialectical laws or to the laws of nature. As I have pointed out,[4] such conceptions are akin to traditional Taoism, which makes the Tao a self-rooted law and an original power of the universe. Taoism also teaches that everything originated from Tao and will return to the same root, so that ultimately human individuals do not count after all. They are destined to be extinguished; then everything returns to the amorphous unity of Tao. Needless to say, such a return to the primeval Tao with its *yin* and *yang* forces is at odds with the tendency of living beings which aspire to higher and differentiated levels of unity.

Fortunately, Mencius (371—289 B.C.), who is generally considered the second most important Confucian teacher after Confucius, took the *jen*, the interhuman affection, as essentially appertaining to human nature, so that any action against *jen* was judged to be subhuman and unworthy of a human being. In spite of this, a self-proclaimed superman could choose to be subhuman and antihuman, as the tyrannical insurgent Chang Hsien-chung (1606—46) showed so clearly. For such eventualities Chung-yung, the thirty-first chapter of the Li-chi, made a peremptory statement: "The ordinance of Heaven is called nature" (天命之謂性). Metaphysically established, this statement would inflict a blow on every theory of a sovereign superman: you must follow your human *jen* nature, because it is the ordinance of Heaven. Heaven appears in the final analysis the origin of human nature. But could Heaven also become the ultimate goal through which man's tendency to develop toward higher levels of unity could be fulfilled?

As I tried to show in a recent paper,[5] the ancient Chinese considered their fate (personal and political) and morality as completely dependent from the ordinance (ling) of Shang-ti, who was conceived as the supreme emperor of the universe. The T'ien (Heaven) of both Confucius and Mencius meant the same Shang-ti. In the Book of Odes, which was written long before Confucius, there is talk about the personal encounter of man with the supreme Deity.[6] In this context the *jen* relationship and societal union must be open to an interpersonal relationship between God and human beings. Thus the hypothesis that "man is the encounter of spiritual beings" (人者 . . . 鬼神之會) would have a very appropriate meaning. And if God is the ultimate goal with whom man should be united on the highest level of societal union, then the norm of morality is also the norm of life development. Moreover, if man is destined to be the partner in this intimate societal union with an immortal God, he must also take part in this immortality, as Teilhard de Chardin remarked: "*Pour s'aimer, il est essentiel de coexister.*"[7] I am fully aware of the fact that in a country where for more than sixty years the main preoccupation has been technical progress, and where religious thought is systematically discarded as an escape from the hard life of everyday, even to think in terms of these concepts may be anathema. Nonetheless, it is an undeniable fact that the Chinese of ancient times had religious insights which were very much in tune with our concept of life.

There is another Confucian interpretation of "Heaven" which Chung-yung spoke about. As we have seen, early Confucians such as Confucius himself and Mencius believed in Shang-ti. But since the period of the Warring States Taoism won more and more the head and heart of the intellectuals. Since the Confucians were not on a par with them in metaphysical speculation, their ideas were strongly influenced by Taoism. Hsün-tzu, the most influential Confucian thinker at that critical period, was typical in his attitude toward Taoistic metaphysics. This approach to metaphysics posited the Tao as the ultimate reality and thereby replaced God: the Tao was the universal law and at the same time the primordial cosmic force with its alternate powers of *yin* and *yang*. Hsün-tzu accepted this Taoistic metaphysical view and combined it with the Confucian will to action. He once observed that T'ien-Ti, i.e., the whole cosmos, is the origin of all productivity; but that it cannot distinguish the things produced, and that, accordingly, it is the privilege of the sages (Chün-tzu) to distinguish and to bring order to the cosmos.[8] In this connection man is said to be the "cosmic heart" in the sense of an exclusive thinking power in the cosmos; he is also said to be the "cosmic virtue" which can assist "T'ien-Ti" and bring order to it. It is noteworthy to see how this form of Confucianism is in tune with the practical-technological thinking of today.

But another kind of Confucian scholar like Tung Chung-shu (179—104 B.C.) was not ready to accept this view. Tung's Heaven and Earth (T'ien-Ti) were naturalistic forces like Tao, and at the same time intentionally operating; they were God. Of course, the God of Tung Chung-shu did not transcend the world but was one with Heaven and Earth. Nonetheless, his world-immanent God works intentionally.[9]

Later, in the Sung and Ming Dynasties (960—1279, 1368—1644) the so-called Neo-Confucians tended more and more toward assimilation of Taoistic metaphysics. But, at the same time, they allotted to Heaven and Earth a benevolent character, and recognized them as the sources of human *jen* nature and consequently of *jen* morality. However, Chu Hsi, the most representative of the Sung Neo-Confucians, did not recognize any conscious thinking of Heaven. Intellectually, he was more akin to Hsün-tzu than to Tung Chung-shu.

At any rate, for all Confucians who assimilated Taoistic metaphysics, human individuals are destined to return to their primeval root of the Tao in its powers of *yin* and *yang*. As stated above, such a conception is at odds with the tendency of human beings who aspire to higher levels of unity rather to an amorphous unity of the Tao and its powers of *yin* and *yang*.

National Chengchi University, Taipei

NOTES

[1] There are two versions of Li-chi, both compiled at the time of the West-Han Dynasty (206 B.C.—8 A.D.). The better known one comprises forty-nine chapters, which were written by Confucian scholars during the period of the Warring States (480—222 B.C.).

[2] The full title of the paper is "Toward a More Comprehensive Concept of Life," in *Analecta Husserliana*, vol. 17 (Dordrecht: D. Reidel, 1984), pp. 21—30.

[3] *The Works of Mencius*, ed. James Legge, bk. 7, pt. 2, Chap. 16.

[4] *Jen chih Che-hsüeh* (Philosophy of Man) (Taipei: Chung-yang-wen-wu-kung-ying-she, 1982), pp. 192—3.

[5] The full title of the paper is "Vom himmlischen Mandat zum Fatum — Aspekte der chinesischen Religiosität." The condensed form of this paper is published in *Philosophy of Religion: Proceedings of the 8th International Wittgenstein Symposium, 15th to 21th August 1983* (Vienna: Hölder-Pichler-Tempsky, 1984), pp. 202—4.

[6] *The Book of Poetry*, ed. James Legge, vol. 4, pt. 3, bk. 1, pp. 427—36.

[7] Pierre Teilhard de Chardin, *Le phénomène humain* (Paris: Editions du Seuil, 1955), pp. 299, 337.

[8] Hsün-tzu, *Hsün-tzu-chi-chieh* (Taipei: World Press, 1961), bks. 17, 19.

[9] Tung Chung-shu, *Ch'ün-ch'iu-fan-lu* (Taipei: He-lo Press, 1974), bks. 48—52, 69.

ALPHONSO LINGIS

MASTERY IN ETERNAL RECURRENCE

Every effort to find in the core of the human being something that could function, epistemologically, as the original identity-pole, the original unity term, presupposed by every act of comprehension, that is, every act of unifying the diversity of experience, or that could figure, metaphysically, as a substance, a self-subsisting entity, or that could be conceived, existentially, as a destiny, the unitary plot of the temporal phases of a life, is radically undermined by Nietzsche's supposition that identities, enduring things, things that are the same, are but logical constructs, correlates of the logical, anthropological, will to power. The ego, he says, is a grammatical fiction.[1] The logic of identities has only the value of a system of simplifications, of falsifications, which has functioned to enable this human species to survive. But — man is no argument.[2] The very concepts of identity, of essence, of underlying unity, of causal source could not then be used to conceive of the sovereign and master form of life Nietzsche's thought is wholly devoted to announcing.

Yet no doubt Nietzsche's work is concerned with defining a certain structure of life — the "most comprehensive soul" — that is capable of affirming what is in the universe, that is capable of being veridical. And that is sovereign — master — capable of commanding itself, an end unto itself, and a generating source of power. And that is utterly individuated, a singular individual, and singular in each of its states and moves.

This sovereign individuality, however, has to be conceived not with the logico-metaphysical concepts, but with Nietzsche's own vision of nature in aimless recurrence, and with Nietzsche's own notion of what recurrence makes the culture of man, his self-overcoming, possible. We will then see that the strange concept, or pseudoconcept, phantasm, of eternal recurrence also diagrams the inner structure of the oversoul of the overman.

A-T. Tymieniecka (ed.), Analecta Husserliana, Vol. XXI, 89—101.

I. THE DISCOVERY OF PURE NATURE

Nietzsche's thought is essentially marked by the conviction that nature is, now, about to be disclosed. The disintegration of the moral and aesthetic structure of our interpretation of nature is the negative condition for this discovery. The critique of the concept of causality will have to be extended to pain, whose causal explanation is the foundation of the moral vision of reality. Pain will no longer be interpretable as the effect of designatable acts, of the essentially guilty attachment to the terrestrial, the transitory, to what passes away because — Platonism had taught — it deserves to pass away.[3] This metaphysics has made of the universe a penal colony. And the critique of the concept of teleology requires that we concede that the universe "is neither perfect nor beautiful, nor noble, nor does it wish to become any of these things."[4] But also the concepts of law, purpose, order, arrangement, form, beauty, wisdom, and the models of organism or machine in which they fit,[5] are to be reduced; with them nature has been idealized — divinized. For us the discovery of nature will have to proceed through a demoralization, a de-aestheticization, a de-deification of nature.

Positively, nature is unveiled in the modern physics of force, its essence in the most scientific thought — the law of eternal recurrence. The conception of infinite force, a force that would bear against nothing, is self-contradictory: a universe of forces, however immense, is finite. Only the time in which modern science situates the universe is infinite. The universe is observably in a state of becoming. But if ever it had a goal, if it could ever reach an equilibrium, that state would have already been reached — since it has already had an infinite time to reach it. Consequently no such goal is possible. The universe, having played out all its cards, having gone through all its possible combinations, can only recommence.

This doctrine, already known, already woven into the structure of all modern scientific knowledge, is yet to arrive into our lives. Another doctrine — that of being, of the logic of identity, the ontology of subsistent terms — has proved necessary in order that this species survive. With a logic and ontology of being, goals, final states, ends, were conceivable in nature, and life, purposive action, was possible. The doctrine of eternal recurrence, formulating the dissolution of all forms of identity and essentiality, declares the fundamental logico-ontological axioms errors.

If Zarathustra is the teacher of the doctrine of eternal recurrence, he is not the teacher of its truth; all modern learning is that, and in fact the coming formulation of this doctrine out of the development of modern science will be but its recurrence — for this newest doctrine is also the oldest doctrine there is, found already taught wherever the first doctrines of ancient mankind were recorded. Zarathustra's mission is rather to teach how to live with this doctrine, how to bear this "heaviest burden," how to conceive not a new conceptual system, but a new form of life, a new nature, capable of enduring truth: "Let us think this thought in its most terrible form: existence as it is, without meaning or aim, yet recurring inevitably without any finale of nothingness: '*the eternal recurrence*.'" "Duration 'in vain,' without any end or aim, is the most paralyzing idea. . . ."[6]

On Zarathustra's mountain, the old soothsayer formulates the ancient understanding of the doctrine as the condemnation of all action: "All is empty, all is the same, all has been!" "All is the same, nothing is worth while, the world is without meaning, knowledge strangles." "Anybody coming to this height . . . would come in vain: caves he would find, and caves behind caves, hiding-places for those addicted to hiding, but no mines of happiness or treasure rooms or new gold veins of happiness . . . . But all is the same, nothing is worth while, no seeking avails, nor are there any blessed isles any more."[7] This paralyzing contemplation has its reflection in Zarathustra's own dark side. His shadow sees the repudiation of teleology as evacuation of reality and growing desert. "O eternal everywhere, O eternal nowhere, O eternal — in vain!"[8]

The thought of eternal recurrence could only come to crush every life in which any hope, any expectation, any unrequited demand put on reality and on time yet subsisted. Only a life already utterly fulfilled, "favorably disposed to itself," already sealed in its own unrestricted self-affirmation could bear the weight of this thought, could even come to "crave nothing more fervently than this ultimate eternal confirmation and seal."[9]

Such a life — such a joy — has once existed: in a culture prior to historical, scientific culture, in the ritual, tragic age. Nietzsche discovered it in the Dionysian epoch in Greek antiquity. Its culminating institution, the ritual, sacred theater, brews the orgiastic ecstasy in which each actor, each participant dreams up a satyr-identity and projects himself wholly into this appearance, knowing not only the joy of producing a hieratic form, a sublime identity, a sacred destiny, making it exist with his own visionary, Apollonian compulsions, his gestures, movements, utterances, his own flesh and blood — but leading it to perfection, to completion and death, with the sentiment of exalted force, force undiminished by this death, as powerful still, capable of bringing the mythical identity back to life again — the same or another form. Such is the specific nature of the actor's pleasure. Artist pleasure, creative pleasure, pleasure of feeling oneself an undetermined fund of force capable of producing an utterly significant identity-form, destined movements, heroic forms, and not afflicted when these phantasms are led to their sublime deaths on the brink of dawn, for feeling oneself undiminished by this death, knowing oneself capable of producing them anew, feeling a force of life capable of producing and destroying innumerable forms of life. Such pleasure is the "metaphysical consolation" the tragic ritual produces — consolation for our grief, finally for our mortality. For here the sense of the end of a life-form, of a destiny, is not wretchedness and desolation, for the creative joy centers the sentiment in the subterranean fund of life-force that is undiminished by that end. Only in living in that subterranean level is it possible to overcome the misery of sensing every life-form, every form of being — and one's own life — destined to be destroyed by the tide of universal Becoming. And this joy is true! This theatrical intoxication, this most superficial pleasure, that of the most superficial man, who is himself nobody, who represents everybody and anybody, in fact communicates with the ontological pleasure of the universe, whose force continually creates forms, forms of being, and is contented, exhausted in that creation, but does not will any of them to survive — "I believe the old Teutonic saying," Nietzsche wrote in his adolescence, "'all the gods must die.'" — a joy then not in the setting up of forms of being, but in the continual creation and destruction of forms of being. The Dionysian ritual is a way to participate in, to feel, that very joy. It is a religious, sacred, experience, enlightenment, for the joy it imparts has the ontological value of insight into the ultimate will of the eternal ontological process.

The tragic ritual has disappeared, the whole tragic culture has long since come to an end, supplanted by the historical, scientific, optimist, Socratic culture set up in order to suppress it. When now, after twenty-five hundred years, out of this rational culture itself the doctrine of eternal recurrence emerges, Nietzsche's task, the Zarathustran mission, can only be to find, now, in the midst of rational culture itself, the form of that ancient joy that might still be possible.

II. THE HISTORICAL SENSE

This joy is itself natural; it is the very depth of joy in nature felt in our nature when our nature is "'*naturalized*' in terms of a pure, newly discovered, newly redeemed nature."[10] Yet this discovery of our own — joyous — nature is itself a cultural task, task of the will to power, of the self-overcoming of man. But can there be a culture of man in a scientific epoch that lays nature bare beneath all moral and aesthetic finalities? For Nietzsche the problem of the possibility of a culture today is not that of whether the lessons of the creative epochs of culture can be recalled today, but whether the forces of the creative ages of culture could recur today, whether the very feelings, the very dreams, from which the cultural forms which stand issued as monuments of the great festive moments of humanity, could recur in the late-born, civilized, rationalized life of today, that is, in the philistine life alone produced by our civilization which only answers to the need for comfort and for security.

The relationship to culture — the "historical sense" — has then for Nietzsche a structure essentially different from that which it has for Hegel. Culture is not maintained in a comprehension that retains the meaning of the deeds that have come to pass and that are past, that retains them not by simple accumulation, but by comprehensive integration — a comprehension that is then systematic and dialectical, engendering the new forms of spirituality out of the integration of divergent and contradictory forms. For Nietzsche the monuments of spirituality have arisen not out of a dialectical fecundity of meaning, but out of forces that are not bound to meaning, not teleological — forces in a region beneath consciousness, Apollonian and Dionsysian compulsions, visionary or oneirotic forces, nonteleological dance movements.

The "most comprehensive soul," that "which can run and stray and roam farthest within itself; the most necessary soul, which out of sheer joy plunges itself into chance; the soul which, having being dives into becoming; the soul which *has* but *wants* to want and will; the soul which flees itself and catches up with itself in the widest circle; the wisest soul, which folly exhorts most sweetly; the soul which loves itself most, in which all things have their sweep and countersweep and ebb and flood —"[11] is one then to be produced by "something altogether new and strange in history," the distinctive virtue and disease of the present age — the "historical sense." The historical sense is the highest form of our spirit; it is our participationist ritual, our orgy and our form of ecstasy. It is not the work of the most comprehending soul, that which retains from all the deeds and misdeeds of history, from "the oldest, the newest, losses, hopes, conquests, and the victories of humanity" their meaning; it is that in which the very griefs and the hopes, the loves and the woundings — the feelings, the very forces — recur.[12] For Nietzsche to believe in the recurrence of time in the present, the representation of the past in the present, is to believe that all the possibilities, all the powers, that were once possible in humanity are still valid, and in force in each individual, and at each moment of history. The historical sense is a moment that wants eternity, an eternity in depth and not in prolongation, wills eternity by willing itself in all its depth, feeling itself living a life that has already been lived thousands of years ago, feeling there recur within itself all the possibilities of life that had ever been known and felt and guessed at and lived through. Feeling in the exorbitance of its feeling the possibility of dreaming once more as once they dreamed those white marble dreams of the Acropolis, dark voluptuous dreams of the Tantric temple of Konarak, the possibility of dancing once more as once men knew how to dance in the Orphic nights in antiquity, the possibility of dancing as the Shiva Nataraj himself dances on ice-covered Himalayan summits, his four arms blessing and cursing, beating the drum of creation and lighting the fires of destruction.

This state Nietzsche calls humanity; it is the possibility that in the individual not the abstract form of humanity, all the affects, all the intensive forces of all individuals recur. The Nietzschean absolute Mind is not contemplative and stabilized in its integral articulation. It is produced not by an assembling of the meanings of all human experiences in a coherence, but by the accumulation of all the feelings and forces of human prostrations and triumphs, a compacting of their contradictions. What is thus produced is a "godlike feeling," which is not blissful, but ecstatic and explosive, which has in the tensions accumulated and produced in itself by virtue of these contradictions a source of ever-excess energies. The action of the most godlike state of life is then not a labor of negativity, an appropriative and reappropriative enterprise, whose exteriorizations are destined for reinteriorization at the end of history. Rather the most divine form of life is equated with nature, with the sun, hub of nature, under whose compulsion Zarathustra's orbit turns. The explosive ecstasy of the most godlike feeling, "the happiness of a god full of power and love, full of tears and laughter," is "a happiness that, like the sun in the evening, continually bestows its inexhaustible riches, pouring them into the sea, feeling richest, as the sun does, only when even the poorest fisherman rows with golden oars!"[13]

III. THE COMPREHENSIVE ARTISTRY

It is then in this nature — demythologized, dedeified, newly discovered, newly redeemed — and in this culture, to which there is access through the distinctive virtue and power of our age, the historical sense, that Zarathustra calls not the higher men, but the overman, the child to come, "innocence and forgetting, a new beginning, a game, a self-propelled wheel, a first movement, a sacred Yes."[14] He is also designated by the name master.[15] Master, that is, one who gives order, one who ordains, one that first orders, and ordains, himself, one that can crown himself, can consecrate himself, that can stand over himself as his own heaven, his azure bell of blessing.

He is also designated by the name noble. The first characteristic of the nobles, the very meaning of the word, is truthful.[16] The nobles are those who are capable of living by the truth. Their vital structure is such that they can affirm all that is and becomes — not only their own ideas and ideals, but all things, all things that are and that become and that become nothing, honey and lees, drunken midnight, tombs and tomb-tears' comfort, gilded evening glow, woe, hell, hatred, disgrace, the cripple, world — "this world, oh you know it!"[17] They can "carry together into One what is fragment and riddle and dreadful accident."[18]

The condition for the possibility of such a most comprehensive soul "which can run and stray and roam farthest within itself . . . in which all things have their sweep and countersweep and ebb and flood"[19] is not the most inordinate dialectical cleverness, which could, like divine reason, abominable spider, weave all things together in rational connections, finding or inventing by ruse a goal and purpose for each thing within the whole: "A *little* reason, to be sure, a seed of wisdom scattered from star to star — this leaven is mixed in with all things: for folly's sake, wisdom is mixed in with all things."[20] But "Gods are shrouded by their beauty; thus you — O heaven above me, pure and deep! You abyss of light! — thus you conceal your stars."[21] "I," says Zarathustra, "am one who can bless and say Yes, if only you are about me, pure and light, you abyss of light; then I carry the blessings of my Yes into all abysses. I have become one who blesses and says Yes; and I fought long for that and was a fighter that I might one day get my hands free to bless. But this is my blessing: to stand over every single thing as its own heaven, as its round roof, its azure bell, and eternal security; and blessed is he who blesses thus .... Verily, it is a blessing and not a blasphemy when I teach: 'Over all things stand the heaven Accident, the heaven Innocence, the heaven Chance, the heaven Prankishness. ... O heaven over me, pure and high! That is what your purity is to me now, that there is no eternal spider or spider web of reason; that you are to me a dance floor for divine accidents, that you are to me a divine table for divine dice and dice players."[22]

Zarathustra teaches then a new way to "carry together into One what is fragment and riddle and dreadful accident."[23] The way of metaphysics, the way of comprehension, is to make the elements, unknowns, contingencies of the universe into one by composing into a totality, a system, their meanings. The way to affirm what is is to affirm its meaning. Metaphysical comprehension is that vast enterprise of exhibiting the finality of all the accidents or happenings, the system composed by all the fragments, the positivity composed by all the negativity, tears and death. "Death ... is the most terrible thing, and to keep and hold fast what is dead demands the greatest force of all," Hegel wrote. "But the life of the mind is not one that shuns death, and keeps clear of destruction; it endures death and in death maintains its being .... [M]ind is this power only by looking the negative in the face, and dwelling with it. This dwelling beside it is the magic power that converts the negative into being."[24]

Nietzsche for his part does not believe in this magic power of the dialectical mind to convert the pain, transience, and death of all things in the universe into something affirmative; this very magical conversion into meaning and ideality was in truth the systematic devaluation of transience and of all transitory things, the will's ill will against time and against all that passes away.[25] Nietzsche does not come to teach a more clever way still, to contrive meaning for all things, and thus to affirm them. He is seeking a mind that can affirm unconditionally, that can love, the fragmentary, the ephemeral, the fortuitous. This affirmation consists not in subordinating them, through meaning, through significance, to one another and to the whole; it consists in affirming the worth of what has no significance, in affirming as an end unto itself what does not serve another. What he has found is a new way to affirm all things through affirming their value. Here value and meaning are no longer identical; to affirm something in its value is not to grasp its referentiality. Value is no longer utilitarian or teleological; to affirm value absolutely is not to affirm that something is an end relatively, in view of something else, or in view of man; it is to affirm that it is an end unto itself, and not for us. "For all things have been baptized in the well of eternity and are beyond good and evil; and good and evil themselves are but intervening shadows and damp depressions and drifting clouds."[26]

Artist-metaphysics, then; for we can say that it is an aesthetic and not a comprehending mind that Nietzsche finds most comprehensive, open to all things, affirmative of all things. The distinction is made in Hegel. The Hegelian, metaphysical, understanding demands that all that is negative — the fragmentary, the false, the evil, the ugly, the moribund — be magically converted into being, and this magic, Hegel declares, the aesthetic mind cannot perform. "Beauty, powerless and helpless, hates understanding," he wrote, "because the latter exacts from it what it cannot perform."[27] Here Hegel takes art — classical art — to be metaphysical; it seeks the same thing as the metaphysical understanding. The essential of the artistry lay in the composition, where the brute contingencies of the human anatomy come to incarnate spirituality, the colors found on the skin come to visualize character, the shapes given by chance in nature come to form a significant event, where each fortuity of color, of density, of brute opacity becomes expressive.

But in the dark splendor of the Dionysian rituals Nietzsche discovered another art.

For us the category of collage gives a postclassical idea of artwork. The first thing it says is that it represents nothing and means nothing. It no longer constitutes significance as it composes materials; it sets forth, without transfiguration, the merely found, haphazard and disproportionate forms, indecisive and murky colors, materials already in an advanced state of decay, noises, that is, sounds that pass without anywhere even sustaining a tone. Here the contingent, the insignificant, the ephemeral, the broken, the random is not composed but just affirmed, set forth. Salvaged, framed, protected by glass and the dehumidified temperature-constant air of museums, exhibited in public places. Here the operation of artistry does not consist in giving it meaning, but in something else — giving it worth. When Breton called marvelous the chance encounter of an umbrella and a flatiron on an operating table, what delighted him was not the hitherto unnoticed relationship the encounter exhibited, but the chance of the encounter between the unrelated things. The given non-sense not given sense, but giving pleasure.

This would be the sort of artistry that the most comprehensive soul works. Its comprehensive artistry is Apollonian: it consists in enhancing, glorifying, stamping each moment of the universe with the seal of eternity, with the worth of eternity. Its comprehensive artistry is Dionysian: it is by its own dance, nonteleological, nondialectical, nondiscursive movement, movement that is at each moment its own end, that it joins all things and joins itself with all things: "O Zarathustra," the animals said, "to those who think as we do, all things themselves are dancing: they come and offer their hands and laugh and flee — and come back. . . . In every Now, being begins, round every Here rolls the sphere There. The center is everywhere. Bent is the path of eternity."[28]

The most noble, the most truthful soul would have to be itself through and through affirmative. For Nietzsche connects the inability to comprise and comprehend the universe and rancor. The soul vast as the universe would have to be able to affirm itself absolutely, affirm all that is fragment, riddle and dreadful accident in itself, affirm all its impulses, affirm all its impulsive nature, all that arises and dies away, discharges, in itself. It would have to be the soul that does not regret or feel remorse or despise any will in itself, or anything it has ever willed, that does not feel discontent with any present will or long for any will to come. Such a mind is no longer a portentous power of negativity, but utterly affirmative, wholly self-affirmation: "This ultimate, most joyous, most wantonly extravagant Yes to life represents not only the highest insight but also the deepest."[29]

IV. THE SIGN OF THE CIRCLE

This form of life is marked with the sign of the great circle, the *circulus vitiosus deus.*[30] It is not only a power of life that is able to endure the idea of eternal recurrence. The law of the circle is its ownmost thought, the thought that arises out of its own nature, the emblem of its own structure, of its own joy. For the perfect joy wills eternity, wills deep, deep eternity, eternity in depth and not in length, eternal recurrence and not eternal endurance:

Have you ever said Yes to a single joy? O my friends, then you said Yes too to *all* woe. All things are entangled, ensnared, enamored; if ever you wanted one thing twice, if ever you said, "You please me, happiness. Abide, moment!" then you wanted *all* back. All anew, all eternally, all entangled, ensnared, enamored — oh, then you *loved* the world. . . . All joy wants the eternity of all things, wants honey, wants lees, wants drunken midnight, wants tombs, wants tomb-tears' comfort, wants gilded evening glow. *What* does joy not want? It is thirstier, more cordial, hungrier, more terrible, more secret than all woe; it wants *itself*, it bites into itself, the ring's will strives in it; it wants love, it wants hatred, it is overrich, gives, throws away, begs that one might take it, thanks the taker, it would like to be hated; so rich is joy that it thirsts for woe, for hell, for hatred, for disgrace, for the cripple, for world — this world, oh, you know it![31]

Joy wills infinitely this moment, just as it is, a moment, already passing, wills its ceaseless passing, its ceaseless movement on the eternal cycles of time. Joy is a will for the now such that the now is not viewed as compensation for the moment past, a will that has no ill will against the past, that wills it still, without remorse or regret, that wills still all it has ever willed, a will that has no ill will against the future, that wills it already, without anticipation or apprehensiveness, that wills already all it can ever will, including its own expiration. Each will past recurs already in the will of the present moment, and each will to come, in coming, recurs, for it is willed already.

The Dionysian joy then is not only exultant, expansive force, but a power that holds itself together, a force by which a life holds itself together. This possession or appropriation of itself, by which it is a self, a singular individual, must not be conceived as a view upon the whole circle of its metamorphoses, an integral or integrating grasp of the diversity of moments through their common meaning, destiny, or law. Joy knows itself inscribed on the infinite cycles of Becoming, knows its infinite repercussion on all the cycles of the universe, but does not view this circle or comprehend it, in the sense that diverse events of physical nature are comprehended through their common formula, or in the sense that the moments of a life are comprehended through the understanding that understands the destiny they realize. The Dionysian soul will die, has already lived innumerable lives and died innumerable deaths, and each time its whole identity has died. Its destiny is the succession of identities, which are engendered and brought to their conclusion one after another, not by a force subsisting in itself, a depth ego, but by the entire force of the universe. The sole form in which all these identities, these personages, these destinies are held together is a recognition of the form: "I have lived this moment already, innumerable times!" This recognition does not base itself on resemblances, it is not a recognizing suddenly something in this life that I had seen before, elsewhere, and which is therefore not *the* same, but similar. The recognition of eternal recurrence could not have that form, for it recognizes that it is the same life, it recognizes identity, the identity of what has no apparent similarities. The recognition is not an induction from moments compared; it imposes itself as a phantasm, taking form in the core of the Zarathustran nature, diagram of its own sovereign will.

V. THE IMPERATIVE PHANTASM

In the servile life this phantasm would come as the most terrible, most goal-annihilating, hope-annihilating of all possible doctrines, radically excluding any possibility of progress or consolation or escape. But in the sovereign life, the one who would have become so well disposed to his life so as to crave nothing more ardently than this ultimate eternal confirmation and seal, this phantasm imposes upon each act of will, of life, an infinite determination of that will. One could henceforth will something only on condition that one could will it all the way, without any hesitation or scruples whatever, will it utterly without anything held back in anticipation of another moment, will it infinitely, will its infinite inscription on all the cycles of Becoming. Each of those moments now becomes an absolute end in itself, loved without any regret or remorse, without any memory of moments past it would terminate or requite, and without any anticipation of what hope or fulfillment might yet be possible, without any apprehensiveness about or apprehension of the imminent end to which it is destined.

Such a life would be, by this phantasm, delivered over to itself absolutely, and to its singularity, to all the contingencies, gratuity, chances and mischances that its birth and the birth of each of its moments have been. The essentially obsessive image of eternal recurrence functions to bind one to this life, to oblige one to one's own life.

And it singularizes that life absolutely. Each hazard, each enigma, each fragment of that life in its singularity, in this very succession and sequence, become destined.

The Pennsylvania State University

NOTES

[1] Friedrich Nietzsche, *Beyond Good and Evil*, trans. Walter Kaufmann (New York: Vintage, 1966), §17.
[2] Friedrich Nietzsche, *The Gay Science*, trans. Walter Kaufmann (New York: Vintage, 1974), §121.
[3] Friedrich Nietzsche, *Thus Spoke Zarathustra*, trans. Walter Kaufmann in *The Portable Nietzsche* (New York: Viking, 1976), I, §20.
[4] *The Gay Science*, §109.
[5] Ibid.
[6] Friedrich Nietzsche, *The Will to Power*, trans. Walter Kaufmann (New York: Random House, 1967), §55.
[7] *Thus Spoke Zarathustra*, II, §19; IV, §2.
[8] Ibid., IV, §9.
[9] *The Gay Science*, §341.
[10] *Ibid.*, §109.
[11] *Thus Spoke Zarathustra*, III, §12, 19.
[12] *The Gay Science*, §337.
[13] Ibid.
[14] *Thus Spoke Zarathustra*, I, §1.
[15] *Beyond Good and Evil*, §260.
[16] Friedrich Nietzsche, *On the Genealogy of Morals*, trans. Walter Kaufmann and R. J. Hollingdale (New York: Vintage, 1969), I, §5.
[17] *Thus Spoke Zarathustra*, IV, §19, 11.
[18] Ibid., II, §20.
[19] Ibid., III, §12, 19.
[20] Ibid., III, §4.
[21] Ibid.
[22] Ibid.
[23] Ibid., II, §20.
[24] G. W. F. Hegel, *Phenomenology of Spirit*, trans. A. V. Miller (Oxford: Oxford University Press, 1977), p. 19.
[25] *Thus Spoke Zarathustra*, II, §20.
[26] Ibid., III, §4.
[27] Hegel, *Phenomenology of Spirit*.
[28] *Thus Spoke Zarathustra*, III, §13, 2.
[29] Friedrich Nietzsche, *Ecce Homo*, trans. Walter Kaufmann (New York: Vintage, 1969), p. 272.
[30] *Beyond Good and Evil*, §56.
[31] *Thus Spoke Zarathustra*, IV, §19, 11.

ANGELA ALES BELLO

THE DEVELOPMENT OF THE SCIENCES IN RELATION TO HUMAN LIFE

Existence Irreducible to Scientific Vision

Also erste Epochè: 'Abstraktion' hinsichtlich aller Wissenschaften bezw. ihres 'Vorurteils' der Objektivität. Die von diesem Vorurteil geleitete Wissenschaft gehört selbst zur Konkretion meiner Lebensumwelt (*Europäische Wissenschaft*).

These words of Edmund Husserl, taken from a manuscript,[1] would seem to turn the customary image of his phenomenological enquiry upside down. Here the image I am referring to is that of systematic, rigorous and therefore 'objective' research with a view to establishing 'how things stand'. The affirmation just quoted may therefore be the cause of scandal, because it seems to cast doubt on the whole of Husserl's programme as stated in *Philosophie als strenge Wissenschaft*.

There is, however, a general remark I should like to make before going any further with the analysis of the passage with which we are here concerned. All the great thinkers have had a series of interpreters, they created philosophical schools and movements that tried more or less explicitly to "conserve" the basic ideas of the master, to 'keep them alive'. In fact, however, interpretations are valid only inasmuch as they are also "betrayals" and none, not even the most faithful, has ever succeeded in bringing out the "true" Plato, the "true' Descartes, the 'true' Hegel, and so on.

I am well aware of this, and it is not my intention to expound "what Husserl really said"; rather I here propose to give my own reading of his interpretation of the sciences, an interpretation that to me seems to be a convincing one because it enables us to understand the process that formed the sciences and brought them into being; it thus establishes their value in relation to human existence. We shall therefore have to use the instruments offered by Husserl's phenomenology to interpret the development of the sciences, or at least those instruments that *to me* seem to be present in it after a systematic study of his published works and a large part of the manuscripts in which he returns to the argument and analyzes it. It seems to me that it is precisely the manuscripts that place greater emphasis on certain aspects that are

A-T. Tymieniecka (ed.), Analecta Husserliana, Vol. XXI, 103–116.

already contained in the published works, but are not treated there in such an explicit manner.

Having explained this, we can now resume our analysis of the passage quoted above. What appears most obvious in it is the relationship — already proposed in the first part of *Crisis* — between the sciences and history, that is to say, the link between the elaboration of the sciences and their historical progress as peculiar formations of a certain civilization, namely European civilization or, as one might say in more general terms, Western civilization.

In parallel with the texts of *Crisis*, the manuscripts especially those written by the mature Husserl, bring out with *ever greater* clarity the need for subjecting the products of culture to a historical and gnoseological analysis. I place the emphasis on the words 'ever greater' because in actual fact one can note even in his earliest studies that his enquiry into the structure of the theoretical sciences, arithmetic and geometry for example, is always accompanied by an altogether particular attention that the author pays to their historical development (see the manuscript K I 28 of 1889/90 containing *Geschichtlicher Überblick über die Entwicklung der Geometrie* and *Geschichtlicher Überblick über die Philosophie der Arithmetik und Analysis*.

Undoubtedly, however, what Husserl was trying to do was to establish the theoretical statute of these sciences, especially after he had abandoned the psychological interpretation of 'number' that he had given in the first part of *Philosophie der Arithmetik*. Rather, his parallel studies of logic tended towards a systematic arrangement of theoretical knowledge by means of a *mathesis universalis*. The problem that arises here is therefore that of establishing the relationship between this theoretical project and the phenomenological analysis that Husserl was developing at that time. To what extent could the phenomenological method adopt the project of the *mathesis universalis*? The relationship between these two aspects is ambiguous and problematical, because in certain respects it drove phenomenological analysis towards the sciences, while in others the move was in the opposite direction, attributing to phenomenology a field of its own. I believe that this problem finds a solution in the text in which Husserl sustains that phenomenology is a *mathesis universalissima* situated beyond any and all scientific formations.[2] If phenomenology is a science, it is so in the etymological sense of the word *Wissenschaft*, i.e. as a search for an originary knowledge that represents the ultimate attempt of 'foundation' understood as legitimation of our human way of speaking of the world.

This seems to me to be the profound message that I have received from Husserl's phenomenological enquiry; in this way a connection is established between science, establishment of knowledge, and the transcendental dimension of consciousness. It is precisely the analysis of the manner in which Husserl understands the sciences, i.e. ever more explicitly as cultural formations, that contributes to clarifying the very sense of phenomenological analysis. Indeed, it leads to a revision of this method, a revision that, while maintaining the basic inspiration of Husserl's proposal, may develop certain aspects and indicate some new perspectives.

1. UTILIZATION OF THE INSTRUMENTS OFFERED BY HUSSERL'S PHENOMENOLOGY FOR INTERPRETING THE DEVELOPMENT OF THE SCIENCES

My own research into the significance of the sciences, carried out in my two books *Husserl e le scienze* (1980) and *L'oggettività come pregiudizio – Analisi di inediti husserliani sulla scienza* (1982), aimed at examining Husserl's point of view about the theoretical and applied sciences and to investigate the underlying structure of all scientific theorization. As regards the first aspect, there emerged an interesting panorama of the evolution of Husserl's interpretation of mathematics, of his evaluation of the natural sciences (*Naturwissenschaften*) and the human sciences or sciences of the spirit (*Geisteswissenschaften*), while the second enquiry brought out the fact that all the sciences were underlain by a double process of idealization and objectivation that constitutes categorization in the true sense of the term.

Idealization presents itself as a gnoseological fact that can be traced back to the variance-invariance relationship that Husserl discovered in the ambit of mathematical calculus, a relationship that is extended to the comprehension (with an invariant as the starting point) of the entire system of possible variations as the totality of concordant possibilities. This operation lies at the basis of the elaboration of the sciences: the variance-invariance pattern serves to understand how there came into being the idea of the world *in se*, an 'exact' world that has to be discovered. In fact, however, this was a case, at least 'at the beginning', of highlighting an identity in the life of experience, the life of the senses, that became idealized. In other words, through the idealization of every partially invariant structure (space-time, substantiality, etc.) there occurred a passage from a typical invariant to the systematic construction of a mathematized nature that was considered unconditionally valid irrespective of the presence of factual residues: relationships, links, conditions, objects, etc., that have to be assigned a place in the idealized tissue.

The expression 'at the beginning', which (together with all the other considerations made above) can be found in manuscript A IV 7,[3] enables us to understand that the gnoseological process has taken place in history and therefore reveals itself as a historical process. In fact, factual being became gradually replaced by the invariant, supposed to be the structure of factual being itself, and it is in this substitution that one has to seek the mechanism that underlies categorization, which — as I have already pointed out — is based on idealization and generates objectivation. And this is also the reason why in the quotation of the beginning of my essay, Husserl writes that the first epochè must consist of the reduction of the prejudice of objectivity that characterizes all the sciences.

There can be no doubt that objectivation also underlies the constitution of our day to day intersubjective world, but the objectivation underlying the formation of the sciences has the connotation of the gradual mathematization of the surrounding world that we see realized, above all, in physics; rather, the specificity of scientific operations leads increasingly to a process of formalization bound up with a particular language, namely the language elaborated with a view to a rigid and univocal utilization of certain terms.[4]

"*Also erste Epochè*": elimination of the prejudice of objectivity that characterizes the exact sciences and forms part of the *Konkretion* of my and our surrounding world, *ours* inasmuch as we are Europeans. One may therefore ask whether this type of objectivation is inevitable or represents merely a possible cultural fact. Husserl replies that the sciences are a possible fact, but not a necessary one, even though extraordinary. Historically this possibility was realized in Europe during the Renaissance and was the work of Galilei but, even though it could not be realized either earlier or in other civilizations, it immediately assumed an extraordinary universality: it became valid for all and at all times.[5]

That it is a possibility capable of being realized is brought out from the gnoseological point of view by the very fact of experience; indeed, this possibility is implicit in the *Stil der Erfahrungswelt*, characterized as it is by a to-and-fro motion between greater and lesser determination in open and indeterminate horizons and by a vague causality that is once again indeterminate. None of this causes difficulties in everyday life, where determination is attained through the permanence of what is typical in a large part of experience, so that what remains indeterminate is practically irrelevant and, as far as practical aspects are concerned, one thus obtains exact knowledge of things or, at least, the kind of knowledge that serves in life. But as soon as in history there came to the fore a theoretical interest and assumed the configuration of philosophical speculation, attention became concentrated on the 'invariant' aspect of experience, on the universal causal style of this world, on the universal structure of the vague and indeterminate knowledge of this world. Reality was thus split into what is subjective and relative and into what is invariant and universal: in other words, the *doxa* was separated from the *episteme* and ever since the latter has represented the idea of rational knowledge that pervades reality *in se*.

According to another manuscript,[6] in this way there takes shape "exact objectivity" which, although at first valid only from a philosophical point of view, was later to become the guiding idea of science as the outcome of a method introduced by the consistent practice of highlighting in each and every experience some exemplary aspect that could be taken as universal. In this way the quality of a thing taken as example becomes idealized and is then considered valid in an unconditional and general manner. This process of idealization oversteps even the limits of the finiteness of experience, thus producing the "infinity" of the world of experience.

As one then seeks to attain to a more thorough understanding of the question, one notes that the gnoseological problem becomes transformed into a historical one or, better, that the gnoseological *possibility* becomes a historical *possibility*. This possibility has been fully realized through the objectivation peculiar of geometry and, more generally, of pure mathematics, where manipulation is not concerned with *things* but with ideas, the end being achieved by means of a twofold process of fixing an idea and deriving one idea from another.

In physics, on the other hand, the process of idealization found its material in the *Ding-Erscheinungen* and the *Ding-Vorstellungen* and the mathematical process did not go beyond the space-time forms, so that in this case objectivation concerned itself with the world as the world of bodies, abstracting it from everything that is not corporeal.

The result of this process, taken as a whole, has become an "a priori" that is connected with the factual being of man, with his particular way of experiencing, thinking, handling and considering things. It is here that the double valence of the "a priori" begins to adumbrate itself, since it is not only an 'object' with respect to man — like any ideal in general — but also a *Gebilde* of man himself, its constructor.

2. WHAT IS BROUGHT OUT BY THIS INTERPRETATION: ONE HAS TO GO 'BEYOND SCIENCE'

"Das Erste ist das Problem der vorwissenschaftlichen Welt".[7] An important consequence derives from the interpretation of science that emerges from the texts here considered: one cannot limit oneself to the scientific point of view but, rather, once one has discovered the mechanism that determines it, it becomes essential to 'go beyond' it if one wants to understand what the prescientific world is like. And thus we have the second aphorism: "First there is the problem of the prescientific world".

If science is a historical cultural formation and does not have the configuration of an authentic instrument of knowledge, the desire to understand what the prescientific world is like makes it essential to examine aspects of life and history that have not been touched by science and we are thus driven to making rapid excursions into the world of primitive people or the so-called uncultured. But to some extent even the uncultured of our own days live within a world marked by science and technology and for this very reason an analysis of their way of being becomes all the more important, for it could provide indications regarding their attitude to things such as to enable us, having noted the difference in approach, to carry out an enquiry of the structural type into the life world.

Continuing the analysis begun by Husserl not only in the manuscripts here mentioned but also in others,[8] being cases in point, one could trace the life world back to its experiences and thus analyze which experiences are peculiar to cultured man and which are peculiar to primitive man. At this point we are faced with two possibilities, for we can either affirm one to be superior to the other, or we can simply describe them as different forms of life. The former possibility, in turn, breaks down into two opposed attitudes, for we can affirm the superiority of either contemporary civilization or of the life of primitive man. But perhaps, given the historical succession, that characterizes them, we could simply describe them without establishing any comparative judgments?

In one manuscript[9] Husserl tends to put them on one and the same level as different forms that can be noted from the experience point of view but which have not yet been submitted to phenomenological analysis. In fact, a true phenomenological analysis of their 'relationship' and their success on, the latter not only in a temporal sense but also in a qualitative sense, does not appear at all. Nevertheless, Husserl asks a number of questions: If science is concealment of life, does life manifest itself in all its fullness before science? And what, in the last resort, is the relationship between life and science?

One thing seems clear in classical phenomenology, namely that the analysis of life must not be limited to a 'natural' level: *vorwissenschaftlich* and *vortheoretisch* do not suggest that one should pinpoint a natural life understood in some ingenuous sense, but rather the intimate and profound roots of life itself. Both the scientific attitude and the natural attitude must be put into brackets and when this is done, it will become possible to grasp the structures of the life world as a prescientific world and thus to understand the deeper significance of the world that surrounds us day by day. This Husserlian project assumes the concrete form of a regressive enquiry, the so-called *regressive Frage* that has been raised by classical phenomenology; but finding an answer to this question remains the goal of our further research.

3. WHAT "GOING BEYOND SCIENCE" MEANS

It was Husserl himself who defined the dimension uncovered by him as originary, and it is at this point that we must gain further insight into Husserl's reflections and, indeed, remeditate them.

The 'deconstruction' of the scientific attitude (which could be defined as 'categorial') and of the natural attitude of everyday life implies a recovery of the 'precategorial' dimension, but what configuration must the latter assume? Is it a dimension? In what sense is it a dimension? Is it the recovery of a fundus or base (*Grund*) or, better, an *Abgrund* that would repropose the search for some unifying thing or quality that lies 'beyond' and thus for a new kind of metaphysics, new on account of being of a sign different from that of traditional metaphysics at least as regards the instruments used for arriving at this base? In this connection I am thinking of a number of developments that have been triggered by the position assumed by Heidegger.[10]

Categorial and precategorial, constructivity and originariness, etc., are but modalities. This statement, of course, calls for further clarification. And here it becomes evident that, even though the distinction between categorial and precategorial was originally proposed by Husserl himself, he does not make this distinction explicit in all its consequences and in some way even tends to conceal it. Even though it is precisely in his examination of the sciences that he highlights this distinction, his own phenomenological analysis still oscillates between the categorial and the precategorial.

In other words, the problem of the significance of the originary also leads one to enquire into the value of phenomenological research itself. As far as Husserl is concerned, indeed, both the history of science and the history of philosophy bear witness to this process of categorization. How can one exit from this process? Is it possible to make such an exit? Husserl proposed a methodological enquiry as a preliminary to a new philosophy that, by very virtue of this enquiry, would come into being in a rigorous manner. But even Husserl himself confused rigour with quest for the essential, which — even though it was no longer of metaphysical value — maintained the characteristics of conceptual rigidity peculiar of the earlier tradition.

On the other hand, Husserl's attempt to reach stable ground, to arrive at a structural moment, must be considered valid, for otherwise there would be no sense in the search itself. And yet one may ask oneself what this 'stability' consists of, i.e. how the analysis of the 'a priori' of the life world may be configured. It seems to me that Husserl suggests a very interesting way out in one of his unpublished manuscripts: indeed, in one manuscript,[11] which is dedicated to the variance-invariance relationship, he highlights the genesis of the invariant and therefore also the genesis of the *eidos* with real and possible perceptions as his starting point. In other words, he there describes the modes of variation of the *Wahrnehmung* and, after having analyzed them, passes on to the *eidetische Einstellung*, the *Wesensschau*, which represents the invariant moment. What is particularly interesting here is the fact that this passage is mediated by the possibility of transformation in action and the change therefore occurs along the road of practice. Before being *theoretical*, therefore, the change presents itself as *practice*: it is the transformation of the pregiven world that drives speculation to seek invariant moments in the variations. Transcendental reduction, rather than utilizing eidetic reduction within its own confines (as Husserl suggested in *Ideen*),[12] here actually unmasks eidetic reduction.

Another fundamental question comes to the fore as soon as we consider the genesis of this categorization process: Are idealization and objectivation perennial modes of organizing knowledge or are they the result of cultural transformations? On the basis of Husserl's own texts regarding the sciences it would seem that on the one hand idealization and objectivation are spontaneous and characteristic processes of a substantially human attitude, while on the other they are special modes of cultural elaboration that in Western civilization generated geometry and its application to nature by means of the mathematization of the *plena* and subsequently led to formalization. Even though in some respects there undoubtedly are profound differences between generalization, typification and formalization, they can yet be traced back to a single process of idealization[13] that underlies even the human sciences. As far as the latter are concerned, Husserl really tends to theorize an epistemological status different from that of the natural sciences, but one can note that the process of 'scientization' has been both massive and capillary ever since the various disciplines in the field of the human sciences began to organize themselves during the course of our century.

What does all this imply? Firstly, that both the categorial attitude and the precategorial attitude are nothing but characteristic modalities of particular cultural formations and, as regards an answer to the question about the significance of the originary, we can therefore postulate that this will not present itself as a metaphysical dimension but rather as a "mode" of life.

This is where one of the most conspicuous results of phenomenological analysis acquires its full significance, namely the fact that pinpointing experiences, together with the study of their modalities and components, makes it possible to understand the attitudes that are inherent in the very cultural formations that draw their origin from them. The tension that exists between intentionality and its fulfilment, the presence of both an iletic moment and a noetic one in experiences themselves, something that was already highlighted by Husserl, can be used for a further analysis of cultural formations. As regards the former, intuitive and adequate fulfilment are characteristic of a precategorial attitude, while in the latter case one notes that iletic connections predominate in some cultural expressions and noetic connections in others.[14]

These brief remarks, which cannot be further developed here, adumbrate the hypothesis of a comparative analysis of the various cultural products that would eventually show up the deeper underlying structures; moreover, since in this way one would be carrying out a systematic study of experiences, the analysis would also maintain the rigour and univocity of Husserl's programme. Such an enquiry, which at first sight seems to pinpoint a particular sector of research, is really characterized by a quite extraordinary radicality and, seeing that it cannot be identified either with the sciences or with philosophy, at least if these terms are understood in the traditional manner, cannot be called by any other name than "phenomenology".

4. FORMATION OF THE SCIENCES AND THEIR INCIDENCE IN HUMAN LIFE

A comparative analysis of the various civilizations and cultural expressions enables us to throw even better light on the significance of the sciences. In this way, indeed, it becomes possible to understand what happened in Western civilization as the very result of the coming into being of the sciences and the effect they have on human life at various levels, be they existential, cultural, social, or moral. Hitherto these levels have been examined only separately and always on the basis of enquiries that, to say no more, concerned themselves with these aspects without ever linking them with the profound transformation of the view of the world that only phenomenology has indicated and identified in its ultimate significance, namely the passage to a view of the world that is characterized by the categorial attitude and which makes the life of man of contemporary Western civilization seems transformed not only with respect to the life of other civilizations, but even with respect to his own past.

This implies that different levels have to be identified even within the historical perspective: there is a level of 'manifest' history, and there is also a level of "latent" history. If one does not penetrate to this latter level, one simply does not grasp the *Auslegung der Welt*, which *für uns alle ist*.[15]

The need for penetrating to this deeper level also transforms our view of human history itself. If an analysis of essences of the type that has been carried out by classical phenomenology enables us to identify a number of invariant aspects of humanity, one may well ask oneself whether, when all is said and done, this analysis does not itself remain within the limits of Western philosophical culture.

The objection we have just made immediately raises two questions, one of which concerns the relationship with interpretations of history of the historicist type, the other the teleological view of history as such and therefore also of the ethical problem.

(1) Abandonment of the essentialist point of view does not imply a lapse into the very historicism that Husserl deprecated; precisely because phenomenology is a rigorous science, it enables us to understand the different modalities with which history unfolds itself and by means of reduction, indeed, helps us to step out of the categorial perspective. This operation is possible because it takes place at a precategorial level and thus despoils itself of categorial incrustations to grasp the specific connotations of the various cultural expressions. Its insistence on understanding 'things themselves', taken as the programme and object of the entire research, should find effective realization in the description of the profound differences of cultural attitudes, thereby recognizing these differences, but without following in the footsteps of historicism by theorizing the impossibility of an observation point that would enable us to grasp them. This viewpoint remains 'structural' and 'philosophical', even though it does so in a manner that is rather different from the one offered by tradition; classical phenomenology is well aware of this, but at the same time declares that, inasmuch as it is *Erste Philosophie*, it remains within the "furrow" of philosophy. But it is precisely when one looks into the significance of this definition that one should become aware of the novelty of phenomenological research: even though it shares the motive of radicality with tradition, it no longer avails itself of the tools and instruments offered by tradition and proposes a new approach.

From the point of view of an analysis of essences one could well raise an objection to what has been said so far and wonder how cultural products with their multiplicity of expression are linked with what is irreducible in man as man. The most recent anthropological studies tend to underscore that, even in the forms considered particularly primitive, cultural expressions already appear in their full form, the "...

dì che nozze tribunali ed are . . .", as Ugo Foscolo puts it[16] in reproposing G. B. Vico's intuition,[17] becomes further and further removed as man proceeds on his road. One may therefore wonder whether it would not be possible to analyze, for example, sentiments, emotions and sensations taken as such, and whether this would not reveal itself to be an essential analysis that affects the essence of what is human? And in what way could laughter and weeping as human manifestations be subjected to an analysis that is not essential? But it is precisely an examination of the cultural products of the past and the present that brings out not only the 'continuity' of humanity but also its 'creativity' as A-T. Tymieniecka has very effectively underscored.[18] Consequently, even though one can stress that creativity is peculiar to man (and all the positions of the essentialist type, be they metaphysical or otherwise, are derived from this observation), this very creativity also leads one to a view of man that considers him to be "mutable" in different spaces and different times, which thus become "his" spaces and "his" times.

In overcoming the contraposition of essentialism and historicism, phenomenology thus helps us to understand that highlighting the invariant moment, considering it to be the 'true' moment of reality, leads to the philosophical positions that see this moment as the ultimate answer to their questions, but that at the same time there are also various and mutable modes of expression.

(2) Even the teleological view of history understood in an immanentist sense undergoes a profound transformation that also involves the ethical question.

The crisis of essentialism, a crisis that is to all intents and purpose characteristic of our times, is also a crisis of values understood in a metahistoric sense, of absolute value judgments, of a rationalist vision that sets itself ideal goals. One can therefore understand the need felt by hermeneutics (thinking of Ricoeur, for example) for recovering 'more than conceptual' dimensions, those of the metaphor being a case in point,[19] or even — notwithstanding the profound differences between them and the former — the presence of irrational tendencies. New spaces are now being sought, because the weight of a reduction to scientism or of a rigidly rational interpretation is making itself felt.

Can one say, then, that there are no longer any ideal goals? Does history have no meaning? Is Husserl's appeal in *Crisis* destined to disappear? In this case, once again, the difficult relationship between the things that seem 'irreducible' because they are essential and the things that change comes to the fore in all its drama.

From the ethical point of view, one certainly notes profoundly different customs, usages and convictions in various 'spaces' and in various times, but at the same time one finds anxiety for happiness, complete fulfilment and overcoming of pain and distress in all men. Every cultural form is in the last resort an attempt to placate these longings, to respond to demands that in certain civilizations are becoming increasingly pressing and pondered. And man responds by means of his creativity, shaping objects, transforming himself, adorning himself, raising temples, but always in different ways, following his phantasy, his instinct, his rationality, thereby creating the common world with its rules and its rites.

Rarely, however, his anxiety for searching becomes placated by a "finished" product. Here we have his ever-present quest for doing better, which can assume concrete form in trustful belief in the perpetuity of his descendence as a challenge to decay or in an incessant and dramatic personal search; which it is to be depends on the particular historical and cultural context.

Contemporary man of European civilization, by very reason of the fact that he has lived right through the illusion created by the sciences that an enlightened and positivist age was within reach, no longer believes in an automatic historical realization of his ideals. These ideals come into being as needs for the fulfilment of perspectives, tensions, but when the ideal is not realized, it gives way to sceptical and nihilist closures.

Although Husserl had indicated the full realization of humanity as an ideal goal and an "infinite" task, he nevertheless seemed to "hope" against "all hope" that mankind would yet reach this goal. All this now seems very problematical. Phenomenology helps us to understand the formation of ideal goals and ideal values as personal and collective needs for adequate fulfilment of our expectations and supplies the instruments for understanding our illusions and possibly also for maintaining our hopes.

"Who does not put his hope in the unhopeable will never find it," declared Heraclitus in oracular form; but in what way must we hope and what is the dimension of hope? Let us therefore understand why we are always driven back to a metahistorical vision of good and of happiness, to a realization that transcends our temporal existence:

inasmuch as we are phenomenologists we can describe how this comes about, inasmuch as we are men we can remain satisfied by this fulfilment.

(English version by Herbert Garrett)

NOTES

[1] MS. trans. A V 15, *Lebenswelt vor der Wissenschaft* (1930, 1932), p. 9.
[2] *Erste Philosophie* II, Husserliana, vol. VIII, p. 249.
[3] *Die physisch-naturale Struktur der Welt . . .* (1931).
[4] An interesting study in this sector has been carried out by D. A. Conci in *L' universo artificiale — Per un'epistemologia fenomenologica*, Rome 1979.
[5] MS. trans. B IV, *Zur "6" Meditation* (1933, 1934).
[6] MS. trans. K III 22, *Paragraph 8a — Arbeit an dem geplantem Paragraph vor Galilei.*
[7] MS. trans. B 10 XIII without title, 1930—1932.
[8] MS. trans. E III 7, *Der Mensch im Schicksal, Religion und Wissenschaft* (1934) and MS. trans. A VII 9, *Horizont* (1933).
[9] MS. trans. B 10 XIII.
[10] E. Nicoletti's enquiries are tending in this direction, especially his essay entitled "Per una fenomenologia ermeneutica," in *Aquinas*, 2—3, 1980, and his contribution "Religione e fenomenologia ermeneutica," in A. Molinaro (Editor), *Le metodologie della ricerca religiosa*, Rome 1983.
[11] MS. trans. A VII 13, *Vorgegebenheit-Wissenschaft* (1921, 1928, 1930).
[12] *Ideen zu einer Phänomenologie und phänomenologischen Philosophie*, Erstes Buch, Husserliana, Vol. III, par. 75.
[13] As regards the process of categorization between the identification of the pure *eidos* and the method of idealization peculiar to mathematical speculation, see *Ideen II*, par. 10, *Logische Untersuchungen VI*; for the differences within this process between generalization and formalization see *Ideen I*, par. 13, and *Erfahrung und Urteil*, par. 90.
[14] Thoroughly revised applications of the phenomenological method have been undertaken by D. A. Conci; in this connection see especially his essay entitled "Le ragioni degli altri. Idee per una metamorfosi antropologica," in *Il Contributo*, No. 2, 1979.
[15] MS. trans. A VII 5, "*In der Welt leben*" (1933).
[16] *I Sepolcri*, Verse 91.
[17] *La Scienza Nuova*, Naples, 1744.
[18] Among the numerous contributions that stand out on account of their originality one might mention: "Man the Creator and His Triple Telos," in *Analecta Husserliana*, Vol. IX, 1979, and "Natural Spontaneity in the Translacing Continuity of Beingness," in *Analecta Husserliana*, Vol. XIV, 1983.
[19] Cf. *La Mètaphor vive*, Paris 1975. But this was also brought out very clearly by P. Ricoeur's own interventions on the occasion of the congress dedicated to his thought and organized by the World Phenomenology Institute: *P. Ricoeur ou l'herméneutique de l'action*, Paris, 16—19 June 1983.

CHANG WING-CHEUK

THE PROBLEM OF THE AUTONOMY OF HUMAN EXISTENCE IN HEIDEGGER'S LATER PHILOSOPHY

Heidegger's most explicit discussion (*Erörterung*) of the relationship between human existence (*Dasein*) and Being after the "turn" (*Kehre*) perhaps lies in his interpretation, in *Identity and Difference*, of Parmenides' fragment 5: "to gar auto noein estin te kai einai."[1] This thesis of identity between human existence (as thinker) and Being, however, gives rise to a question: does this identity eliminate the autonomy of human existence? Does it reduce human existence to Being? It is the object of my paper to offer an answer to this question.

Taking into consideration Heidegger's position in his later writings, the appearance of this question is by no means accidental. First, in "On the Essence of Truth" Heidegger explicitly declares that "die ek-sistente Freiheit als Wesen der Wahrheit [ist] nicht eine Eigenschaft des Menschen . . . sondern der Mensch nur als Eigentum dieser Freiheit ek-sistiert."[2] Second, in "Letter on Humanism" Heidegger writes that "das Wesen der Ek-sistenz ist existenzial-ekstatisch aus dem Wesen der Wahrheit des Seins."[3] Similarly, in *Nietzsche II*, "Das Wesen der Menschen bestimmt sich aus dem Wesen (Verbal) der Wahrheit des Seins durch das Sein selbst."[4] Finally, in *What is Called Thinking*? Heidegger characterizes human thinking as "Geheiss."[5]

These announcements force us to ask: How can human existence maintain its autonomy in the face of Being? As pointed out by Ruprecht Pflaumer, "Die alte theologisch-metaphysische Frage nach der möglichen *Freiheit* des Menschen angesichts der allmächtiger *Wahrheit* des Seins (Gottes) drängt unabweislich hervor."[6]

To discover an answer to the above question, we may begin by clarifying the nature of the identity-relationship between human existence and Being. From the standpoint of the history of philosophy, we can see that during the period of German Idealism both Schelling and Hegel developed their respective philosophies of identity through their different interpretations of Parmenides' fragment 5. But insofar as the philosophies of identity of the German Idealists belong to the tradition of the metaphysics of subjectivity and hence are to be overcome (*überwunden*), Heidegger certainly does not understand the concept of

A-T. Tymieniecka (ed.), Analecta Husserliana, Vol. XXI, 117–124.

"identity" in the same way as German Idealism. Indeed, in *Identity and Difference* Heidegger points out that "identity as it is thought of in metaphysics is represented as a characteristic of Being."[7] Since the German Idealists are imprisoned in this metaphysically represented conception of identity, they cannot apprehend the essence of identity.

What, then, does the term "identity" mean for Heidegger? According to Heidegger, "identity" means "belonging together." It is important to note that "belonging together" is not to be understood in the sense of *nexus* or *connexio*. Rather, Heidegger points out that "the 'together' is now determined by the belonging."[8] Clearly, "identity" in the sense of belonging together is not an external relationship. As a belonging together, identity is internal to human existence as well as to Being. On the one hand, "A belonging to Being prevails within man, a belonging which listens to Being because it is appropriated to Being"; on the other hand, "Being is present and abides only as it concerns man through the claim it makes on him."[9]

From this clarification of the nature of identity as belonging together we can see that the thesis of the identity between human existence and Being implies two points. First, it shows that there is an inseparability between human existence and Being. Second, it thematizes the response-character of human existence to Being. Human existence is visualized "as the being who thinks, is open to Being, face to face with Being; thus man remains referred to Being and so answers to it."[10] However, this merely means that the human being as listener is essentially inseparable from Being; it does not mean that human existence can be totally absorbed into Being and hence lose its autonomy. Indeed, the "mutuality" of belonging definitely prevents human existence from being reduced to Being.

Moreover, in claiming that human existence "*is* essentially this relationship of responding to Being, and he is only this," Heidegger himself admits that "This 'only' does not mean a limitation, but rather an excess."[11] That is, in explicating human existence as correspondence to Being, Heidegger actually tries to reveal the distinctive feature of human existence in the face of Being. Clearly, understanding such a distinctive feature can enable us to determine the autonomy of human existence even in the face of Being. In short, the resposive ability forms the first major characteristic of the autonomy of human existence in the face of Being.

In fact, Heidegger not only thematized the essential inseparability of human existence from Being, he also emphasized the essential inseparability of Being from human existence. He explicitly delcares: "Being is present to man neither incidentally nor only on rare occasions. Being is present and abides *only* as it concerns man through the claim it makes on him."[12] For it is human existence, among all beings, "who *alone* lets Being arrive as presence."[13] This capability of "letting Being arrive as presence" constitutes the second major characteristic of the autonomy of human existence in the face of Being. To be sure, as has already been pointed out by Heidegger, this "does not at all mean that Being is posited first and only by man."[14] For human existence lets Being arrive as presence *only* when it is open to Being. Indeed, the response-character of human existence prohibits any possibility of subjectification of Being. Nevertheless, this prohibition does no harm to the autonomy of human existence in the face of Being. For the autonomy of human existence has no necessary connection with human existence's being as subject (*subjectum*). What the announcements we cited above are meant by Heidegger to deny is only the master position of human existence over Being. But the denial of the master position of human existence and that of the autonomy of human existence are fundamentally two different things. Certainly, these announcements also imply that the autonomy of human existence is basically granted by its appropriation (*Ereignis*). But this only means that human existence cannot by itself posit its own autonomy, and that the autonomy of human existence must be determined in cooperation with Being. For, according to Heidegger, the appropriation is nothing but the letting belong together.[15] Thus, it is incorrect to say that Heidegger sees human existence as a kind of puppet which is totally controlled by Being.

In fact, a brief discussion of the semantical behavior of the word "letting" will be helpful. The meaningfulness of an expression like "X lets B happen" presupposes that X has a kind of autonomy. In other words, despite the response-character of the word "letting," its active character cannot be overlooked. Accordingly, in letting Being arrive as presence, human existence clearly shows its own autonomy.

It is true that Heidegger also tells us that this letting-be is not possessed by human existence as a *property*.[16] However, this merely implies the following three things. First, the letting Being arrive as presence is not an ontical property of human existence, but rather constitutes its Being. Second, as far as the source of this capability of human existence is concerned, human existence only receives it from the appropriation but does not posit it at its own disposal. Finally, human existence can achieve its own autonomy only when it cooperates with Being. In other words, the letting Being arrive as presence is not an arbitrary activity of human existence. Clearly, this says nothing more than what we have already pointed out in the above discussion. Accordingly, they cannot be used as evidence to refute the thesis that human existence has its own autonomy in the face of Being.

In fact, Being itself differs from the letting Being arrive as presence. That is to say, Being in itself cannot replace this letting Being arrive as presence. It is only human existence, among other beings, that can take this role of letting Being arrive as presence. This irreplaceability of human existence's letting Being arrive as presence not only demonstrates human existence's autonomy in the face of Being, but also shows in what way Being needs human existence in the appropriation. Thus, the "needed" character of human existence once again guarantees its autonomy in the face of Being. That is to say, since the appropriation of Being without the appropriation of human existence would be impossible, the autonomy of human existence in the face of Being is undeniable.

In *Sein und Mensch im Denken Heideggers*, Pflaumer discusses the possibility of finding an answer to the question of the autonomy of human existence within Heidegger's later philosophy. However, Pflaumer seems to conclude that one cannot find a positive answer to this question within the ontological dimension allowed by Heidegger. Hence, in defending the autonomy of human existence, Pflaumer appeals to a quasi-ontological solution. The key point of his attempt is to define the autonomy of human existence in terms of "living body."[17] He explicitly declares that "auch der Leib des Menschen ist 'Haus des Seins.'"[18] For, according to him, "das Sprechen der Sprache kann er nur durch seinen aus der Sprache und für sie gebauten Leib vollbringen."[19] However, Pflaumer has to "ontologize" that which Heidegger originally takes to be ontical, or if we insist that our living body is essentially ontical, the kind of autonomy Pflaumer attributes to human existence is merely ontical.

Taking the first alternative, Pflarmer's solution goes beyond Heidegger's own position and hence becomes problematic from a Heideggerian standpoint. However, the second alternative is not only too weak, but also distorts Heidegger's original position. Heidegger himself clearly writes: "But we also say *too much* when we mean Being as the all-encompassing and thereby represent man only as a particular being among others (plants, animals) and put both into the relation."[20] Indeed, insofar as the relationship of identity between Being and human existence is concerned, "what is at issue here is certainly not the relationship of Being to 'ontical' man" — as it is aptly termed by Werner Marx.[21] Thus, instead of bringing in the problem whether one should broaden Heidegger's ontological position, we can provide a purely ontological solution by concretizing our above discussion of the identity between human existence and Being in terms of Heidegger's analysis of the nature of language.

It is important to note that when Heidegger declares that language speaks, he does not deny that man is the speaker, but rather asks, "How does man speak?"[22] In answering this question, Heidegger writes that "speaking, as the listening to language, lets Saying be said to it, this letting can obtain only in so far — and so near — as our own nature has been admitted and entered into Saying."[23] That is, man speaks only insofar as he, according to his Being, belongs to language and is capable of responding to it. Although human speaking is needed by language, language "is not merely of the making or at the command of our speech activity."[24] Accordingly, human speaking has its own autonomy in the face of language. Moreover, since speech (*Rede*) is an *existential* of human existence, here human speaking is something more than an ontical phenomenon. Indeed, speech belongs in an essential manner to man's appropriation. One can also clearly see that speaking shows the *ontological* autonomy of human existence in the face of Being in a concrete manner. To be sure, since human speaking is essentially listening, it echoes again the response-character of human existence in the face of Being. However, one should note that this role as listener merely prevents human existence from becoming a master over Being; it is not incompatible with its autonomy in the face of Being. For Heidegger, the autonomy of human existence in the face of Being lies exactly in its role as a listener to the saying of language.

Moreover, one should note that when Heidegger says that the ability to listen to Saying . . . lies in Appropriation," he does not want to deny the autonomy of human existence either.[25] Rather, he wants to emphasize that the autonomy of human existence is only shown in appropriation: "Appropriation grants to mortals their abode within their nature, so that they may be capable of being those who speak."[26]

This clearly shows that in what way human existence can gain its autonomy in the ontological sense. Accordingly, only when human existence speaks appropriately, can it demonstrate its autonomy in a proper manner. Certainly, this also tells us that human existence can never "produce" or "posit" its own autonomy; rather, its autonomy is granted by appropriation. However, this does not mean that the granting of autonomy to human existence is accidental to appropriation. According to Heidegger, "appropriation" means essentially that "man is delivered over to the ownership of Being and Being is appropriate to the essence of man."[27] Thus, if human existence had no autonomy in the ontological sense, appropriation would become impossible.

From the later Heidegger's influence on the development of ontological hermeneutics, we can discover that the relationship between the interpreter and the text is structurally similar to that between human existence and Being. Accordingly, the kind of autonomy human existence has in the face of Being is analogous to that of the interpreter facing the text. Following Heidegger, Gadamer thematizes the relationship between the interpreter and the text in terms of the dialogical model rather than by appeal to Hegelian dialectics. By means of a mapping we can also discover that such a dialogical relationship exists between human existence and Being. Indeed, as pointed out by Werner Marx, man's "essence must be defined in terms of the 'dialogue' with the nonhuman 'claim' and 'appeal' of Being, the dialogue which man *is* in his essence."[28] Accordingly, the answering role, as well as the necessity of *subtilitas applicandi*, helps us in an analogous manner to understand the nature of the corresponding autonomy possessed by human existence in the face of Being.

It is well known that Heidegger, in developing his fundamental ontology, adopts Dilthey's view of the hermeneutical circularity between individual existence and history (or tradition) as the model for explicating the relationship between human existence and Being. Clearly, as history or tradition is inseparable from individual existence, Being needs human existence. Indeed, this circularity does not mean that human existence is reducible to Being. Rather, since this circularity emphasizes the "interactive" character of the interplay between these two partners, it demonstrates the autonomy of human existence in the face of Being.

Thus, we can now conclude that when the later Heidegger stresses the responsive character of human existence, he is not attempting to eliminate the autonomy of human existence. He instead merely tries to prevent us from any attempt to reduce Being to human existence. Indeed, in characterizing human existence as a "listener," Heidegger is appropriataing rather than denying the autonomy of human existence in the face of Being. As is insightfully pointed out by Werner Marx, "all of these seemingly 'passive' characterizations of human existence should not blind us to the fact that man is allotted the role of co-creating creatively in the occurrence of Being. Man is thought by Heidegger as a necessary and creative *co-player* in the play of the world."[29]

Tunghai University

NOTES

1 Heidegger, *Identity and Difference*, trans. Joan Stambaugh (New York, 1969), pp. 27ff.
2 Heidegger, "Vom Wesen der Wahrheit," in *Wegmarken* (Frankfurt, 1978), p. 186.
3 Heidegger, "Brief über den Humanismus," in ibid., p. 330.
4 Heidegger, *Nietzsche II* (Pfullingen, 1961), p. 194.
5 Heidegger, *Was Heisst Denken*? (Tübingen, 1971), pp. 153ff.
6 Ruprecht Pflaumer, "Sein und Mensch im Denken Heidegger," *Philosophische Rundschau* 13 (1966): 178.
7 *Identity and Difference*, p. 28.
8 Ibid., p. 29.
9 Ibid., p. 31.
10 Ibid.
11 Ibid.
12 Ibid.
13 Ibid.
14 Ibid.
15 Cf. Ibid., p. 39.
16 Cf. "Vom Wesen der Wahrheit."
17 Cf. Pflaumer, "Sein und Mensch," pp. 229, 232—33.
18 Ibid., p. 232.
19 Ibid., p. 233.
20 Heidegger, *The Question of Being*, trans. William Kluback and Jean T. Wilde (New York, 1958), p. 75.
21 Werner Marx, *Heidegger and the Tradition*, trans. Theodore Kisiel and Murray Greene (Evanston, 1971), p. 213.
22 Heidegger, *Poetry, Language, Thought*, trans. Albert Hofstadter (New York, 1971), p. 198.
23 Heidegger, *On the Way to Language*, trans. Peter Hertz (New York, 1971), p. 124.

[24] Ibid., p. 125. The German original reads: "Die Sprache braucht das menschliche Sprechen und ist gleichwohl nicht das blosse Gemächte unserer Sprechtätigkeit" (*Unterwegs zur Sprache* [Pfullingen, 1979], p. 256).

[25] Ibid., p. 129.

[26] Ibid., p. 128.

[27] *Identity and Difference*, p. 36.

[28] *Heidegger and the Tradition*, p. 225.

[29] Ibid., p. 227.

PART II

NATURE, SPIRIT, SOUL

MARIO SANCIPRIANO

DE LA NATURE À L'ESPRIT PAR L'EXPÉRIENCE HUMAINE

1. L'HOMME ET LA NATURE

L'analyse du rapport de l'homme avec la nature peut commencer par la considération des actes internes qui réalisent tout d'abord l'auto-connaissance de l'homme. En fait, la perception immanente (*immanente Wahrnehmung*) de Husserl est la seule qui peut être vraiment adéquate à son objet. Par elle, la conscience réfléchit sur la *cogitatio* et la saisit intimement dans sa plénitude; tandis que, par la perception transcendante (*transzendente Wahrnehmung*) dirigée sur le monde, la conscience, bien qu'orientée à l'être originaire des choses, est soumise à la fuyante apparition d'événements qui lui sont étrangers.[1]

On peut dire que, par l'expérience immanente, la *cogitatio*, elle-même, devient un objet intentionnel et que la conscience trouve en elle-même l'accomplissement (*Erfüllung*) de sa propre activité; tandis que, par l'expérience transcendante, le monde (*Welt*) devient notre monde (*Umwelt*) par une sorte de *diffusion* de l'activité intentionnelle de la conscience, qui lui donne sa signification humaine. Ainsi les phénomènes de la nature aboutissent finalement dans le *forum internum* de la conscience humaine par une sorte de *concentration*, car ces phénomènes y trouvent leur centre. Certainement ce n'est pas la conscience pure, mais c'est sa vision intentionnelle qui se diffuse dans l'univers; et ce ne sont pas les phénomènes externes, mais leurs représentations qui se concentrent dans la conscience.[2]

En effet, pour la phénoménologie, la nature n'est pas le "tout autre", ni le totalement *inadéquat*, par rapport à l'esprit, car elle n'est pas tout à fait extérieure à la vie intellectuelle, comme si elle en était tout entière "*aliénée*", selon l'expression hégélienne, ou comme si elle était un "*prope nihil*", selon les théories les plus strictement idéalistes. Ainsi l'esprit n'est pas le "tout autre" par rapport à la nature, comme s'il était fermé en soi et pour soi: cela signifie qu'il n'est pas tout "intérieur" à lui-même. On peut le constater dans l'*action* de l'homme, qui envahit le domaine des choses. L'action, en effet, est déterminée par la *volonté*, qui est un objet primaire de la réflection; mais telle action aboutit enfin à un resultat qui agit en dehors de l'ego, dans le monde.

A-T. Tymieniecka (ed.), Analecta Husserliana, Vol. XXI, 127–134.

Donc la vie du *Kosmos* se “concentre” et se conforme (lat. *adaequatio*) dans l’homme en tant qu’il est un *microcosme*; et la vie de l’homme se “diffuse” dans l’univers qui, en tant que *macrocosme*, en reproduit à son tour l’image, quoique renversée qu’elle soit, par un changement de perspectives. La nature se donne à l’homme dans sa naïveté originaire (*natura naturans*); et l’homme en reçoit les ressources auxquelles il s’est disposé par la constitution de son être moral (*natura moralis*). C’est toujours l’homme qui réunit, par son expérience la plus complète, la nature à l’esprit. Il a son propre corps (*Leib*) enraciné dans la nature, il a sa forme spirituelle (*Seele*) en lui-même; il a enfin la nature (*Kosmos*) comme théâtre de ses actions et de sa vie.

Mais l’homme peut améliorer les conditions de la nature, au fur et à mesure qu’il se perfectionne lui-même, à partir de son auto-interprétation, comme l’a bien affirmé Mme Tymieniecka, dans la Leçon inaugurale du VIIme Congrès International de Phénoménologie (1978).[3] Il y a une *donation de sens* pour les événements de la nature, c’est-à-dire une action significative de la part de la conscience: c’est une action qui dérive, comme une conséquence immédiate, de la faculté qu’on a de faire un libre choix des choses, en tant qu’on fait un libre choix de soi-même. Sens et non-sens de la vie personnelle se réfléchissent dans la vie de la nature, dont le déterminisme est mitigé par l’action de l’homme.

2. LA VIE SPIRITUELLE

Dans l’activité de l’ âme, on reconnaît éminemment les liaisons qui sont possibles entre la nature et l’esprit. On peut commencer à resoudre ce dualisme, par l’analyse “intentionnelle” de celle des “facultés” spirituelles qui est la plus proche des impressions de la nature: c’est-à-dire l’*imagination.* En effet, ce sont les phénomènes de la nature qui s’impriment dans l’*imagination reproductrice*, et dans la mémoire imaginative (*imaginatio*); mais c’est la conscience créatrice qui s’épanouit dans l’*imagination novatrice* (*phantasia*).[4]

À ce propos, il n’y a pas toujours une opposition entre le *monde*, qui est reproduit dans la perception, et l’*imagination novatrice*, qui produit des formes nouvelles et qui s’épanouit dans la vie de l’esprit. Il y a une ligne de continuité et aussi de transcendance, de la nature à l’esprit, à partir des données naturelles pour arriver à leur transfiguration créatrice, c’est-à-dire aux formes “croyables” de la vie spirituelle; mais il y a aussi une réalité "incroyable", dans la nature et la vie, qui dépassent toute prévision et toute imagination de l'homme.

Par la voie des sens, la nature s'offre aux impressions et à la mémoire: l'esprit n'est pas l'anti-nature et l'imagination n'est pas toujours la "folle du logis", car elle a aussi le don, tout raisonnable, de la composition libre et créatrice des données immédiates de la conscience, dans toutes les formes — contemplatives, actives et mystiques — de la vie spirituelle.

Outre cela, les *sentiments* peuvent offrir plusieurs liaisons, infiniment nuancées, entre l'esprit et la nature. Il est vrai que les émotions se réalisent parfois par l'impulsion de forces débordantes de la nature qui, pour ainsi dire, pénètrent dans l'âme (*commotiones animi*); mais les *passions de l'âme* peuvent s'accorder avec des fins raisonnables. Si cela n'est pas toujours possible, s'il s'agit de passions indomptables, celles-ci peuvent toutefois s'équilibrer entr'elles-mêmes.[5]

De la philosophie grecque, nous avons appris que l'univers est réglé par les forces de l'*union* et de la *séparation*, c'est-à-dire de l'*amour* et de la *haine* (Empédocle).[6] Mais seulement l'homme en a la conscience: en lui la *force cosmique* de l'amour se transforme, pour devenir une *énergie morale*. En lui se "concentre" l'amour qui est "diffusé" dans l'universe (Aurobindo). Le monde extérieur, par le sentiment, se fait interne et "adéquat", dans un certain sens, à mesure qu'il partage les dimensions humaines. Avec sa force de pénétration, la nature est très *proche* de la vie sentimentale. Pour cela elle est plus immédiatement connue par les émotions de l'homme que par sa vie rationnelle.

Du naturalisme de la Renaissance nous avons appris que la chaleur vitale contribue à la formation des "passions" dites "chaudes", telles que l'amour et la colère, et que le froid est une propriété de la haine: "*Amor ex calore . . . odium ex frigidis est* (J. L. Vivès). Est-ce là un symbole, une métaphore, une expression figurative? Je le crois bien, mais la chaleur physique devient un ingrédient naturel des émotions et donc une qualité expansive, pour l'homme qui l'utilise et qui enfin la dépasse dans l'"émergence" de sa vie spirituelle.[7]

De plus, on peut faire encore allusion à la *volonté* et à son action, qui dépasse l'intimité de la conscience et se manifeste en dehors. Les impulsions de la nature instinctive donnent bien des motivations émotionnelles aux actes qui seront décidés enfin par la *volonté*, c'est-à-dire par une fonction qui peut resulter inaltérée et parfois indifférente, au regard de ces motivations mêmes. En fait la volonté, réduite à ses actes du *propos* et du *choix*, dispose de chances très considérables pour diriger les instincts de la vie vers des buts préférables, orientés aux valeurs humaines.

A propos de la volonté pure (inaltérée) et des "passions de l'âme", on pourrait se demander encore s'il y a vraiment dans l'homme une énergie spirituelle qui émerge de la force de la nature, comme on l'a vu dans la *Consciousness-Force* d'Aurobindo, que nous avons déjà discutée (note 2), ou plutôt s'il est possible une initiative tout à fait originaire, dans le cadre de la raison pure. Il faut reconnaître que, bien au delà de l'idée de force, qui est contingente, l'*eidos-ego* atteint nécessairement sa constitution dans la forme de la liberté.

Néanmoins, sans perdre sa sobriété habituelle, l'esprit connaît bien les heures de l'enthousiasme qui entraînent la raison dans une sorte d'énivrement ou de *μανία* (Platon).[8] Mais, dans le thème des passions humaines, l'intelligence nous montre, par l'*ἐποχή*, comme l'on peut vivre *dans* la nature sans être *de* la nature et comme l'on peut se regler par la *ratio sobria* et la rendre toujours dominante, même lorsqu'elle agit dans l'*eros*, par la liberté créatrice.

Certainement on trouve aussi, dans le "monde de la vie", la déception et la revolte totale contre les téléologies qui dirigent toute action humaine. C'est là que se pose le problème du mal, c'est-à-dire de l'*imperfection* du monde et de la *corruption* humaine, opposées soit à l' *eros* soit à la *raison* de l'homme. C'est un thème dont le développement peut se rendre utile à ébaucher une "théodicée" phénoménologique. En effet, la phénoménologie, par sa continuelle réference au sujet transcendental, semble se poser exclusivement au dessus de toute imperfection anthropologique, jusqu'à exciter certaines réactions légitimes de la part de la *psychanalyse* et de l'*existentialisme.* On arrive, par l'"instinct de la mort" (Freud), jusqu'aux racines les plus profondes de notre constitution psychologique et par là aux ressources inconscientes de notre présence dans le monde de la vie. On pourrait demander, comme on l'a demandé en effet, si Freud a atteint la dernière profondeur de la conscience ou bien si, tout court, il l'a même dépassée (Lacan). L'*Inconscient* n'est pas dans la profondeur de la conscience: c'est au delà de la conscience même et l'on y arrive, pour ainsi dire, par une transcendance négative.

Mais, bien que l'instinct de la déstruction soit enraciné en nous et bien que le vide interne à l'*Inconscient* soit immense, le germe de la corruption n'atteint pas le sujet, d'où l'intelligence et la volonté puisent leur vie la plus pure. Si cela n'était pas vrai, on ne pourrait pas juger de la corruption même: on trouverait normales la négation de l'être aussi bien que son affirmation et l'on ne pourrait jamais projeter une *destructio destructionis*. Bien que plongés dans les contrastes de l'existence, nous ne sommes pas constitués dans l'absurde. En reconnaissant la contradiction dans la nature "aliénée," nous tâchons d'en sortir et de remonter, plus haut que la contradiction même, à la source, la plus originaire, de notre vie spirituelle. Si l'homme n'a pas encore réussi, à cause de son avidité, à détruire la nature, de son côté la nature n'a pas encore pu, par ses mécanismes, enchaîner la liberté de l'homme.

3. L'ÊTRE POUR LA VIE

La question de l'unité de l'*homme* et de la *nature*, en tant qu'il est possible d'envisager l'intégration (ou même la destruction réciproque) de leur existence, nous conduit à la découverte de l'*être de la vie*, à partir, encore une fois, de la vie de l'homme: c'est-à-dire à la découverte de l'être du sujet (*Subjektsein*).[9] Certainement il ne s'agit pas du seul sujet psychologique (le *je*), mais surtout du sujet *constituant*, ou même *créateur* (le *moi*), qui est absolu en lui-même, avec le *cogito* et toutes ses implications spirituelles, qui lui sont vraiment *adéquates*. Enfin l'unité totale de l'*ego* présume une *unité d'être*, c'est-à-dire une *identité* universelle et non seulement une *unification* particulière des fonctions humaines.[10]

De cette façon, le sujet existant — le même et identique sujet, le "moi fondamental" — est *en-soi* dans le monde, avec sa participation à la vie; mais il est *pour-soi*, lorsqu'il pense la vie. Cette distinction renvoie à une synthèse transcendante, c'est-à-dire à l'unité de l'*esprit*, en tant qu'il est une chose complète en lui même. Ainsi le monde de la vie, dans la multiplicité de ses données, n'est pas simplement le résultat des intentions de la conscience: il est réellement un objet (*vorgegebene Lebenswelt*) ayant sa propre continuité, qui s'entrelace avec celle de la vie humaine. On se rencontre avec la nature, telle qu'elle est, faisant part de son élan vital et on la dépasse dans le moment qu'on la pense. La nature *pensée* n'est pas la nature vécue, car l'*εἶδος* y engage une participation intellectuelle, tandis que le *βάθος* de la nature, sa vie profonde, a toujours quelque chose de sauvage.

Cependant, si nous allons à la recherche de notre vie originelle dans la multiplicité de ses formes, nous pouvons bien la trouver dans notre intégration, bien qu'*inadéquate*, avec la nature originaire; mais si nous allons à la recherche de l'*unité d'être* de la vie même, nous devons partir de l'unité absolue que l'homme reconnaît dans son être, car la nature même y trouve son centre: "*In interiore homine habitat veritas*" (St. Augustin).[11]

S'il m'est permis d'employer un néologisme, dans un langage qui, tout en se conformant à la philosophie de la vie, la dépasse par ses intentions métaphysiques, je voudrais dire qu'il y a une *tension primaire* du vivant vers la vie, car le vivant suit l'instinct de sa conservation, qui combat l'instinct de la mort, avec l'intention, pour l'homme, de le vaincre, jusqu'à l'outre-tombe. A cette tension primaire, nous devons nos préférences, lorsqu'elle s'accorde avec les téléologies rationnelles de la conscience. Dans la nature, il y a donc une tension du vivant pour la vie, selon la *durée réelle*; mais, dans le domaine de l'esprit, on reconnaît une tendance de la vie pour la vie, dans l'*identité de l'être.* Or si la durée réelle suffit aux phénomènes de la nature, entre les termes de la naissance et de la mort, l'esprit de l'homme repose sur l'*identité* de l'être (qui est eternelle), car l'*identité* appartient à l'essence de l'esprit individuel, ainsi qu'à l'essence de l'être.[12]

University of Siena

NOTES

[1] Selon Husserl, la perception du monde est toujours inadéquate (*inadäquate Wahrnehmung*), parce qu'elle est contingente (E. Husserl, *Erste Philosophie*, II, Vorles. 33). Ainsi nous trouvons l'opposition entre: 1) la *perception transcendante*, qui remonte même à l'être originaire des choses, mais par *Abschattungen*, et qui est donc relative, et 2) la *perception immanente* (au sujet), qui seule est absolue: "*Das einzige absolute Sein ist aber Subjektsein als für sich selbst ursprünglich Konstituiertsein*". (E. Husserl, *Erste Philosophie*, II, Vorles. 54; *ibidem*, Beilage XXXI. Du même auteur: *Logische Untersuchungen*, V, 1, §2; *Ideen I*, §38 et §45; *Analysen zur passiven Synthesis*, "Husserliana, XI, p. 16. Cf. aussi: S. Strasser, *Das Problem der Leiblichkeit in der phänomenologischen Bewegung*, Analecta Husserliana, vol. XVI, 1983, p. 19 (volume consacré au thème "Soul and Body in Husserlian Phenomenology — Man and Nature"). Dans ce volume cf. P. Trotignon, *L'oeil de la chair* (notamment p. 51). A propos de la traduction de "*Wahrnehmung*" (*transzendente Wahrnehmung* = perception transcendante; *immanente Wahrnehmung* = perception immanente), cf. E. Husserl, *Idées directrices pour une phénoménologie*, traduction par P. Ricoeur, Gallimard, 1950, p. 121) Pour la "perception" (sens correspondant à *Wahrnehmung*) "interne" et "extérieure", cf. A. Lalande, *Vocabulaire technique et critique de la Philosophie*, P.U.F., Paris, 6me édit. 1951, s.v. *Perception*, c, d.

[2] Cf. E. Husserl, *Ideen II*, §22: "Das reine Ich als Ichpol"; *Ideen III*, Beilage I, §4, a: "Das Leib als Localisationsfeld der Empfindungen". On peut reconnaître l'unité et le centre: "In the stream the various experiences of the ego are not merely unified in a serial progression; they are also *centralised* into an ego pole, as Husserl realised in his later *Universal Teleology*" (E. M. Barker, *The Ego-Body Subject and the Stream of Experience in Husserl*, Analecta Husserliana, vol. XVI, cité, p. 183). On peut comparer cette *centralisation* avec les conceptions de la *diffusion* et de la *concentration* de la conscience dans la philosophie indienne. Mais la phénoménologie opère une réduction, par l'*ἐποχή*, qui la sépare de la conception théosophique du monde. Selon Aurobindo, la conscience absolue se diffuse dans l'univers (*Selfdiffusion*) et se concentre en elle-même (*Selfconcentration*). Mais il suppose une *Consciousness-Force* et non une *conscience pure*: pour lui, la matière et l'esprit ne sont que deux degrés ou deux expressions de la même *force* consciente. Au contraire, en opérant la reduction phénoménologique, on possède la conscience dans sa structure originaire. (Cf. Sri Aurobindo, *The Life Divine* (1939), Pondicherry, 1973; R. Shanker Srivastava, *Sri Aurobindo and the Theories of Evolution*, Chowkhamba Publ., Varanasi, 968, p. 27). Selon Aurobindo "the Absolute is both Consciousness and Force". It is *cit* (esprit) as well as *śakti* (énergie). While the dynamic aspect welds into creation, the passive aspect contains the energy in a potential state." (R. Shanker Srivastava, *Oeuvre citée*", p. 63) Ainsi "*l'individuo è un centro della coscienza universale*" (Cf. E. Labanchi, *L'Uno e i Molti: Sri Aurobindo e il pensiero occidentale moderno*, Dissert. doct. Università di Siena, Facoltà di Magistero, 1978, p. 58). — Au contraire, dans l'ancienne philosophie taoïste, Tchouang Tseu (personnage chinois ou personification d'un livre du même nom, IVe siècle av. J.-C.) arrive, pour ainsi dire, à la conscience pure, par une sorte de *ἐποχή*, avec le symbole taoïste de la simplicité naturelle de l'homme, "lorsque son potentiel spécifique intérieur n'a pas été altéré par la sapience ou circonvenu par la morale" Selon Tchouang Tseu, on peut obtenir un état de la conscience, où l'on voit sans regarder, où l'on entend sans écouter, où l'on sait sans penser. Il y a là une vision intérieure, ou perception généralisée, qui ne peut entrer en jeu que lorsque s'anéantit la distinction entre le "dedans" et le "dehors", le "soi" et les "choses", le "ceci" et le "cela". D'où l'immanentisme réaliste qui est spécifique du taoïsme (A. Waley, *Three Ways of Thought in Ancient China*, London, 1940, trad. française, p. 70). Cf. aussi H. Maspero, *Le Taoïsme et les religions chinoises*, Gallimard, Paris, 1971, p. 462. A ce sujet, le voyage récent de Mme Tymieniecka en Chine nous offre de nouvelles occasions (avec sa conférence à l'Académie de Pékin), dont elle a informé le Centre italien de la "International Husserl and Phenomenological Research Society" (Rome, 1982).

[3] Ainsi, dans sa conférence de Rome (1978), A-T. Tymieniecka a expliqué, de façon originale, l'*auto-interprétation* de l'homme, qui entre pour établir sa propre "nature", et l'*auto-règlement* (de·vie) des autres créatures vivantes. Cf. aussi: A-T. Tymieniecka, *Imaginatio creatrix. The Creative versus the Constitutive Function of Man and the Possible Worlds*, Analecta Husserliana, vol. III, 1974, p. 3—41.

[4] A l'époque de la Renaissance, Jean Louis Vivès, l'ami d'Erasme, nous donna une phénoménologie avant la lettre, de la visée cognitive: "Imaginativae actio est in animo quae oculi in corpore, recipere imagines intuendo [. . .]. Phantasia vero coniungit et disiungit ea quae singula et simplicia imaginatio acceperat" (J. L. Vives, *De anima et vita*, édit. M. Sancipriano avec traduction italienne en regard, Gregoriana, Padova, 1974, lib. I, p. 168: "De cognitione interiore").

[5] Sénèque nous a appris que, *si la raison ne suffit pas à modérer les moeurs*, on peut opposer, dans l'intériorité, une "passion" contre l'autre: que l'on peut, par exemple, vaincre la peur en renforçant l'espérance (Seneca *Epistolae ad Lucilium*, XIII, 12). Les mouvements (*commotiones*) contraires dans la nature sauvage se recomposent dans l'équilibre (*tranquillitas*) de l'âme.

[6] Empédocle, *Fragments*, Diels, 177, 17.

[7] "Amor ex calore illo ingenti impellit ad grandia opera edenda atque admiranda, quae vires, ingenium, facultatem excedunt" (I. L. Vives, *Oeuvre citée*, III, p. 524 et 620).

[8] Platon, *Phaedr.*, 244 a; 249 d; *Conv.* 180 n. On peut aussi rappeler la "declamatio" d'Erasme, *Stultitiae laus*, 1509 et la théorie de l'amour chez G. Bruno, *De gli eroici furori*, 1585.

[9] Cf. E. Husserl, *Erste Philosophie*, Vorles. 54 et Beilage XXXI: Transzendentale Aufklärung des Unterschiedes in der Seinsweise: Sein der Natur (und objectiven Welt) — Sein der Subjekte als Menschen-Subjekte — Sein der Monaden-Subjektivität.

[10] Cf. G. Galli, *L'Uno e i Molti*, Chiantore, Torino, 1944, p. 29.

[11] Augustin, *De vera religione*, 39. Husserl précise, dans sa phénoménologie du sujet transcendantal: "*Man musst erst die Welt durch ἐποχή verlieren, um sie in universaler Selbstbesinnung wiederzugewinnen*: Noli foras ire, *sagt Augustin*, in te redi; in interiore homine habitat veritas" (E. Husserl, *Cartesianische Meditationen*, §64). Mais, dans un manuscrit appartenant à son âge avancé, il prélude au fondement absolu: "absolutes in einem neuen überweltlichen, übermenschlischen, übertranszendental-subjektiven Sinn" (Ms E III 4, 1930—1934), cité par I. Kern, *Husserl und Kant. Eine Untersuchung über Husserls Verhältnis zu Kant und zum Neukantianismus*, Nijhoff, Den Haag, 1964, p. 300.

[12] Pour d'autres détails, on peut voir: M. Sancipriano, *L'evoluzione ideale. Fenomenologia pura e teoria dell'evoluzione*, Morcelliana, Brescia, 2me édit. 1961; *Teleology and the Constitution of Spiritual Forms*, Analecta Husserliana, vol. IX, 1979; *La pensée anthropologique de J. L. Vivès: L'Entéléchie*, "Arbeitsgespräch in der Herzog August Bibliothek, Wolfenbüttel", Vorträge herausgegeben von A. Buck, Hauswedell, Hamburg, 1981.

LI ZEHOU

THE PHILOSOPHY OF KANT AND A THEORY OF SUBJECTIVITY

I

Crocé said that all history is contemporary history.[1] We can also say that all history of philosophy is contemporary philosophy. It might be interesting to look at Kant's philosophy from this perspective to see whether it can contribute anything to contemporary Marxist philosophy. In my book, *Critique of Critical Philosophy, A New Key to Kant*,[2] I have presented my ideas on these matters, and here I would like briefly to outline some of them.

In China there are heated debates over human nature, and I would like to start with the problem, What is human nature? People now agree that human nature is not equivalent to class nature. This is clear since class societies have occupied only a short span in the history of human society. Even without class nature, human nature still exists.

Is human nature then equivalent to animal nature? Does human nature merely consist in eating, sleeping, sex, and such? Many people equate human nature with this sort of animal nature or with man's primitive desires. I do not agree with this. Human nature is precisely those particular characteristics that are specific to man and distinguish him from animals. These human characteristics have been seriously alienated in class society, and it is exactly this alienation that tends to make human nature appear as animal nature. As Marx pointed out in his *Economic and Philosophic Manuscripts of 1844*, the alienation of labor has caused man to feel free only in such activities of an animal nature as eating and drinking. Although there is a connection between freedom and human nature, this sort of animal freedom has nothing to do with human nature.

Can human nature, then, be equated with social nature? Many theories have argued this, but the concept "social nature" is hardly clear. Does it mean group nature? Animals also have a group nature. In many animal groups there can be found certain types of organization and division of labor, even a sort of "moral behavior" and "altruism" with individual sacrifice for the preservation of the group. Human

A-T. Tymieniecka (ed.), Analecta Husserliana, Vol. XXI, 135–149.

nature certainly cannot be equated with this instinctive type of "social nature."

Is social nature just social consciousness? We know that social nature, in this understanding, has often been explained as a thing of purely rational character which both is detached from and dominates sensuality and emotion. Thus it frequently possesses the character of alienation. Rather than say that this is human nature, we should call this some sort of divine nature, something of supernatural character which has been thrust upon man. This sort of alienated rationality, or "social nature," forms a counterpart to the alienated sensuality, or animal nature, discussed above. The reduction of human nature to things of an alienated social nature, for instance, to religion, asceticism, subservience to computers, or man's slavery to production, is no different from understanding human nature as alienated animal nature, as primitive desires, hedonism, pornography, or man's slavery to consumption.

In fact, none of these is human nature. Human nature is the opposite side of alienated sensuality and alienated rationality. It is the unification of sensuality and rationality, the unification of animal nature and social nature. This unification is not a mere adding, combining, or mixing of the two contrary things, not "half angel, half devil." Rather, it is rationality, social character, in the midst of sensuality, animal nature, or the sedimentation of rationality within sensuality, with the two dissolved together into a single whole. This is just the humanization of nature or humanized nature of which Marx spoke in the *Economic and Philosophic Manuscripts of 1844*.

In this sense human nature is subjectivity. There has already been much discussion of subjectivity. I think we should distinguish between the subjectivity of the individual and that of mankind. The latter precedes and is a presupposition of the former. In contrast to the whole objective world, mankind (not the individual) creates for itself a structure of subjectivity. This structure is not only concretely sensuous but also superbiological, having a universal necessity. Marx was right when he said that with animals there is no distinction between subject and object. Man is different. Through prolonged historical practice, he has constructed a complete system of superbiological subjectivity which is distinct from the natural world and can act upon it. This is what I understand by human nature.

Thus subjectivity in Marxist theory cannot be explained as only class consciousness or moral will but should be understood as a psychological structure which has a foundation in the practical, material activity of the whole of mankind. Class consciousness and moral will have only temporary and limited concrete contents, but the psychological structure is something sedimented through the entire history of mankind. Thus it has a universal form. In this sense, the subjectivity of Marxist theory entails not only a theory of revolution but also a theory of the construction of mankind's spiritual civilization.

Mankind has not only made, accumulated, and developed the material civilization of the external world, from primitive pottery to today's spacecraft, but is also continuously creating, accumulating, and developing the spiritual civilization of the internal world. Man is different from animals not only due to his material civilization but also due to his structure of psychology, which is both social and built on animal, physiological mechanisms. This cultural-psychological structure, which is the concrete embodiment of human nature, distinguishes man from animals.

This structure can be analyzed as having at least three aspects, the structures of intelligence, morality, and aesthetics (knowledge, will, and feeling). All kinds of ideas in knowledge, moral standards, and artistic taste are invariably in constant change; each have a particular, changing content that corresponds to a given society, time, and class. But at the same time they have constantly sedimented structural results, which are more continuous and unvarying than the former. It is precisely in this constant movement of change that the cultural-psychological structure, both ability and form, is ceaselessly accumulated, consolidated, and continued. This structure is psychological and yet is built on a physiological basis. It is not the structure of man's brain alone in isolation from the history of mankind. It is the merging and unification of physiological and social aspects. Therefore, insofar as it is manifested as the universality of sensibility, this universality is not biological but social. It belongs not to certain classes but rather to the whole of mankind. It belongs not to all kinds of animals but to a particular species, that of human beings.

Kant's great accomplishment lay in raising the problem of subjectivity in a comprehensive manner. Here I would like to use Marxist philosophy to consider the questions raised by Kant. Kant's philosophy consists of three main sections, epistemology, ethics, and aesthetics and teleology, which correspond to his three *Critiques.* What follows will be a simple discussion of these three areas.

II

(A) In his epistemology Kant delineated a series of a priori forms of knowledge, from the pure intuition of time and space to pure concepts of understanding. If man did not have these a priori forms, it would be impossible to gain scientific knowledge of universal necessity. And it has already been demonstrated by modern science that cognition is not a gradual, sequential process, passive and still like reflection in a mirror as traditional materialism understood it. In fact, at the very beginning of the process of cognition, sensation contains a whole set of subjective elements. For animals, the reception of and reaction to the outer world by the sense organs is intertwined with the whole activity of their biological processes. The central nervous system frequently has a dominating or even determining function in the regulation of the sense organs. If animals are already this complicated, man is even more so. The practical activity of human labor and the whole history of society have had a great impact on the dominating function of the central nervous system, including its biologically inherited factors. Marx has stated that man's five senses are the fruits of all of world history. Konrad Lorenz pointed out that, from animal to man, from sense perception to forms of thought, all are formed through the adaptation of organisms to the environment in mutual interaction.[3]

In *Critique of Critical Philosophy*, I repeatedly emphasized the determining function of human practical activity in molding man's whole psychological structure and processes. While practical activity progressively enlarges its field and content following the advance of history, the foundation, though not the totality, of this practical activity is in the use and making of tools. The fundamental line of demarcation between man's practical activity and the life activity of animals lies in the use and making of tools.[4] Tools extend man's limbs artificially, and man's two hands become organs specialized in the use of tools. The human biological body together with its tools forms an super-biological body, and its activity is super-biological behavior. It is precisely this kind of behavior which is the true foundation of man's super-biological existence.

Thus we can see that super-biological existence is prior to super-biological sight or hearing; practice is prior to sensation and perception. Man's organs which move, his hands and feet for instance, perhaps were the very earliest to possess a humanized nature, and only later did the sense organs like eyes and ears become humanized. Later also was the humanized content of language and the linguistic organs, the emergence and development of real language and thought. Therefore it is practical activity and not sense data or language which is the starting point of epistemology. Man's entire subjective psychological structure of cognition, from sense perception to conceptual thought and language, is erected upon the extremely prolonged span of man's laboring activities in the use, creation, improvement, and coordination of tools. All sorts of knowedge and forms of natural laws are first preserved and accumulated within this sort of human practical activity, then turned into the information systems of languages, symbols, and cultures, and finally are sedimented into the structure of man's psychology. Not language but the practice of using and making tools is the link between the natural world and human beings. It transforms the natural laws and natural struggles into the forms of human society, human nature, even the human brain. In this way there emerged a human subjectivity of cognition which enables man to know the world in his super-biological modes.

Thus man's learning behavior is different from that of animals. The structure of knowing which man learns beginning in his childhood cannot be equated with Pavlov's system of secondary signals. This structure of knowing is a kind of ability to symbolize which includes an intellectual intuition neither inductive nor deductive. (Kant thought that only God has intellectual intuition; in fact this is not so.) Signals of signals in the end are merely signals and are fundamentally distinct from this sort of cognitive structure or ability.

For example, mathematics is a tool of knowledge peculiar to man. Kant connected mathematics with a priori sensibility and, in his table of categories of understanding, put quantity before quality. This order is the opposite of Hegel's, and Kant's, I think, is the more correct one. Regardless of whether one looks at man's primitive knowledge or at modern science, man's use of mathematics to abstract and grasp the objective world is a clear demonstration of the special characteristic of subjectivity of human cognition. This is an important indication of the fundamentally different forms of knowing of man and animals.

In *Critique of Critical Philosophy,* I wrote that the origin of mathematics is not analysis or induction but rather the basic practical activity of man. Thus certain elements and forms of primative operations, like counting, relating, or ordering, give rise to such abstractions as the natural numbers, plus, minus, equation, and the associative, commutative, and transitive properties. The establishment and development of mathematics is inseparable from the ability for intellectual intuition which man has gained through his basic, practical activity. Human intellectual intuition, which is neither separable from sensation nor yet equivalent to it, has played a crucial role in the discoveries of mathematics and the other sciences. Piaget has also pointed out that logic and mathematics originate in certain practical operations. Yet he did not investigate the problem from the perspective of the totality of the history of mankind's practice. He did not connect psychology with anthropology to raise the problem of human nature from the perspective of a philosophy of subjectivity.

(B) In sum, practice is not equivalent to ordinary sense perception, nor can practice be regarded as arbitrary subjective activity. We must return to practice its characteristic of regularity as a concrete structure, its historically concrete, objective social character. Only then will we have a true understanding of practice. This is the reason why I feel it necessary to repeatedly emphasize the use and creation of tools. Today there are many people who discuss the Marxist theory of practice or praxis, but few stress this point.[5]

What finally is practice? No one has really explained it clearly. Despite all the discussion about practice and praxis, most people still overlook the fact that the activity of employing tools is man's fundamental practice. Yet there is a consistant thread, the concern to arrive at an objective understanding of practice, which runs from Marx' early *Theses on Feuerbach* through Engels' later writing, *From Ape to Man*.[6] It is just this which constitutes historical materialism and is the central theme of Marxist philosophy.[7] Historical materialism is precisely a theory of practice. The content of the theory of practice is the activity of subject on object, while historical materialism, using productive force as a criterion, deals with the control and transformation of the objective world by man. The two are the same thing, and theories which part them are all contrary to Marxism. In my opinion, not the philosophy of praxis but the philosophy of practice is the real philosophy of Marx with which he wanted to change philosophy and reality.

Separated from the theory of practice, historical materialism easily becomes an ordinary sociological principle, a vulgar determinism. Apart from the real, material actions of the human subject, both individual and group, "social existence" would loose the lively, active content which it originally possessed. It would loose its inherent, practical nature. "Social existence" would become instead a sort of objective, "environmental" existence, and man would turn into something passive, manipulated, and controlled, an insignificant grain of sand or a tiny gear in the midst of an alienated social structure. This kind of historical materialism is really fatalism or mechanical, vulgar determinism. In the Soviet Union, the Stalinist bureaucratic system's "orthodox" theory falls into this category.

On the other hand, if the Marxist theory of practice departs from historical materialism, it also moves away from the historically concrete, objective laws of mankind's material production. If practice loses its basic meaning of the use and creation of tools in actual history, it can be equated with ordinary empiricist sensation or lead toward a theory of voluntarism. Understood as a simple subjective force, practice frequently comes to mean theoretical activity, critiques of culture, moral consciousness, and revolutions of the superstructure. Certain Western Marxist theories of praxis are of this type,[8] and Mao Zedong in his later years also had such a voluntaristic tendency.

> If one were to talk of human practice in a vacuum isolated from historical necessity, one might easily ground practice in unhistorical psychological needs. Then the motive force of history becomes not economics, or the modes of production, but rather psychology, subjective needs.[9]

This sort of theory is exactly what I do not mean by the human subjectivity which I am emphasizing here.

III

(A) In certain respects, Kant was more perceptive than Hegel. Kant recognized that epistemology does not exhaust philosophy, but Hegel equated almost the whole of philosophy with logic or with the logical progress of the absolute Idea. In fact this amounts to a sort of pan-rationalism. However it leaves open the question of why man is alive. What is the real value and meaning of human life? In what do the content, depth, and richness of existence consist? Life, death, anxiety, loneliness, terror, these are not necessarily problems of logic, though they are profound philosophical problems. Their character of immediate reality, under specific conditions, makes them more profound than problems in logic and epistemology. When one turns everything into a question of rational logic, the profound, immediately real character of man's existence is frequently overlooked or erased. Man becomes an inconsequencial link in the historical workings of logic or the rational machine, and the significance of being, with its ability to create history through subjectivity, is concealed and deracinated. This sort of pan-rationalism of Hegel's has had its unhealthy influence on Marxism. It neglects man's real, immediate existence and ignores ethics. Ethics have no position in Hegel's philosophy, being merely a single link in his logic. Thus the subjectivity of individual existence is lost.

From the perspective of what is universal in mankind, practice sediments and then transforms itself into the human structure of logic and knowledge. On the other hand, practice is always individual. It is formed, realized, and completed in the practice of individuals. This sort of immediacy of individual practice is precisely the concrete, real character of individual existence, behavior, emotions, will, and desires. This sort of reality is prior to and superior to the universal character of knowledge. The nature of man should not be separated from his concrete existence to become some sort of external dictator.

Thus philosophy includes logic and epistemology, in other words theories of scientific method. Philosophy of science, analytic philosophy, and dialectical materialism all fall within its scope. Yet philosophy cannot be equated with these.

(B) In the realm of ethics, Kant also gave prominence to the problem of subjectivity, in a way which far surpassed superficial French utilitarianism. Let us take an example. Suppose a great scientist, Einstein for instance, died in order to rescue a crippled, old man of no consequence from a burning house just before he would have completed his theory of relativity. Is this in accordance with morality? From the perspective of utilitarianism, there is no need to exchange the life of a man capable of making great contributions to mankind for the life of a man who has nothing to contribute. This kind of sacrifice would be valueless and immoral.

Yet from the point of view of Kant's ethics, it would be necessary to act in this way. To act this way is a sort of "categorical imperative", and it is just in this kind of "categorical imperative" that morality expresses its dignity and its tremendous strength. In terms of an anthropological ontological philosophy of practice, the value and meaning of Kant's "categorical imperative" lies in the following. It demands the establishment of subjectivity with regard to individual practice. It demands that individual practice have the duty and the sense of responsibility to bear the burden of building the principles for the existence and development of the whole of mankind. This sort of sense of responsibility and moral behavior transcend any concrete interests of a time, society, or group, just as in the Old Chinese saying, "To establish virtue is of the first importance," before achievements or writings. Individual existence, being unrepeatable, expresses in this its incomparable greatness.

In this manner, the individual and the whole of mankind are completely united. And thus, this responsibility transcends any laws of cause and effect or consequences for the interests of any individual or group. It is just for this reason that it possesses such a lofty and stirring nature. Mankind alone possesses this kind of moral law. Such ethical behavior is distinct from animals' instinctive sacrifice of the individual for the group.

But the individual's morality only exists by virtue of man's self aware, conscious, rational control, and this belongs in the realm of the establishment of man's subjectivity. Just as there is rationality sedimented in the sense intuition of epistemology, so there is rationality sedimented in man's sensibilities of emotion, will, and wish. This finally is true free will, corresponding to free creation (intellectual intuition) in epistemology. Thus, though seemingly a purely formal principle, in comparison with utilitarian ethics, Kant's ethics more penetratingly come in contact with the nature of morality and with the core of mankind's subjectivity.

There is a great deal of discussion of "moral heritage" in China. But moral continuities in history cannot be the inheritance of concrete content, as there are great differences, even oppositions, in the content of morals with differing times, societies, or classes. Neither is it the borrowing from previous times of formal terms or concepts of morality. What is actually handed down is the form of human psychological structure. Even though the content may be concretely determined by history in a society, epoch, or class, still it is exactly this kind of formal structure which links them all, constituting what is human ethics itself.

(C) Subjectivity is not just individual morality. To look on subjectivity as merely the moral strength of the individual is idealism. In *Critique of Critical Philosophy,* I distinguished the "big I", the human collectivity, and the small I", the individual self. I pointed out that the individual self, including both the individual's material needs and spiritual wishes, is always concretely connected to and restricted by history within specific social conditions and surroundings. Thus the seemingly concrete passions and will if separated from history are in fact abstract and even nonexistent, while the seemingly abstract social modes of production and productive relations, on the contrary, are concrete, historical realities. What is of a changeless, commonly shared character perhaps belongs only to animal nature. Differing lives, forms of marriage, aesthetic tastes, and affections, all are concretely restricted within and determined by the social environment and history. Both Kant and Hegel pointed out that the individual in isolation has an animal nature and that objectivity and rationality all derive from collective society. The value, meaning, and richness of the human individual's life vanish when he is looked at as an animal. All of these are rather the riches and products of human history. Thus the individual subjectivity discussed in philosophy and ethics is not animal individuality but instead is the individual as a member in society's collective existence. In this sense, the subjectivity of human ethical beings also has objective elements.

However, up until the present there are still severe contradictions and conflicts between the individual and the group, the "small I" and the "big I". The phenomenon of alienation in which the individual self is obliterated in the whole of the collectivity constitutes a necessary phase in the history of mankind. Thus, on the one hand it is quite shallow to reduce Marxism to humanism or individualism. And on the other hand it is also a mistake to take historical materialism to be mechanical determinism and man a mere tool. A Marxist ethics cannot evade or depreciate these problems. Rather it should recognize that in the future the immense meaning and value of individual existence will become an increasingly prominent and important topic.

> The individual, as a natural, existential body of flesh and blood in a given state of conditions has an awareness of particular force of his own existence's uniqueness. . . . Actually, existentialism precisely demonstrates . . . the rich social meaning of the individual.[10]

In terms of the broad sweep of history, to stress individual practice is to emphasize the accidental in historical development. From Hegel to contemporary Marxism, there has been an emphasis on historical necessity which is inappropriate, almost fatalistic. This overlooks the individual and the self's freedom of choice and thus overlooks the great effects of accidents including the actions, wills, abilities, and personalities of the different individuals in history. On the one hand voluntarism is incorrect. In broad historical terms, from primitive society to contemporary society, from small scale, agricultural production to large scale, industrial production, material civilization has made progress and has had a great impact on man's life and consciousness. In these processes, there are in fact objective regularities and historical laws which are beyond redirection by men's wills. On the other hand however, I am also opposed to vulgar determinism. Within mankind, the practice of every individual actively creates history, and the process of history is full of a great quantity of accidental factors.

Another aspect of subjectivity as a part of ethics is that the value and meaning of individual existence reside in the relationships of human association and harmonious coexistence. The nature of human ethics lies in what Marx, in his *Theses on Feuerbach*, called a "socialized mankind". In the contemporary world, alienated by power, money, and technology, the lonely individual longs for a return to human nature. This is a desire to go back to ethical relationships which are replete with the flavor of human feelings; it is not a wish to regress to the primitive world of animals. In this respect, the teachings of Confucius, retaining vestiges of clan society's humanism, practical rationality, and ethical emotions, provide a valuable historical legacy. Up to the present, this legacy is still partially preserved in the cultural-psychological structure of the Chinese people. Perhaps in the future it may yet give play to its own contribution in the formation of mankind's ethical subjectivity.[11]

IV

The third section of Kant's philosophy concerns aesthetics and teleology. At the very beginning of his "Critique of Aesthetic Judgment", Kant distinguished the pleasure of aesthetic feeling from animal sensory pleasure and conceptual intellectual cognition. Sensory pleasure takes something as beautiful only on the basis of pleasure received by the sense organs. The pleasure of aesthetic feeling, an achievement of sensibility enjoyed by all mankind and possessing the character of a subjective universal necessity, is of a different nature. Pleasure is derived only after an object has first been judged beautiful. In other words, aesthetic pleasure is not passive sensation or reflection on the part of sense organs, nor is it the intellect's abstract, logical act of knowing. It is the result of the activity of many different psychological functions including the sense perception, imagination, emotions, intentions and passions, and understanding of the human subject. Kant called this aesthetic judgement. The concrete structure of this type of "judgement" awaits the further research and discoveries of a future psychology.

Here the important point is that aesthetic judgement forms the third aspect of subjectivity, which completes the humanization of man's internal, biological world and is counterposed to the humanization of the external, objective world of nature through material civilization. As free form, beauty is the unity of the two characteristics of regularity and of teleology. It is the external humanization of nature or humanized nature. As the psychological structure which corresponds to this free form, aesthetic feeling is the dissolving together of sensation and rationality. It is mankind's internal humanized nature. It is the final achievement of man's subjectivity, the clearest and most distinct manifestation of human nature because herein are the sedimentation of mankind in the individual, the sedimentation of rationality in sensuality, and the sedimentation of society in nature. The sense organs, originally of animal nature, have been humanized. The animal structure and quality of psychology have been transformed into something with a human nature.

Its scope is exceedingly broad. Eating is not merely due to hunger but becomes dining. The relationship between the two sexes is not merely one of copulation but becomes love.[12]

In the *Economic and Philosophic Manuscripts of 1844*, Marx wrote:

This direct, natural, and necessary relation of person to person is the *relation of man to woman.* In this *natural* species relationship man's relation to nature is immediately his relation to man, just as his relation to man is immediately his relation to nature — his own *natural* destination. In this relationship, therefore, is *sensuously manifested,* reduced to an observable *fact,* the extent to which the human essence has become nature to man, or to which nature to him has become the human essence of man. From this relationship one can therefore judge man's whole level of development. From the character of this relationship follows how much *man* as a *species being,* as *man,* has come to be himself and to comprehend himself; the relation of man to woman is *the most natural* relation of human being to human being. It therefore reveals the extent to which man's *natural* behavior has become *human,* or the extent to which the *human* essence in him has become a *natural* essence — the extent to which his *human nature* has come to be *nature to him.* In this relationship is revealed, too, the extent to which man's *need* has become a *human* need; the extent to which, therefore, the *other* person as a person has become for him a need — the extent to which he in his individual existence is at the same time a social being.[13]

Sexual passion becomes love, natural relationships become human relationships, natural sense organs become the sensory organs of aesthetic feeling. This is the final aspect of sediment subjectivity. Add it together with the structure of the intelligence of epistemology and the free will of ethics, and they form the three important aspects and main content of subjectivity.

Just as I mentioned above that some aspects of the teachings of Confucius may be helpful for a future philosophy of ethics, so too it is worthwhile attending to the traditional Chinese aesthetic. The saying "The movement of Heaven is vigorous; the gentleman ceaselessly strengthens himself," displays the Confucian spirit of constantly striving forward. The philosophy of the Taoist Zhuang Zi (c.369—286 B.C.) took an aesthetic attitude toward life. Chinese, Chan Buddhism was not opposed to involvement in the daily affairs of life. Qu Yuan (c.340—278 B.C.), the greatest poet of China, represents a tradition of unyielding persistance for a beautiful ideal. Together these make up the soul of Chinese aesthetics. They should be capable of their own contribution toward the establishment of a philosophy of beauty in a new era.

The nature of beauty is the nature of man in its most complete unfolding, and the philosophy of beauty is the highest peak in the philosophy of man. In philosophical terms, this is a question of subjectivity, while in terms of science, it is a problem of the cultural-psychological structure. If for his own times, the branch of learning related to historical materialism which Marx emphasized in his researches was political economy, in current society, with its high level of technology, the problems of the cultural-psychological structure are increasingly prominent and pressing. It is not economic poverty, but spiritual impoverishment, loneliness, isolation, and boredom which will become increasingly serious problems in the world of the future.

Philosophy is not revolutionary prescriptions. It cannot solve problems of reality directly and quickly. It would rather gaze forward, probing and understanding the special characteristics of the sedimentation of the cultural-psychological structure of subjectivity. It should pay heed to the melding together of science and morality, of material civilization and spiritual culture, of the restraints of collective universality and the many faceted, latent potential of the individual. In this may lie the future direction of development of historical materialism and the philosophy of practice. Then the science of education, the study of man's whole psychological growth and development, will become the most crucial discipline for future society. All of these are precisely the combination of naturalism and humanism which Marx expected and for which he hoped, his great ideal that natural science and humanistic studies should be fused into a single science. And here too, after ridding it of its passive elements which derived from small scale, agricultural production, the important concept in traditional Chinese philosophy of the "unification of Heaven and man" could prove useful for future philosophy. The ideas raised in this article, due to limitations of space, provide only a brief outline of the directions such a philosophy might take.

Chinese Academy of Social Sciences, Beijing

(Translated by Catherine Lynch)

NOTES

[1] This essay has been translated, after revisions by the author, from Li Zehou's "Kande zhexue yu jianli zhutixing lungang," in *Lun Kande Heige-er zhexue* (On the philosophies of Kant and Hegel) (Shanghai: Renmin chubanshe, 1981), pp. 1—15.

[2] Li Zehou, *Pipan zhexue de pipan — Kande shu ping* (Beijing: Renmin chubanshe, 1979).

[3] Konrad Lorenz, *Behind the Mirror: A Search for a Natural History of Human Knowledge*, trans. Ronald Taylor (New York: Harcourt, Brace, and Jovanovich, 1977).

[4] Although some animals, chimpanzees for instance, can use and make tools, only for man is the use of tools as an activity universally necessary for fundamental existence. This kind of activity in animals only serves to demonstrate that the premise and potential for the specifically human activity of using tools does actually exist in biology. So too the "moral" behavior of animals in groups exhibits the biological basis of human ethics, although with animals these biological foundations are not joined with true social existence. I do not agree with the ideas of socio-biology.

[5] Li Zehou, *Pipan zhexue de pipan*, p. 195.

[6] I do not agree with the exaggeration of theoretical differences between Marx and Engels. Although the two do have their differences, still their commonalities are the more important.

[7] The soul of Hegel's dialectics is formed from the historical sense of the whole of mankind. Hegel's logical movement of "Spirit" is merely an upsidedown, idealist

reflection of the development of man's history. Thus the result of inverting Hegel's idealist dialectics could not be ordinary "dialectical materialism", which is unrelated to human history. Rather it can only be historical materialism, which takes the development of the history of mankind as its central theme.

[8] This is why I do not agree with the theories of early Lúkacs, the Frankfurt school, and the Yugoslavian school.

[9] Li Zehou, *Pipan zhexue de pipan*, p. 352.

[10] Ibid., p. 407.

[11] See Li Zehou, "Kongzi zai pingjia" (A reassessment of Confucius), *Zhongguo shehui kexue* (Social science in China) **2** (1980), 77—96.

[12] Li Zehou, *Pipan zhexue de pipan*, op. cit., p. 401—402.

[13] Karl Marx, *Economic and Philosophic Manuscripts of 1844*, Dirk J. Struik, ed. (London: Lawrence and Wishart, 1970), p. 134.

CLYDE PAX

HUMAN NATURE AND MIND IN MARTIN HEIDEGGER

When Professor Tymieniecka asked me to present a brief paper at this Third Symposium on Phenomenology and Oriental Philosophy, she said that what was needed for the occasion was a straightforward presentation of some of the central ideas of Heidegger on the questions of human nature and mind. The reason for her request, no doubt, is her experience that too frequently discussions of the thought of Heidegger quickly reach a point where argument about what a particular text means pushes aside fruitful dialogue. I shall make every attempt to respond to the request of Professor Tymieniecka. However, the thought of Heidegger is itself not always as straightforward as it is hermeneutical. Moreover, I cannot claim to have mastered Heidegger's thought to the extent that I can claim to be able to give a definitive interpretation of his meaning of human nature and mind. With these two caveats in mind I will make my presentation as clear as I am able; I hope it will be suitable to enable us to carry on a fruitful dialogue.

All of the work of Heidegger emerges from and remains centered upon a single question, namely, the question of the meaning of Being. The question of the meaning of Being, he tells us, must be distinguished from questions about the meaning or meanings of beings, from questions about the ground of beings, and from questions about beings as a whole. Thus the question about the meaning of Being is no ordinary question, but one that is totally unique. He describes this question as the first of all questions, not chronologically to be sure, but first in that it is the broadest, the deepest, and the most fundamental. It is the question which in its uniqueness makes possible all other questions and all other concerns.[1]

Failure to keep this question about the meaning of Being centrally before us while we ask about human nature and mind can only result in a vast misunderstanding of Heidegger's thinking. Such a misunderstanding would be present, for example, if one were to read his long and careful analysis of human existence (*Dasein*) simply as a study of philosophical anthropology. It would be another and related misunderstanding to think of Heidegger as a subjectivistic thinker because of his insistence

A-T. Tymieniecka (ed.), Analecta Husserliana, Vol. XXI, 151–158.

upon the interpretive character of all experience and all understanding. Neither of these views could be farther removed from the genuine concerns of Heidegger. His primary concern began as and remained a concern with the question of the meaning of Being, not with the issues of human nature or of mind. This Heidegger makes explicitly clear in the *Letter on Humanism.* There, in response to the question of Jean Beaufret, he asks whether there is any reason to retain or to attempt to recover a meaning for the term humanism; he argues that there is reason to do so only if the term can be placed back into the context of the question of Being.[2]

Having insisted upon the uniqueness and primacy of the question of Being, it is necessary to say, however, that the questions of what it means to be human and what it means to think cannot properly be designated as secondary questions. Heidegger tells us in the work *On the Question of Being* that we always say too little if we say Being without saying man.[3] The reason for this is that the question of the meaning of Being gets raised only if and only when man raises the question of Being. Only in this questioning is Being (in distinction from beings) *there*; and only in this same questioning is man there in his most proper reality. In this questioning man is the *there* of Being; in this questioning man is authentically who he is as the coming out of concealment, that is, as the self-disclosure of Being.

In his own proper being, man is an ecstasis. He is relational toward Being in such a way that, as Gadamer comments, the relation must be given priority over the members of the relation. Insofar as he is the questioner of Being, man is not appropriately thought of as one of the beings that are, nor does he belong to a species called human beings; he does not have an essence if that term is meant to designate some inner core which either statically or dynamically exists as what it is and thereby causes men to be essentially. If we speak of essence in relation to human existence we must, Heidegger insists, say that the essence of man lies in his being toward Being.[4]

The insistence that man is properly *da-* and *zu-Sein* means that there is a real and far-reaching sense in which Heidegger wishes to distance his work from all talk of human nature and of mind. The "destruction of the old ontologies" is at the same time a destruction of all theories of human nature and all theories about mind insofar as nature and mind are taken in an essentialistic or substantial sense.

This destruction or overcoming does not, however, deny the uniqueness of man; nor does it remove or lessen Heidegger's interest in the matter of thinking. Quite the contrary. The uniqueness of the human person, as Dasein, is severely heightened and the question of thinking is placed at the center of concern. Heidegger refers approvingly to the position of Parmenides that thinking and Being belong together. His life work is nothing else, in fact, but the effort to think thoroughly this belonging together of thinking and of Being. The issue of Dasein and his "nature" gets drawn into this issue of the belonging together of thinking and Being because Dasein is ontically who he is by being the one who is standing in or existing in the ontological question. The human person in his or her authentic life is the one who lives his ontic being by being ontological as well as ontic.

This standing in the question of Being means for Dasein that his ownmost ontic being gets thrown into question; his ownmost being is always already and totally at issue. For this reason it is not enough to say that Dasein's way of being is to be ontological; it is necessary to say further that this way of being is ontological *as care*. Dasein is there, is existing, always in such a way that the meaning of his own being and the meaning of Being are at issue, that is, are in need of being questioned. The need of questioning arises because clarity is lacking; but it arises much more profoundly because what is at issue is what is most worthy to be questioned.

Because the question of what it means to be Dasein and the question of how it stands with Being can be neither reduced to one question nor separated from one another, a radical difficulty and a radical ambiguity pervade all of Heidegger's thinking on the issues of human nature and mind. The proximity of the question means, on the one hand, that all authentic questioning about man and about mind must remain faithful to the primordial and abiding ecstasis which constitutes man. On the other hand, the proximity is a denial of the identity of the two questions. Dasein is not Sein, but is always radically other than Sein. Dasein is ontic, and is ontological only according to the manner of its being ontic. There is always possible a double consideration regarding Dasein: one consideration is ontic and the other is ontological. Heidegger fixes this difference by using the term "existentiell" to designate the ontic consideration and the word "existential" to designate the ontological consideration.[5] To uncover the meaning of Dasein one must think both the existentiell and the existential aspects of the way of Dasein's comportment. It seems to me that Heidegger's insistence upon the necessity of keeping both aspects in view provides at least a region for bringing Heidegger's thinking into fruitful discourse with some central concerns and central ideas of Chinese philosophy. I refer, for example, to the Confucian concern with *jen*, which was to catch both an inward greatness and a greatness of manifestation; I refer to the question of whether *jen* is primordially good, bad, or indifferent — a question that has troubled Confucianism from the beginning and which contains in its history the question of how one is to begin to genuinely experience *jen*. Of particular interest also are the meaning of the five great relations of Mencius, and the tremendously fascinating issues of *wu-wei, wu-yu, and wu-hsin* as these issues were raised in Taoism and in the long development of Buddhism in China.

Dasein emerges and is present neither as sheer fact nor as necessity, but rather as question and as care. Dasein normally finds itself in the world and at work with all of the things in the world. But because Dasein is not fact but question and care, this normal situation can be a situation in which the whole issue of the meaning of Being is forgotten and covered over by the concern that Dasein has (the fascination that Dasein feels) for the beings and the events of the world. When the question of the meaning of Being is forgotten, Being is no longer *there* (except in the mode of being forgotten); in the same act of forgetfulness Dasein as Dasein is also overlooked, that is, is no longer authentically existent. It is at this point, in this situation of inauthentic existence and forgetfulness, that arise various theories of human nature and of mind and of natural objects. Heidegger's thinking here is not at all foreign to the thought expressed in the verse of the *Tao te-ching* which reads: "When the great Tao declined the doctrines of *jen* and *li* emerged."[6]

As, that is, because, Being manifests itself by coming out of concealment Being is there and there-being (Dasein) stands forth as who he is. But also as, that is, in the manner that, Being manifests itself there are all the beings that are; and there-being (Dasein) by being ontological is there as the open region for all the beings that are. By being ontological in being ontic, Dasein is also ontic-ontological. Heidegger expresses this way of being of the human person as "being-in-the-world."

It is of primary importance to remember that "being-in-the-world" is not another name for Dasein. Rather it is only by being toward Sein that Dasein is being-in-the-world. Failure to keep it clearly in mind that the human person is, for Heidegger, a being in the world only within the context of being-toward-Being can only give rise again to massive misunderstandings of Heidegger's thought as it touches upon the issues of human nature and of mind. The most severe misunderstanding would be to think of man simply as one of the beings in the world, alongside of other beings. Such a way of understanding the human person is possible, perhaps is very common; it is even a way that is revelatory, namely, revelatory of man as the one who can forget and fail even to ask authentically who he is. But it is a way of understanding that is withdrawn from our most primordal experience both of other beings and of ourselves as being-in-the-world. In Heidegger's descriptive interpretation we meet other beings as "ready-to-hand," that is, as always already involved with and involving our own comportment. From the perspective of this primordial experience, it is only as another and specifically identifiable interpretation that other things can be seen as objects. The further interpretation is one that separates the "ready-to-hand" from the original experiencing and places it "on its own footing," as something that is no longer involved with Dasein but something that is present, outside of, independent of Dasein. In this same act of interpretation Dasein is also further interpreted and is no longer "being-in-the-world" but becomes a subject before the present and presentable object. Dasein becomes presently capable of being thought of as having a (human) nature. Thus it becomes possible to ask for a self-understanding by asking about human nature precisely within the confines of an interpretation that has forgotten the experience of being human. The distinction and play between the interpretations of "readiness-to-hand" and "presence-to-hand" relates closely, it would seem, to the effort in Neo-Confucian thought to discover which way of asking about a thing or an event is the *best* way to ask about a thing or an event.[7]

The failure to keep "being-in-the-world" in the context of the question of Being can result in a second misunderstanding. This misunderstanding is the one which would genuinely address Dasein as being-in-the-world and not as one of the beings who are in the world (as in the objectivistic interpretation), but would take the human situation of being-in-the-world simply as a fact, rather than as a wonder that needs to be thought. And the proper phrase here is "needs to be thought," not "thought about."

To accept Dasein's being-in-the-world sheerly as a fact is to see the human person as existential to a degree. Such an understanding, however, would not reveal Dasein as open to Being. It would therefore fail to appreciate the radicalness of Heidegger's thinking regarding what it means to be human and what it means to think. Such a limited meaning to the term existential is more in accord with the thought in Sartre's *Being and Nothingness* than with Heidegger's thought. It is perhaps for this reason that Heidegger, more than Sartre, is brought into communion with both Taoism and Zen Buddhism.

The whole of Heidegger's life work is absorbed in the question of thinking the meaning of Being and the meaning of the difference between Being and beings. Since this difference, the ontological difference, arises in and as there-being or Dasein, it could correctly be said that Heidegger's thought never withdraws from the questions of who man is and what it means to be called to thinking. To be human person, however, and thinking are not two of the things that are. Rather, being human person and thinking, like change in the words of the *I-Ching*, penetrate all of the things that are.

In a number of his writings Heidegger refers to the pursuit of the question of the meaning of Being as following a Way. This way is one that is not visible and laid out before the traveler who enters upon it. Whether it leads anywhere is not clear except in the following of it. Most importantly, however, it is a way for Heidegger which like the Tao of Lao-tze takes precedence over those who are on the way.

It is neither possible nor appropriate here to detail all of the movements of Heidegger as he seeks to move along his way. He himself tells us that the task is not to seek to record and to clarify (by constant textual study) the way of Heidegger, but to do our own thinking along the way. Here I would like to suggest and only briefly discuss three or four ways of the being of Dasein and of thinking which are suggested by Heidegger's word Care — which he refers to as the being of Dasein.[8]

Understood as the being of Dasein, care has nothing to do (initially at least) with an emotional concernfulness. Nor is care some action that Dasein does. Care is rather the way in which Dasein exists. Dasein finds itself always already away from its origin (*geworfen*) and always not yet at its own wholeness; from birth Dasein is on the way to wholeness (being-unto-death). Because Dasein exists as the one who is always away from origin and unto death, the meaning of Dasein as Care is temporality. But to say that Dasein is temporal is not to place Dasein within cosmic time; it is rather to say that cosmic time is real, is passing, is present, is scarce, is precious, or is boring and heavy by reason of the temporality of Dasein. It is because the temporal presence of those things that are present is such within a more primordial meaning of time which is not defined by presence, that Heidegger feels compelled to criticize a metaphysics of presence. This theme has obvious kinship with much of both Chinese and Indian thought inasmuch as these latter frequently feel less than comfortable with the attitude which merely accepts the "ten thousand things" without questioning the meaning of their presence.

There is a Buddhist text which suggests the exercise of looking into one's own face before one is born. As far as I know there is no similar exercise suggested by Heidegger. There is, however, a description of an experience of self-discovery that is every bit as startling as the exercise suggested in the Buddhist text. This experience is the experience of self-finding, or perhaps of being found as oneself, which Heidegger calls *Befindlichkeit* and which is sometimes translated as fundamental mood or state-of-mind. It does not seem to me that these translations convey the startling character of the experience, and perhaps the German word does not either. At any rate Heidegger goes on to use the more dramatic word *Angst* and takes pains to disassociate his meaning from fear and to draw the word closer to the dreadful, the alarming. The experience which these words point toward is the experience of standing in the radical possibility that I myself might not be rather than be. An even more startling dimension of the experience is that Dasein is thereby placed in the nearness of Sein, but we do not need to follow this aspect of the issue at the present moment.

For a discussion of what it means to be human person and what it means to think, this discovery and unconcealing of Dasein as possibility is of tremendous significance. To say that Dasein is revealed as a possibility means that in his actuality, in his facticity, and throughout his existence, he remains primordially possibility. To say this is to say a great deal. It is among other things to separate Heidegger's understanding of human nature and of thinking from those understandings which start with the given presentness of man as the point of departure for thought rather than the subject matter which needs to be thought. It is to separate him also from those understandings in both East and West which see man as fundamentally negative or fundamentally empty, and from those understandings which seek to reduce human persons to a part of a system of either nature or spirit.

This there-being who is in-the-world and who is revealed as possible-being is, Heidegger tells us, equiprimordially understanding and discourse. Here, as in the discussion of the existentialia above, it is imperative to think both understanding and discourse not as actions of Dasein but as constitutive of the way of being of Dasein. If we can keep our thinking about understanding and discourse clearly within the horizon of the ontic-ontological mode, it is clear that understanding and discourse both are *there*, are present, in the self-disclosure of Being. Hence it becomes thinkable why Heidegger approves so deeply the statement of Parmenides that thinking and Being belong together. It becomes thinkable also how the thinking that the human person does within this primordial thinking might be called a thanking. Also thinkable within this, or a similar, context are the many Taoist texts which on their surface seem to disparage knowledge. On a deeper level the intent of the texts may not be at all to keep the people ignorant but to keep them from forgetting. Perhaps thinkable also is the *hao jan chih ch'ih* of Mencius, that difficult phrase which is translated in such different ways but which is said to fill the space between heaven and earth.[9]

I would like to close with only one more effort of thinking. It is this. None of the above has anything to do with mysticism, unless that word is used to designate a very appropriate act of human persons who are seeking to think what it means that they are human persons.

College of the Holy Cross, Worcester, Mass.

NOTES

[1] Martin Heidegger, *An Introduction to Metaphysics*, trans. Ralph Manheim (New Haven, 1959), chap. 1.

[2] Heidegger, *Letter on Humanism* in *Basic Writings*, ed. David F. Krell (New York, 1977), pp. 189ff.

[3] Heidegger, *The Question of Being*, trans. William Kluback and Jean T. Wilde (New Haven, 1958), p. 75.

[4] Heidegger, *Sein und Zeit* (Tübingen: Niemeyer, 1967), sec. 9.

[5] Ibid., sec. 4.

[6] *Tao-te Ching*, chap. 18, in *A Source Book in Chinese Philosophy*, trans. Wing-tsit Chan (Princeton, 1963), p. 148.

[7] Cf. selections from sho Yung and Chang Tsai in *A Source Book in Chinese Philosophy*, chaps. 29—30.

[8] Heidegger, *Sein und Zeit*, pp. 39ff.

[9] *The Book of Mencius*, 2A:2, in *A Source Book in Chinese Philosophy*, p. 63.

CHUNG-YING CHENG

THE NATURE-BEING-PRINCIPLE: A CONSIDERATION FROM CHU HSÏ

What is the nature of mind? Mind is conceived to perform the function of perception and knowing (*chih-chüeh*[a]) and understanding (*li-hui*[b]), among other things. Chu Hsi[b′] says the following: "The knowing and perception of mind is such that it [mind] has this *li*[b″] and performs this action"[1] "Investigation of things (*ko-wu*)[b‴] is to understand *li* (*li-hui*) in things. To reach knowledge is to clarify the mind[2] (*tźu-hsin-tóu-ché*[c])."[3] One shuld note that for Chu Hsi understanding as a function of mind is to have *li* as its object, and understanding in the sense of the Chinese term *li-hui* would simply mean "meeting the *li*." *Li* is also held to be the object of knowing and perception by Chu Hsi. He says: "What mind knows and perceives is *li*. *Li* is not separate from knowing and perceiving; and knowing and perceiving is not separate from *li*."

If *li* is the object of perception, knowledge, and understanding, is *li* external or internal to mind? With this question the double meanings of *li* become evident. For Chu Hsi, *li* is both internal and external: it is internal in mind in the sense that mind has all *li*, and it is external to mind in the sense that *li* resides in all ten thousand things as it should. To understand, know, and perceive is therefore to do so in a double sense: it is to know, perceive, and understand *li* in the mind and *li* in the things. It is indeed when the *li* of the mind and the *li* of things *coincide* that one may say that understanding, perception, and knowing take place. This peculiar theory of perception, knowing, and understanding is based on the ontological conception of mind as embodying all *li*, and indeed embodying *li* as its nature and substance.[4] The ontology of *li* in this sense is essential for understanding Chu Hsi's notion of mind and his theory of knowledge and understanding.

To explain that mind has *li* as its substance, he speaks of the *nature* (*hsing*[d]): He says: "*Hsing* is all the *li* the mind has: mind is where all *li* meet."[5] "*Hsing* is *li*; mind comprehends, holds, manages, and operates the *li*"[6] He explains: "Mind takes *hsing* as substance, mind treats *hsing* as if it is the inside of a dumpling. For the reason why the mind has *li* is that it has nature."[7] From these statements, how are we to understand the relation of *hsing* to *li* in the mind? In the absence of a detailed

A-T. Tymieniecka (ed.), Analecta Husserliana, Vol. XXI, 159—163.

account of the background of *hsing* in the Confucian classics, we may say that *hsing* is what makes a particular object the type of object that it is. Specifically, *hsing* is the fundamental organizing principle for human existence and therefore the ultimate ground of *humanity*. In light of our analysis of the metaphysics of *li-ch'i*[d'] and dialectics of *t'ai-chi*[d"], *hsing* can be said to represent the refined result of the dialectical development of the *t'ai-chi*, and constitute a specific configuration and unity of *li-ch'i* in this process of development. This conforms to the *Chung Yung*[d'''] statement: "What heaven ordains is *hsing*," and this alludes to the dialectical presupposition of the notion of *hsing*. Chu Hsi recognizes this significance of *hsing* as the ontological basis of humanity, but does not seem to be aware of the dialectical presupposition of *hsing* as he is generally unaware of the model of the dialectical development of *t'ai-chi* which exist as the guiding principle of his thoughts on mind and learning, epistemology and methodology. He did recognize *hsing* as reflecting the ultimate reality of the *t'ai-chi* in mind. Thus he strongly approves the following statement of Shao Yung[e]: "*Hsing* is the shape and body of the *tao*, and mind is the walls containing the *hsing*."[8]

As the *tao*[e'] (or the *t'ai-chi*) is a unity of *li* and *ch'i*[e"] and not *li* alone, *hsing* should have *ch'i* also. To simply identify *hsing* with *li* is apparently an oversimplification and perhaps a mistake. That *hsing* has *ch'i* as a component is shown by the fact that there is movement of the *hsing*, namely, the existence of feelings (*ch'ing*[f]). Chu Hsi recognizes this: "*Ch'ing* is the activation (*fa*[g]) of *hsing*,"[9] "*Hsing* is root, *ching* is the seeds."[10] To illustrate this, he says: "Benevolence (*jen*[h]) is *hsing*, love is feeling. The feeling is the result of functioning (*yung*[i]) of nature and feelings is the state of *hsing* unactivated (*wei-fa*[j])."[11] In light of this it can be said that mind is the unity of *hsing* and *ch'ing* in the dialectical sense, and the unity of *li* and *ch'i* in the ontological sense. Chu Hsi recognizes this by saying: "The mind contains the *hsing* and *ch'ing*; *hsing* is the substance and *ch'ing* is the function."[12] The image of "containing" (*pao*[k]) and the paradigm of substance-function, however, do not really explain the dialectical nature of *hsing* and *ch'ing* in the mind, although they do lead one to see mind as functioning itself of *hsing* and *ch'ing* and therefore ultimately a function of *li-ch'i*.

This functioning of *hsing* and *ch'ing*, insofar as it is an activity that can be identified as *ch'i*, is the source and momentum of activity and energy. This is perhaps why *mind* is said to be a matter of *ch'i*. To say that mind is a matter of *ch'i* is to point to mind as an activity and process, but in light of the dialectical ontology of the *t'ai-chi*, mind is a unity of *li* and *ch'i* which is capable of realizing *li* and *hsing*. This mind-realization is possible strictly because of the ontological nature of mind and the dialectical creativity of mind qua the *t'ai-chi*, not simply because of mind as *ch'i*. This subtle distinction is important for understanding the possibility of *ko-wu* and *han-yang*[l] (perserving and nourishing mind) as two aspects of the methodology of mind-realization (*chin-hsin*[l']) and nature-realization (*chin-hsing*[m]).

The ontology of mind can therefore be described in a threefold direction. In the first place, mind contains *li* and *hsing*, and in knowing, perceiving, and understanding will realize or manifest *li* or nature. The second dimension of mind is the *ch'i*-activity of mind. In characterizing mind in this light, mind is called "the five elements of *ch'i*"[13] by Chu Hsi. As five elements of *ch'i*, mind is conceived as vacuid (*hsü*[n]) and most subtly active (*ling*[o]), most clear and sensitive (*pu-mei*[p]). He says: "*Hsü-ling* (vacuity and activity) of course is the substance of mind."[14] "Mind is the subtle brightness [of man], therefore it contains all *li* and is able to respond to all things."[15] The *ch'i*-nature of mind makes mind the most important and indeed the only instrument and agent for the transformation of person, and, consequently, the starting point for the transformation of a society and the whole world. This is possible, however, only when the activity of mind assumes the form of perception, knowing, and understanding the *li*, and reaches the point of total clarity and enlightenment which make the mental activity both correct and powerful.[16]

The third dimension of mind lies in its controlling and directing power which influences action and conduct, but which begins in the relation of mind to the dialectical ontology of the *li* and *ch'i*. Chu Hsi recognizes this dimension of mind and stresses the following expression of Chang Tsai[p'] as important: "The mind command (*t'ung*[q]) *hsing-ch'ing*."[17] In light of the ontological composition of mind, the "command" should mean both "unify" and "direct or control." Chu Hsi says: "*Hsing* is the *li* of mind, and *ching* is the moving of *hsing*. Thus, mind is the master (*chu*[r]) of *hsing-ch'ing*[r']."[18] Apparently, mind is recognized as a free agent that can make decisions and choices. But the reason why mind can direct and control itself is because it unifies both *hsing* and *ch'ing* ontologically and thus creates and moves dialectically. The statement "The mind commands *hsing-ch'ing*" has to be again seen in reference to the dialectics of unity and division of the *t'ai-chi*. It demonstrates how *li* and *ch'i* together at the level of mind may reach an accord that will both transform and influence. But again one has to see how mind may actually exercise its commanding power through a methodological cultivation of itself in terms of *ko-wu, ch'ih-ching*[s] (holding central-mindedness), and *ch'ü-ssu*[t] (removing selfish desires), etc. This latter point leads to the final characterization of mind as capable of self-cultivation. The self-cultivation of mind is the beginning of cultivation of a person, a family, a community, a society, a nation, and the whole world. For Chu Hsi, this cultivation will take the from of an epistemological quest for *li* together with an ontological reflection and nourishment of mind itself. These two constitute the two major aspects of his methodology for the cultivation of a person in the direction of achieving sagehood. That these two are possible is due to the ontological function and the dialectical creativity of mind. This constitutes the ultimate concern of his philosophy of man.

Summarizing the above, we may represent the dialectical structure of the ontology of mind by the following diagram:

t'ai-ch i

li *ch'i*

hsing (nature)

ch'ing (*ch'i*) *hsing* (*li*)

hsin (mind)

Tamkang University and University of Hawaii at Manoa

NOTES

[1] See *Chu Tzu Wen Chi*[u], Chüan[v] 55, Answer to Pan Ch'ien-chih[w].
[2] See *Chu Tzu Yü Lei*[x], Chüan 15.
[3] Ibid., Chüan 15.
[4] Ibid., Chüan 5.
[5] Ibid., Chüan 5.
[6] Ibid., Chüan 5.
[7] Ibid., Chüan 5.
[8] Ibid., Chüan 10 and Chüan 4.
[9] Ibid., Chüan 59.
[10] Ibid., Chüan 119.
[11] Ibid., Chüan 20.
[12] Ibid., Chüan 5.

[13] Ibid., Chüan 5.
[14] Ibid., Chüan 5.
[15] See *Commentary on Mencius, Ching-hsin* chapter. See also *Commentary on Ta Hsüeh*[y] *and Ta Hsüeh Huo Wen*[z].
[16] Through practice, it is capable of directing and transforming practice and action.
[17] Ibid., Chüan 98 and Chüan 101.
[18] Ibid., Chüan 5.

CHINESE GLOSSARY

a.	chih-chüeh	知覺	l.	han-yang	涵養
b.	li-hui	理會	l′.	chin-hsin	盡心
b′.	Chu Hsi	朱熹	m.	chin-hsing	盡性
b″.	li	理	n.	hsü	虛
b‴.	ko-wu	格物	o.	ling	靈
c.	tźu-hsin-tóu-ché	此心透澈	p.	pu-mei	不昧
d.	hsing	性	p′.	Chang Tsai	張載
d′.	li-ch'i	理氣	q.	t'ung	統
d″.	t'ai-chi	太極	r.	chu	主
d‴.	Chung Yung	中庸	r′.	hsing-ch'ing	性情
e.	Shao Yung	邵雍	s.	ch'ih-ching	持敬
e′.	tao	道	t	ch'ü-ssu	去私
e″.	ch'i	氣	u.	Chu Tzu Wen Chi	朱子文集
f.	ch'ing	情	v.	chüan	卷
g.	fa	發	w.	Pan Ch'ien-chih	潘謙之
h.	jen	仁	x.	chu Tzu Yü Lei	朱子語類
i	yung	用	y.	Ta Hsüeh	大學
j.	wei-fa	未發	z.	Ta Hsüeh Huo Wen	大學或問
k.	pao	包			

PART III

THE SPATIO-TEMPORAL ARTERIES OF INDIVIDUAL EXISTENCE

G. B. MADISON

DU CORPS A LA CHAIR: MAURICE MERLEAU-PONTY

I

Connais-toi toi-même! Depuis Socrates l'injonction de l'oracle qui est à Delphes est devenue l'injonction de la philosophie elle-même, et elle lui a préscrit sa tâche. Répondre à cette injonction, c'est, pour le sujet philosophant conscient de soi, se demander ce qu'il est est au juste et comment il se peut qu'il est, justement, conscient de soi comme existant. Qu'est que cela veut dire? Comment cela, est-il possible?

Depuis l'antiquité on s'est demandé en particulier comment il se peut que chez l'homme un corps (matière) se trouve comme par miracle doublé d'un esprit (conscience). Ainsi le problème de l'identité personnelle — Qui sommes-nous? — problème fondamental de la philosophie, se pose aux philosophes comme le problème de l'union de l'âme et du corps. Et ce problème, les philosophes l'ont débattu depuis l'aube de la philosophie — sans grand succès, il faut l'avouer.

Au début de la philosophie moderne, quand les grandes synthèses médiévales avaient perdu ce qu'elles auraient pu avoir de convaincant pour les esprits réfléchissants, Pascal écrivait:

> Qui ne croirait, à nous voir composer toutes choses d'esprit et de corps, que ce mélange-là nous serait très compréhensible? C'est néanmoins la chose qu'on comprend le moins. L'homme est à lui-même le plus prodigieux objet de la nature; car il ne peut concevoir ce que c'est que corps, et encore moins ce que c'est qu'esprit, et moins qu'aucune chose comme un coprs peut être uni avec un esprit. C'est là le comble de ses difficultés, et cependant c'est son propre être.

Et il ajoutait, citant S. Augustin, cet individu qui enjambe deux mondes, le monde de l'antiquité et le monde pré-moderne:

> La façon dont l'esprit est uni au corps ne peut pas être comprise par l'homme, et cependant c'est l'homme.[1]

Un état de choses incompréhensible, en effet, plus on y pense. Et le progrès de la philosophie moderne depuis Pascal ne semble pas avoir grandement contribué à sa clarification. On pourrait même dire qu'avec

A-T. Tymieniecka (ed.), Analecta Husserliana, Vol. XXI, 167—188.

le cartésianisme elle a fait de l'ancien problème de l'union de l'âme et du corps un problème tout à fait impossible à résoudre.

Si en effet les termes "corporel" et "spirituel" ou "psychique" désignent deux genres de substances — choses étendues et choses pensantes — qui sont de par leur nature la plus propre — c'est-à-dire, conceptuellement — absolument distinctes l'une de l'autre, comment en effet pourraient-elles jamais s'unir pour former, non pas une simple juxtaposition de deux choses disparates, mais, comme c'est évidemment le cas, une seule chose avec sa réalité propre: l'homme, la personne humaine? Cette union mystérieuse ne peut s'accomplir que par miracle — par le *fiat* divin. Mais accorder le secret à Dieu, c'est le retirer aux hommes, qui ont le plus besoin de le connaître. Le résultat du cartésianisme, c'est de poser, d'un côté, une nature totalement objective et totalement intelligible — n'étant que matière en motion dont les lois sont susceptibles d'être découvertes par les science objectives de la natures — et, de l'autre côtè, des sujets conscients qui ne sont que pure intériorité et qui, puisqu'ils ne sont que des purs regards *sur* la nature n'existant nulle part *dans* cette nature intelligible en soi, ne peuvent ni être intelligible pour soi, ni objets d'aucune science.

Un seul mot peut servir à désigner l'essence de la pensée moderne depuis Descartes jusqu'à Sartre: le *dualisme*. Et le résultat du dualisme, c'est l'aliénation: l'aliénation du sujet de l'objet, du psychique du corporel, l'aliénation donc de l'homme de la nature et de sa propre nature. Une conscience aliénée d'elle-même est une conscience scindée à l'intérieur d'elle-même, et une telle conscience, nous le savons depuis Hegel, est une conscience malheureuse.

Pourra-t-on jamais sortir de ce malheur causé par la rupture entre le subjectif et l'objectif, le soi et le monde, le corporel et le psychique, inaugurant de la sorte une époque nouvelle dans la pensée? C'est, je pense, le défi que notre temps lance à la philosophie.

Quand on se retrouve dans une impasse, c'est parce que quelque part en cours de route on a pris un faux tournant. Le seule remède à un tel moment, c'est de retracer ses pas. C'est ce qu'accomplit la psychanalyse vis-à-vis de l'individu égaré. En le faisant revivre des expérience déformatrices de son passé, elle permet à l'individu de prendre un départ nouveau dans sa vie.

Or, le grand mérite de Husserl, c'est d'avoir entrepris le même genre de traitement vis-à-vis de l'homme de l'époque moderne. Dans l'ouvrage qu'il a commencé à la veille de sa mort, *La crise des sciences européennes et la phénoménologie transcendantale,* Husserl entreprenait une véritable archéologie de la conscience moderne dans le but de clarifier "l'origine de l'opposition moderne entre l'objectivisme physicaliste et le subjectivisme transcendental."[2] "Ce qui est évidemment nécessaire, disait-il, c'est que nous réfléchissons en arrière, d'une manière parfaitement *historique* et *critique*, de sorte à rendre possible une compréhension de soi radicale."[3] Ce "Questionnement en retour" (*Rückfrage*) est une véritable déconstruction de la philosophie moderne; il est la tentative, selon l'expression d'un interprète, "de revivre la tradition dont nous faisons partie avec le but de nous libérer des préjugés qui sont inhérents à cette tradition."[4]

Ce que, de cette façon, Husserl réussit à montrer, c'est comment l'origine du dualisme dans sa forme moderne coincide avec la montée de la science moderne et sa façon de concevoir la nature comme un monde de simples corps fermé sur lui-même. La conséquence de cette conception du monde est, pour ainsi dire, de scinder le monde en deux mondes: la nature physique et l'univers psychique. Cette "rationalité naturelle-scientifique" rend ainsi possible la psychologie au sens moderne du mot, qui s'occupe du psychique, par opposition avec le physique.[5] Le développement de cette psychologie à travers Descartes et les empiristes anglais jusqu'à la psychologie expérimentale d'aujourd'hui est le développement d'une lutte entre l'*objectivisme* et le *transcendantalisme.* Pour Husserl la vocation historique de la phénoménologie à cet égard est de résoudre ce conflit en montrant comment le monde dit objectif ne se suffit justement pas à lui-même, mais est plutôt quelque chose d'entièrement relatif et a en fait son origine dans la subjectivité transcendantale. C'est cette dernière qui elle-même constitue le "sens ontique" (*Seinssinn*) du monde, qui constitue "le monde "objectivement vrai""; le monde de la science objectiviste n'est autre chose qu'une structure de la conscience transcendantale.[6]

Mais on pourrait bien se demander si Husserl a vraiment réussi à dépasser le dualisme moderne ou s'il n'a pas plutôt simplement substitué à un dualisme de substances (matière, espirt) un dualisme épistémologique (objectivisme, transcendantalisme). Bien sûr, l'objectivité naturaliste est chez Husserl absorbé dans la subjectivité transcendantale qui est alors conçue comme *omnitudo realitatis,* le *seul* univers qu'il y a — ce qui constitue une forme de réductionisme (correspondant à la réduction du subjectif à l'objectif dans le physicalisme). Mais tout réductionisme présuppose, conceptuellement, le dualisme, qu'il n'élimine donc pas.

L'idéalisme transcendantal de Husserl qui privilègie la subjectivité transcendantale, est-il, en fin de compte, beaucoup plus satisfaisant qu'un matérialisme objectiviste qui fait de la conscience un simple épiphénomène du corps physique?

Merleau-Ponty ne le croyait pas, et il était rebuté par l'idéalisme transcendantal de Husserl. Ce qu'il ne pouvait pas accepter, c'est la notion d'un Ego transcendantal qui fait tout, qui constitue tout, et en première instance son propre corps comme chose mondaine.[7] La difficulté que Husserl éprouvait dans la Ve Méditation cartésienne à se défendre d'une manière convaincante contre l'objection de solipsisme suffisait, je pense, à convaincre Merleau-Ponty que l'idéalisme transcendantal de Husserl restait toujours une forme de subjectivisme et donc prisonier du dualisme moderne. Ce qu'il disait de Sartre, il aurait pu le dire aussi de Husserl: "Chez Sartre, c'est toujours *moi* qui fait la profondeur, qui la creuse, qui fait tout, et qui ferme du dedans ma prison sur moi . . . ."[8]

Pour Merleau-Ponty il n'était jamais question d'un Ego transcendantalement souverain, ni même, comme Husserl disait, d'une "âme" ou d'un "esprit" directement accessible par la "perception interne". Faisant en fait reférence à Husserl, il disait: ". . . il n'y a pas d'homme intérieur, l'homme est au monde, c'est dans le monde qu'il se connait."[9] Et ce que son premier grand livre, la *Phénoménologie de la perception* avait pour but de démontrer, c'est qu'entre moi et mon corps, entre ce corps que je suis et le monde qu'il habite et anime, et entre moi et autrui il n'existe aucun dualisme.

Si le dualisme moderne n'est en fait que le résultat d'une certaine manière arbitraire de voir le monde, s'il est en fait le produit de l'abstraction et de l'idéalisation scientifiques, la redécouverte du *Lebenswelt* en-deça de la pensée objectivante de la science nous fournit le moyen de dépasser le dualisme moderne. C'est justement la leçon que Merleau-Ponty tirait de Husserl. Mais Husserl voulait montrer ensuite comment le *Lebenswelt* est lui-même le produit d'un Ego constituant, et c'est ce que Merleau-Ponty a refusé d'accepter. Dans cette tentative de surmonter définitivement le dualisme et le subjectivisme modernes, Merleau-Ponty rejoignait le courant dominant de notre temps. Son effort philosophique s'alliait en ce sens avec celui de Heidegger. Mais on pourrait aussi se demander si, en détruisant la primauté de la conscience et en défaisant le cogito, en insistant sur la priorité de l'irréfléchi sur la réflexion et sur le fait que la conscience n'est pas constituante mais plutôt donnée à elle-même à partir de la vie obscure du corps, à partir d'une chair polymorphe qui est anonymat et généralité confuse, on pourrait se demander si Merleau-Ponty n'annonçait pas en fait ce mode de pensée qui a pris l'essort depuis sa mort et qui s'est donné la tâche de déconstruire la subjectivité elle-même, rejetant en cours de route toute la pensée moderne, en premier lieu l'idée même du sujet rationnel et autonome, c'est-à-dire l'"homme".

Ce courant de pensée, c'est ce que des sociologues ont appellé le post-modernisme, et il suffit peut-être pour le caractériser de citer trois de ses porte-paroles:

L'écrivain américain, Norman O. Brown, qui dans ses livres *Life Against Death* et *Love's Body* a célébré la perversité polymorphe et a proclamé la supériorité du principe du plaisir sur le principe de la réalité, de l'instinct sur la raison. L'effort de Brown était de détroner la conscience en faveur du rêve. Pour lui l'essentiel était la dissolution du soi dans le corps, l'éradication de toutes les bornes et de toutes les distinctions — les distinction entre le soi et le corps, entre l'homme et la femme, entre le sujet et l'objet, entre le psychique et le physique. Ces divisions constituaient pour lui la chute originelle, la chute dans la division et le mensonge, et ce qui importait pour lui était l'abolition de "tous les dualismes".

Le philosophe français, Michel Foucault, qui lui aussi a contesté la primauté de la conscience et de la rationnalité — enfin toute cette pensée moderne qui a tant insisté sur la liberté, la dignité, et la responsabilité de l'individu. L'idée de l'homme, de l'homme rationnel et autonome, il faut la déconstruire par — ici aussi — une archéologie du savoir. Foucault, l'historien de la folie, nous suggère que la folie est une forme de connaissance que la bourgeoisie moderne a délibérément supprimée. La morale de son histoire étant que le rationnel ne détient aucun privilège sur l'irrationnel. La folie est libératrice.

Le psychologue anglais, R. D. Laing, enfin, qui a fait l'éloge de la folie et qui considère la schizophrénie comme une forme de "futurisme". Le schizophrène est pour lui un courageux cosmonaute explorant les confins d'un univers encore inconnu.

Ce qui caractérise le post-modernisme en général, c'est le refus de toutes les distinctions et de toutes les bornes — entre, par example, le corps et l'esprit, le physique et le mental, l'irrationnel et la raison, l'intellectuel et le sensuel, le soi et l'autrui, la nature et la culture, la réalité et l'utopie. D'un côté le rejet de l'ordre, des limites, et des restrictions, du contrôle et mesure de la conscience raisonnable, de l'apolinien et en général de ce qu'on avait jusqui'ici coutume d'appeler la civilisation, il est, de l'autre côté, une célébration du dionysien, du délire, de l'onirique, de l'érotique, de la sensualité libinalle, de l'instinctif et de la spontanéite, du narcissisme. Antinomien et anti-institutionaliste, utopique et anarchique, appel à la suppression de la conscience oppressive, libération des instincts, émancipation de la sensualité, il s'allie facilement avec le radicalisme politique.

La question que je voudrais considérer ici, c'est donc de savoir comment il faut interpréter le rejet de la part de Merleau-Ponty du dualisme et du subjectivisme modernes. Comment, vingt ans après sa mort, comment faut-il le situer vis-à-vis de la pensée contemporaine? La compréhension humaine étant toujours rétroactive, on comprend ce qui a été en le situant par rapport à ce qui n'existait pas encore à ce moment-là, par rapport à ce qui l'a en fait suivi dans le temps. Devra-t-on donc voir en Merleau-Ponty comme un précurseur du post-modernisme? Ou devra-t-on au contraire interpréter sa critique du modernisme d'une autre façon encore, de sorte à ne pas faire de lui un simple proto-post-structuraliste?

II

Dans son tout premier livre, *La Structure du comportement*, Merleau-Ponty s' est donné comme tâche "de comprendre les rapports de la conscience et de la nature."[10] Autrement dit, les rapports de l'objectif et du subjectif, de l'extérieur et de l'intérieur, du corporel et du psychique. Cette problématique devait rester la problématique centrale de toute son oeuvre ultérieure. Voyons donc brièvement comment dans ses deux premiers ouvrages le jeune Merleau-Ponty répondait à la question.

D'abord, dans *La Structure du comportement* il soulevait expressément le problème de l'union de l'âme et du corps. Et il rejetait catégoriquement l'idée qu'il s'agit là de l'union de deux substances ou de deux ordres de réalité, de deux mondes, comme le dirait Husserl. C'est dire qu'il relativise complètement le corps et l'esprit, insistant sur ce qu'ils n'ont aucune réalité en eux-mêmes. Ce qu'ils sont, c'est plutôt deux termes variants d'une structure ou d'une structuration unique qu'est l'existence même. Le rapport entre les deux est entièrement dialectique: "Il ne s'agit pas de deux ordres de faits extérieurs l'un à l'autre, mais de deux types de rapports dont le second intègre le premier." (SC, 195)

Il nous est cependant important de noter que si Merleau-Ponty rejette ici tout *dualisme*, il ne préconise pas pour autant un simple *monisme*. L'intégration du corps et de l'âme n'est point une *fusion*. Il n'y a certes pas deux substances en l'homme, mais par contre l'homme n'est pas une entité rigide et monolithe. Il y a bien une "âme" et un "corps", mais le corps n'est corps *humain* qu'en étant la fondation même de l'âme, l'expression visible d'une vie "spirituelle", et l'âme n'est l'âme que par le moyen d'un corps qui est comme sa propre manifestation. Il y a entre âme et corps comme une polarité de tension, et l'homme total n'est rien d'autre que cette tension sans cesse renouvelée. Si ainsi Merleau-Ponty nie ici catégoriquement tout *dualisme* chez l'homme, il ne reclame pas moins une certaine *dualité* à son être. "Il y a une dualité, dit-il, qui réapparaît toujours à un niveau ou à l'autre." C'est dire que l'"intégration n'est jamais absolue et elle échoue toujours, plus haut chez l'écrivain, plus bas chez l'aphasique." (SC 226)[10]

Il est important de noter aussi que si l'âme et le corps sont tout relatifs et ne sont que deux moments d'une seule structure ou *Gestalt*, cette structuration est pourtant *verticale*. Je veux dire par là que l'"âme" désigne ce qui, à n'importe quel niveau, relève du supérieur dans le comportement humain, le "corps" désigne ce qui relève de l'inférieur.[11]

Il importe de noter également que, contrairement à une certaine tendence de pensée, Merleau-Ponty n'essaie jamais d'expliquer le supérieur en termes de l'inférieur. L'avènement du supérieur supprime comme autonome l'inférieur, en l'intégrant dans un comportement élargi et mieux structuré.[12] Comme Husserl qui insistait que l'anormalité n'est qu'une variante de la normalité (et pas l'inverse, comme chez les post-modernistes),[13] Merleau-Ponty maintient que l'anormal et le pathologique doivent se comprendre par reférence au comportement dit normal.

La conclusion qu'il faut en tirer, c'est que Merleau-Ponty ne voulait absolument pas à l'époque abolir "tous les dualismes". Qu'en est-il maintenant de la *Phénoménologie de la perception*?

A lire le compte rendu de la conférence faite par Merleau-Ponty sur "Le Primat de la perception et ses conséquences philosophiques", il est évident que beaucoup de ses auditeurs pensaient que Merleau-Ponty voulait dissoudre le cogito dans l'existence anonyme du corps — dans le

Lebenswelt — et qu'il niait l'existence du sujet rationnel et autonome. Ne voulait-il pas célébrer ce qu'il appellait "la spontanéité enseignante du corps" au grand dépens du dur travail de la réflexion rationnelle?

Le Merleau-Ponty de la *Phénoménologie de la perception* n'était pourtant pas un philosophe de l'irrationnel ou de l'irréfléchi, et il est un peu ironique que certains ont critiqué son projet central, qui était de réfléchir sur l'irréfléchi comme tel, en disant qu'il était une contradiction dans les termes, puisque du moment que l'irréfléchi est saisi par la réflexion il cesse d'être, justement, l'irréfléchi. On ne peut pas retourner à et coincider avec l'irréfléchi comme tel. L'ironie est que Merleau-Ponty n'a jamais proposé de faire coincider la réflexion à l'irréfléchi. Sa philosophie était une philosophie proprement réflexive, au sens le plus propre du mot, et il n'a jamais soutenu que l'important est de retourner à l'expérience simplement vécu afin de s'y installer. Pour lui la philosophie était une entreprise réflexive, et était donc une chose toute différente de l'existence directement vécue d'une façon irréfléchie. Ce qu'il voulait faire était quelque chose de proprement apolinien; il ne voulait pas renoncer à la raison mais voulait plutôt l'élargir. Faisant referençe à Hegel, il disait que "la tâche de notre siècle" est d'"explorer l'irrationnel et l'intégrer à une raison élargie".[14] Ce qu'il voulait comprendre était justement la conscience rationnelle elle-même, en tant précisément, qu'elle ne se possède que sur le fond d'une existence irréfléchie d'ou elle émerge — mais aussi qu'elle transcende (c'est-à-dire, à laquelle elle ne se réduit absolument pas).

Les deux textes suivants disent, je pense, tout ce qu'il y a à dire à ce sujet:

> Assurément, une vie n'est pas une philosophie. Je pensais avoir indiqué . . . que la description n'est pas le retour à l'expérience immédiate; on ne retourne jamais à l'expérience immédiate. Il s'agit de savoir seulement si nous allons essayer de la comprendre. Je crois que la tentative d'exprimer l'expérience immédiate n'est pas une trahison de la raison mais, au contraire, une contribution à son aggrandissement.[15]

> Loin que la philosophie nous apparaisse comme un inutile redoublement de la vie, elle est au contraire pour nous l'instance sans laquelle la vie aurait chance de se dissiper dans l'ignorance de soi ou dans le chaos.[16]

Travaillant ainsi pour l'agrandissement de la raison et étant opposé à la dissipation de la conscience dans l'ignorance de soi et dans le chaos — je soutiens donc que le jeune Merleau-Ponty n'était nullement un prophète de ce Dionysus nouveau, l'homme post-moderne, l'homme enfin libéré de la tyrannie de la conscience de soi.

III

Peut-on en dire autant, cependant, du dernier Merleau-Ponty? N'est-il pas plutôt le cas que dans *Le Visible et l'invisible* tout a été bouleversé et mis à la dérive? N'a-t-il pas lui-même dit qu'il fallait "tout reprendre, rejeter les instruments que la réflexion et l'intuition se sont donnés, s'installer en un lieu où elles ne se distinguent pas encore, dans des expériences qui n'aient pas encore été "travailées", qui nous offrent tout à la fois pêle-mêle, et le "sujet" et l'"objet", et l'existence et l'essence, et lui donne donc les moyens de les redéfinir"? (VI, 172) En reconnaissant la "Nécessite d'un retour à l'ontologie", une "ontologie de l'Etre brut", (VI, 219) au "principe barbare dont parlait Schelling", n'était-il pas tous les dualismes qu'il voulait alors surpasser, et d'une façon décisive?

Il est en tout cas évident que cette nouvelle interrogation ontologique devait avoir pour lui des importantes ramifications sur "la question sujet-objet", "la question de l'inter-subjectivité", "la question de la Nature". (VI, 219) N'est-il pas le cas qu'en s'enfonçant dans "le sous-sol de la phénoménologie",[17] de la conscience, dans l'obscur royaume de l'inconscient (qu'il avait auparavant nié), Merleau-Ponty s'est retrouvé de l'autre côté de la barrière, de l'autre cote de toute philosophie de la conscience et de la subjectivité elle-même? C'était, semble-t-il, un peu l'impression de Sartre. "En un sens, disait-il, rien n'a changé des idées qu'il soutenait dans sa thèse (la PhP); en un autre, tout est méconnaissable: il s'est enfoncé dans la nuit du non-savoir, à la recherche de ce qu'il appelle, à présent, le "fondemental"."[18] Merleau-Ponty, devenait-il à la fin un disciple de ce grand précurseur du post-modernisme, Baudelaire, qui, à la fin de son poème, *Le Voyage*, avait écrit:

Nous voulons, tant ce feu nous brûle le cerveau,
Plonger au fond du gouffre, Enfer ou Ciel, qu'importe?
Au fond de l'Inconnu pour trouver du *nouveau*!

En tout cas, le "fondemental" nouveau qu'a découvert Merleau-Ponty par sa descente au sou-sol de la conscience, il lui a donné un nom: la *chair*. Et il faut bien reconnaître que la chair n'est pas simplement un autre nom pour ce qu'en philosophie traditionnelle on appelle le corps; "l'on sait, disait-il, qu'il n'y a pas de nom en philosophie traditionnelle pour désigner cela", (VI, 183) pour désigner ce que désigne ce mot philosophiquement insolite: la chair. En effet, la chair n'est ni matière, ni esprit, ni substance. (VI, 184) L'exploration de la chair devait donc inéluctablement amener Merleau-Ponty a des "conséquences extravagantes". (VI, 184)

Elle devait en fait l'amener à une sorte de monisme philosophique par delà tous les dualismes. N'étant pas le corps objectif, ni non plus le corps pensé par l'âme comme sien, la chair est plutôt le sensible lui-même, "le sensible au double sens de ce qu'on sent et ce qui sent." (VI, 313) La chair est le milieu formateur et du corporel et du psychique, et de l'objet et du sujet; elle est l'Etre d'indivision d'avant le clivage conscience-objet. Elle est la "généralité du Sensible en soi", l'"anonymat inné du Moi-même". (VI, 183) Dans une note de travail de novembre 1960 il écrivait:

> L'unité *préalable* moi-monde, monde et ses parties, parties de mon corps, unité avant ségrégation, avant dimensions multiples, — et de même l'unité du temps — Non pas architecture de noèses-noèmes, l'une sur l'ature posées, se relativisant l'une l'autre sans réussir à s'unifier: mais il y a d'abord leur lien profond par *non-différence* (VI, 315)

La découverte de la chair n'exige rien de moins qu'une "reconstruction complète de la philosophie". (VI, 246) La philosophie ne peut plus être la simple réflexion psychologique ou même transcendantale. (VI, 209) Tout le vocabulaire de la réflexion philosophique traditionnelle est à rejeter. "Il nous faut renoncer, en commençant, à des notions telles qu' "actes de conscience", "états de conscience", "matière", "forme", et même "image" et "perception"." (VI, 209)

C'est une "reconstruction complète" de notre langage lui-même qui est exigée. Le "langage est une puissance d'erreur, dit-il, puisqu'il coupe le tissu continu qui nous joint vitalement aux choses et au passé, et s'installe entre lui et nous comme un écran." (VI, 166) Ce que la philosophie devrait chercher, il semble vouloir dire dans ce moment tardif de sa vie, c'est "un langage de la coincidence, une manière de faire parler les choses mêmes. Ce serait un langage dont il [le philosophe] ne serait pas l'organisateur, des mots qu'il n'assemblerait pas, qui s'uniraient à travers lui par entrelacement naturel de leur sens, par le trafic occulte de la métaphore — ce qui compte n'étant plus le sens manifeste de chaque mot et de chaque image, mais les rapports latéraux, les parentés, qui sont impliqués dans leurs virements et leur échanges." (VI, 167)

Ce langage libéré de l'emprise de la conscience calculatrice et manipulatrice, de ses significations claires et distinctes, ce langage métaphoriquement sauvage qui dit sans que rien ne soit apparemment dit, qui parle en suggestions occultes, n'est-il pas justement ce langage réalisé par les philosophes et les écrivains qui sont venus occuper l'avant-scène de notre culture depuis la mort de Merleau-Ponty? La "reconstruction complète" demandée par Merleau-Ponty, n'est-elle pas cette déconstruction complète opérée par les post-modernistes du structuralisme?

En tout cas, il faut reconnaître que la découverte de la chair et de l'être brut ou sauvage a amené Merleau-Ponty à dire des choses qui sont effectivement extravagantes d'un point de vue traditionnel. Il dit par exemple que "les choses nous ont et que ce n'est pas nous qui avons les choses . . . Que le langage nous a et que ce n'est pas nous qui avons le langage. Que c'est l'être qui parle en nous et non nous qui parlons de l'être". (VI, 247) Le point de vue du *sujet* parlant et pensant semble bien être dépassé: ". . . les choses se *trouvent dites* et se *trouvent pensées* comme par une Parole et par un Penser que nous n'avons pas, qui nous ont." (S, 27)

L'idée même de l'homme ne semble plus avoir cours, puisque ce n'est pas lui qui a proprement parler *fait* quoi que ce soit. "Ce n'est pas nous qui perçevons, c'est la chose qui se perçoit là-bas, – ce n'est pas nous qui parlons, c'est la vérité qui se parle au fond de la pensée – Devenir nature de l'homme qui est le devenir homme de la nature". (VI, 239) Ainsi, l'homme semble se retrouver réintéré à la nature naturante. L'anthropologie a été déconstruite. "Quand nous parlons de la chair du visible, insiste-il, nous n'entendons pas faire de l'anthropologie . . ." (VI, 179) La philosophie du dernier Merleau-Ponty semble bien donc ne plus être une philosophie de l'homme mais la pensée d'un être polymorphe ou amorphe, un être de promiscuité et de transitivisme.[19] Merleau-Ponty dit que sa philosophie devait se réaliser "sans aucun compromis avec l'*humanisme*" et qu' "Il faut décrire *le visible* comme quelque chose qui se réalise à travers l'homme, mais qui n'est nullement anthropologie". (VI, 328)

Un dépassement de tous les dualismes de ce genre serait de sorte à effectuer la guérison du malheur de la conscience moderne causé par le subjectivisme et la séparation de la conscience de la Nature qui est "la chair, la mère", (VI, 321) "matrice polymorphe", (VI, 274) "Etre d'indivision", (VI, 262), "prégnance de possibilités". (VI, 304) Guérison par le retour au sein maternel et à l'union d'avant la séparation. De cette sorte se trouverait surmontée la séparation la plus douloureuse, la ségrégation du soi et de l'autrui. "Il n'y a pas ici de problème de l'*alter ego*, dit-il, parce que ce n'est pas *moi* qui vois, pas *lui* qui voit, qu'une visibilité anonyme nous habite tous deux, une vision en général . . ."; (VI, 187) "il y a transitivisme par généralité". (VI, 322) Moi et autrui, "Nous fonctionnons comme un corps unique"; (VI, 268) "lui et moi sommes comme les organes d'une seule intercorporéité." (S, 213) "S'il y a coupure, dit-il, ce n'est pas entre moi et l'autre, c'est entre une généralité primordiale où nous sommes confondus et le système précis moi-les autres." (S, 220) Ce qu'il y a au fond est une "corporéité en générale" qui, comme "l'égocentrisme de l'enfant" est "transitivisme et confusion du moi et de l'autre." (S, 220) Adultes civilisés qui ont appris à voir nos vies "comme une serie d'états de consciences privés", (S, 221) nous avons aussi "appris à distribuer entre des vies uniques ce qu'il y a de douleur et de plaisir dans le monde. Mais la vérité est moins simple". (S, 220) La conclusion de Merleau-Ponty: "Il faut donc concevoir, — non pas certes une âme du monde ou du groupe ou du couple, dont nous serions les instruments, — mais un *On* primordial qui a son autenticité, qui d'ailleurs ne cesse jamais, soutient les plus grandes passions de l'adulte, et dont chaque perception renouvelle en nous l'expérience . . ." (S, 221) L'"égologie" et la "sphère d'appartenance" dont parlait encore Husserl se retrouve ainsi dissoutes en faveur des "attaches primordiales". Et ainsi "la comprésence de ma "conscience" et de mon "corps" se prolonge dans la compréscence d'autrui et de moi"; la présence de moi à moi et la présence de moi à autrui sont "substituables dans la présence absolue des origines." (S, 221) La chair n'est donc pas ma chair ou votre chair mais la chair indivise du monde, de l'Etre brut et sauvage, de la Terre-mère qui nous englobe tous.

Le temps lui-même, enfin, se retrouve aboli, le temps de l'homme éveillé et conscient de soi, source de toutes les distinctions et de toutes les douleurs, le "Cela était" de Nietzsche, origine du ressentiment.

Il s'agit de trouver dans le présent, la chair du monde (et non dans le passé) un "toujours neuf" et "toujours le même" — Une sorte de temps du sommeil. . . . Le sensible, la Nature, transcendent la distinction passé, présent, réalisant un passage par le dedans de l'un dans l'aitre Eternité existentielle. L'indestructible, le Principe barbare (VI, 320—1)[20]

Voilà donc la sorte de propos que tenait le dernier Merleau-Ponty. Tout cela, il faut l'avouer, pourrait être assez déconcertant pour une philosophie du sujet qui récuse le dualisme et le subjectivisme modernes et qui pourtant refuse la dissolution post-moderniste du sujet lui-même. Il est donc temps de s'interroger sur la signification précise de ce passage du corps à la chair chez le dernier Merleau-Ponty.

Ce qu'il faut noter toute de suite à cet égard, c'est que si Merleau-Ponty a voulu dépasser résolument tout dualisme, il ne proposait pas pour autant de nier toutes les dualités ni d'abolir toutes les distinctions. La chair n'est pas la notion-clé d'une philosophie de l'Identité par-delà la Différence. La chair n'est pas le corps, il est vrai. Elle contient et le corporel et le psychique; elle est l'unique réalité dont l'esprit et le corps, le sujet et l'objet, ne sont que des "parties totales". Mais si on peut même parler de "parties totales", c'est que les mots corporel et psychique gardent toujours un sens; ils continuent de désigner quelque chose, non pas, bien sûr, deux substances,[21] mais, quand même, deux *dimensions* irréductibles d'un Etre qui est la dimensionalité même. La chair est le fait qu'entre le psychique et le corporel, entre le sujet et son corps, entre lui et le monde, entre le soi et l'autre, il y a rapport de circularité et même de réversibilité. Mais réversibilité n'est pas coincidence. "Il est temps de souligner, disait-il, qu'il s'agit d'une réversibilité toujours imminente et jamais réalisée en fait." Quand une de mes mains touche l'autre en train de toucher les choses, "je ne parviens jamais à la coincidence". Cette coincidence s'éclipse toujours, et il reste comme un "hiatus" entre les deux côtés du corps propre, le côté "sujet" et le côté "objet". (VI, 194—5)

Dans sa dernière philosophie, Merleau-Ponty ne *supprime* donc pas la dualité sujet-objet, mais il la *dépasse* en concevant le rapport intentionnel, conscience-objet, comme un rapport qui se fait à l'intérieur de l'Etre lui-même, un Etre qui transcende la distinction conscience-objet.[22]

S'il n'y a pas, entre le psychique et le corporel, identité, fusion ou coincidence, le but de la philosophie elle-même ne peut pas être de nous faire coincider avec "la présence absolue des origines". Elle ne peut pas avoir pour tâche d'effectuer une fusion avec le sensible, l'Etre d'indivision par delà le rapport sujet-objet. Nous sommes donc peut-être en mesure maintenant de comprendre cet étrange passage dans *Le Visible et l'invisible* où Merleau-Ponty parlait d'"un langage de coincidence" et où il disait que le philosophe devait livrer son langage au "trafic occulte de la métaphore". (VI, 166—7) Et en effet il ajoutait à la fin de ce passage une remarque que je n'avais pas cité et qui change tout. Si le philosophe revendique un langage de ce genre, "il faut bien voir, disait-il, la conséquence: si le langage n'est pas nécessairement trompeur, la vérité n'est pas coincidence, muette." (VI, 167)

Merleau-Ponty n'était donc nullement en train de proposer que le philosophe "devrait se taire, coincider en silence, et rejoindre dans l'Etre une philosophie qui y est déjà faite." (VI, 166) La situation véritable était pour lui tout autre. Le langage du sujet réfléchissant n'est pas *coincidence* avec la chair mais *sublimation* de la chair. La thématisation du sens perceptif ou du *Lebenswelt* — objet de la philosophie — doit être comprise comme un "comportement de degré supérieur". (VI, 230) Entre le silence des origines et le langage de la réflexion, il y a un rapport, dit-il, dialectique: "le langage réalise en brisant le silence ce que le silence voulait en n'obtenait pas." (VI, 230) [Et comme] il disait aussi, d'une façon qu'on ne pourrait vouloir plus nette, la "description même du silence repose entièrement sur les vertus du langage. La prise de possession du monde du silence, telle que l'effectue la description du corps humain, n'est plus ce monde du silence, elle est le monde articulé, élevé au *Wesen*, parlé". (VI, 233) Le silence qui suit le langage n'est pas le silence qui le précède; c'est un produit du travail de la réflexion. La philosophie n'est donc pas inutile, vanité subjectiviste, car sans elles on ne *saurait* même pas *ce que veut dire* notre inhérence à la chair du monde.

Merleau-Ponty ne propose donc pas de supprimer la conscience, la réflexion, la science et la philosophie. Dans une note de travail intitulée, "Science et philosophie", il s'attaque au scientisme moderne et à la phénoménologie des *Erlebnisse* ou de la conscience (disant en passant qu'on ne peut pas comprendre l'histoire en termes de décisions subjectives) — mais il s'attaque aussi à ce "primitivisme" tant célébré par les post-modernistes anti-scientifiques. Il y écrit:

> La recherche de la vue "sauvage" du monde ne se limite nullement à un retour à la précompréhension ou à la préscience. Le "primitivisme" n'est que contrepartie du scientisme, et encore scientisme. Les phénoménologues (Scheler, Heidegger) ont raison d'indiquer cette précompréhension qui précède l'inductivité, car c'est elle qui met en question la valeur ontologique du *Gegen-stand.* Mais un retour à la pré-science n'est pas le but. La reconquête du *Lebenswelt*, c'est la reconquête d'une *dimension* dans laquelle les objectivations de la science gardent elles-mêmes un sens et sont à comprendre comme *vraies* . . . — le pré-scientifique n'est qu'une invitation à comprendre le méta-scientifique et celui-ci n'est pas non-science. Il est même dévoilé *par* les démarches constitutives de la science . . . . (VI, 236)

Ne voulant pas supprimer la réflexion scientifique et philosophique, il ne voulait surtout pas supprimer ce qu'il avait appelé la pensée du subjectif, de rejeter cette subjectivité qui est la grande découverte (devrait-on dire plutôt "invention"?) de la philosophie moderne. Sur ce point il était catégorique. Faisant référence à Heidegger qui, comme on le sait, voulait se débarasser de la subjectivité ou de la conscience de soi parce qu'elle nous cache, pensait-il, l'être, Merleau-Ponty disait au contraire: "la pensée du subjectif est une de ces solides que la philosophie devra digérer. Ou encore, disons qu'une fois "infectée" par certaines pensées, elle ne peut plus les annuler; il faut qu'elle en guérisse en inventant mieux. . . . La subjectivité est une de ces pensées en deça desquelles on ne revient pas, même et surtout si on les dépasse." (S, 194) Chez Merleau-Ponty il n'était aucunement question de la mort de l'homme, et il ne se proposait pas, comme les post-modernistes, d'achever la conscience de soi.

Dans la première partie de ce texte, j'ai dit qu'en matières politiques le post-modernisme s'allie facilement au radicalisme ou au révolutionisme et qu'on peut alors le reconnaître par son antinomisme et son anti-institutionalisme, par sa croyance en l'utopie et l'anarchie. La position politique du dernier Merleau-Ponty devrait alors confirmer le fait qu'il n'est pas à compter aux rangs des post-modernistes. Car à la fin de sa vie il ne croyait absolument pas dans l'Utopie, dans la possibilité d'une résolution définitive des différences et des conflits qui séparent les hommes, et c'est la raison pour laquelle il a nettement pris ses distances vis-à-vis du marxisme. Dans *Les aventures de la dialectique* Merleau-Ponty rejetait d'une façon décisive l'idéologie révolutionniste et affirmait sa solidarité avec la tradition libérale de l'Occident. Il y disait que "la révolution qui recréerait l'histoire est à l'infini",[23] et il qualifiait de "fictions" les notions du pouvoir prolétarien, de la démocratie directe et du déprissement de l'Etat. (AD, 299)

Loin d'être anti-institutionaliste, Merleau-Ponty insistait sur ce que sans les institutions libérales il n'y a pas de liberté, et sans la liberté aucune résolution acceptable et humaine du problème social. Si les problèmes qui nous agitent aujourd'hui ont une solution, disait-il, "elle est libérale". (S, 433) L'important pour lui, c'est qu'on trouve "des institutions qui implantent dans les moeurs cette pratique de la liberté." (S, 434) Car, disait-il, "il faut à la liberté du substantiel, il lui faut un Etat, qui la porte et qu'elle anime." (S, 433) Dans la politique Merleau-Ponty adhérait pleinement à l'essentiel du libéralisme traditionnel de nos démocracies occidentales, basé qu'il est sur les notions de l'individu autonome et responsable, l'importance des limites et des formes.[24]

Rejeter de cette sorte l'utopie revient à réaffirmer l'histoire, cette même histoire que les structuralistes autant que les révolutionnistes et les psychanalistes de la coincidence ou de la perversité polymorphe voudraient achever. C'est réaffirmer les vertus de ce qu'on a traditionnellement appelé la Civilisation. Ce n'est pas une éloge du primitivisme ou du naturel que Merleau-Ponty faisait quand en 1957, de retour d'un séjour à Madagascar, il répondait aux questions d'un journaliste. A la fin de l'interview l'interviewer lui a dit, "Vous semblez croire à une supériorité de nos valeurs, de celles des civilisations occidentales, sur celles des pays sous-développés . . .", et Merleau-Ponty lui a répondu avec les remarques qui suivent et que, pour finir avec cette partie de mon texte, je voudrais citer en entier, puisque c'est rarement qu'on se dérange pour les lire.

> Non certes à leur valeur morale, et encore moins à leur beauté supérieure, mais, comment dire, à leur valeur *historique*. En atterrissant à Orly, au petit jour, après un mois à Madagascar, quelle stupeur de voir tant de routes, tant d'objets, tant de patience, de labeur, de savoir, de deviner aux lumières qui s'allument tant de vies distinctes qui se réveillent dans le matin. Ce grand arrangement fiévreux et accablant de l'humanité dite dévelopée, c'est, après tout, ce qui fera un jour ques tous les hommes de la terre puissent manger. Il a déjà fait que les hommes existent les uns aux yeux des autres, au lieu de proliférer chacun dans leur pays comme des arbres. La rencontre s'est faite dans le sang, la peur, la haine, et c'est ce qui doit finir. Je ne peux pas sérieusement la considérer comme un mal. En tout cas, c'est chose faite, il ne peut être question de recréer l'archaisme, nous sommes tous embarqués et ce n'est pas rien d'avoir engagé cette partie. (S, 418)

Non, certainement pas, Merleau-Ponty ne croyait pas que notre civilisation occidentale avec ses valeurs traditionnelles était quelque chose qu'il fallait déconstruire. Nous sommes tous embarqués, il faut continuer la partie, en faisant, si possible, mieux.

IV

Affirmer l'histoire, c'est reconnaître qu'il n'y a pas de résolution définitive aux problèmes humains; c'est donc rejeter la nostalgie qui nous fait croire au salut historique et au dépassement de toutes les aliénations. La conclusion d'une réflexion sur l'histoire, disait Merleau-Ponty, "ce n'est pas la révolte, c'est la *virtu* sans aucune résignation." (S, 47) La conclusion, c'est une croyance prudente et raisonnable dans la possibilité d'un certain *progrès*. La conclusion, c'est de croire que "toutes réserves faites sur les stagnations ou les reculs, les relations humaines sont capables de mûrir, de tourner leur avatars en enseignements, de recueillir dans leur présent la vérité de leur passé, d'éliminer certains des secrets qui les rendent opaques et de se faire plus transparentes."[25]

Si de même on peut parler de progrès en philosophie, ce n'est pas en ce sens qu'un jour la philosophie aura trouvé la réponse aux problèmes qu'elle s'est posés à son début, de sorte qu'"elle pourra, comme disait Hegel, déposer son nom d'*amour du savoir* pour être savoir *effectivement réel*."[26] S'il y a du progrès en philosophie, c'est plutôt en ce sens qu'elle arrive parfois à mieux poser ses problèmes, à mieux comprendre le vrai sens de ses problèmes, et à éliminer des faux problèmes, de sorte qu'il n'est plus besoin d'y revenir ou de les traiter tel qu'ils avaient été traité dans le passé. Le progrès de la philosophie consiste, précisément, à réussir parfois à tourner ses avatars en enseignements.

Cela s'applique en particulier au très vieux problème de l'union de l'âme et du corps. L'echec de la philosophie moderne à résoudre ce problème est de sorte à nous enseigner que c'était là un faux problème, ou du moins un problème mal posé. Ce que nous avons appris de l'échec de la philosophie moderne, c'est que c'était une erreur même de parler de l'âme et du corps, comme si c'était deux choses dont l'existence était évidente. En fait, ces notions ne correspondent à rien d'absolu et d'évident, et certainement pas à des réalités de notre expérience — Husserl déjà nous a fait voir autant.

Mais peut-on en dire autant des notions modernes de subjectivité, de conscience? N'étaient-elles que des pures inventions qui aujourd'hui nous masquent l'être et qui sont donc à défaire? Est-ce qu'il n'y a aucune vérité dans ces notions du passé qui puisse être receuillie dans notre présent? Sont-elles des noms pour des problèmes qui ne sont pas tellement à reprendre dans une *Aufhebung* mais plutôt a déconstruire dans un *Niedergang*?

Si cela était le cas, et si les grandes découvertes de la philosophie moderne — la subjectivité, la conscience — étaient abandonnées, la philosophie aurait fait tellement de "progrès" qu'elle n'existerait même plus. Ce serait la déconstruction de la philosophie elle-même. Et, en effet, on parle de plus en plus de la fin de la philosophie.[27]

La dissolution de la subjectivité et la "mort de l'homme" seraient alors la fin de la philosophie, car depuis au moins Socrates l'objet principal de la philosophie a été l'homme lui-même. "Le philosophe, disait Merleau-Ponty, est l'homme qui s'éveille et qui parle."[28] La philosophie est l'interrogation réflexive que poursuit l'homme éveillé au fait de son existence. La philosophie surgit an moment où un sujet devient conscient de soi et retrouve devant lui comme son vis-à-vis (*Gegen-stand*) un monde opaque qui, comme un miroir obscur, lui renvoie une image qui est l'énigme de sa propre existence. Nous sommes nous-mêmes, Merleau-Ponty écrit, "une seule question continuée, une entreprise perpétuelle de relèvement de nous-mêmes sur les constellations du monde, et des choses sur nos dimensions." (VI, 140)

Qu'est-ce que c'est, enfin, que l'homme dans la nature? Cet être qui, comme Pascal observait, est un tout à l'égard de rien et un rien à l'égard de tout, un milieu entre tout et rien. Qu'est-ce que c'est qu'un objet qui est aussi sujet, sujet qui est aussi objet, un visible qui est aussi voyant, un voyant qui est aussi visible, ni simple partie du monde (en soi) ni pur regard sur le monde (pour soi), mais les deux à la fois, puisqu'il ne pourrait pas voir s'il n'était pas lui-même visible? Qu'est-ce que c'est que ce sujet qui est la métamorphose du visible lui-même? Qu'est-ce que c'est que cet étrange entrelac ou chiasme entre le sentant et le sensible dans la chair du monde qu'est l'individu humain? Comme Merleau-Ponty disait du sujet perçevant dont le corps est un pouvoir d'expression spontanée, "c'est là une grande merveille dont le mot d'*homme* ne doit pas nous masquer l'étrangeté." (S, 82)

Toutes les questions de la science ou de la simple curiosité sont animées intérieurement, dit Merleau-Ponty, par "l'interrogation fondamentale qui apparaît à nu dans la philosophie." (VI, 140) L'homme est pour lui-même une question, et le "Qui suis-je?", le "Que suis-je?", est la question centrale de la philosophie. La philosophie n'est que la réponse faite par l'homme à la claire *conscience* qu'il a de son obscure *existence*. Ce que, au fond, l'individu se demande, surtout dans les moments exceptionels de sa vie et surtout le malade (mais l'homme lui-même n'est-il pas en un sens l'animal malade, comme disait Hegel?), c'est, dans la formulation que lui donne Merleau-Ponty: ". . . pourquoi suis-je moi? Quel âge ai-je vraiment? Suis-je vraiment seul à être moi? N'ai-je pas quelque part un double, un jumeau?" (VI, 141) "Toute question, dit-il, même celle de la simple connaissance, fait partie de cette question centrale qui est nous mêmes . . ." (VI, 141) La question de l'identité personnelle est en effet la question centrale de la philosophie.

Si Merleau-Ponty a fait quelque apport à la pensée de notre époque et quelque contribution au progrès de la philosophie, c'est, je pense, en ce qui concerne cette question, ce problème de l'identité personnelle. Certes, il ne nous a pas légué une solution au problème de la subjectivité et du rapport entre le psychique et le corporel, mais du moins il n'a pas essayé de simplement esquiver le problème en niant toutes les dualités. L'important pour lui, ce n'était pas de dissoudre la subjectivité, mais de la reconcevoir, d'une façon moins métaphysique que dans le passé. Ce qu'il a fait, en fin de compte, c'est peut-être d'avoir réussi à tourner l'échec du problème moderne de l'union de l'âme et du corps en enseignement. Husserl avait déjà montré ce qu'il a d'artificiel en concevant le rapport entre le psychique et le corporel comme le rapport entre deux mondes, mais Merleau-Ponty, qui a toujours voulu rester fidèle à Husserl en dépit de sa critique très poussée de ce philosophe de la conscience, a avancé la phénoménologie de quelques pas en rejetant carrément toute "égologie" et la notion husserlienne d'une "sphère d'appartenance". Il nous a ainsi débarassé de tout ce bric-à-brac, comme il disait, de la réflexion psychologique ou même transcendentale — états de conscience, actes de conscience, noèses, noèmes, images, hylé, etc., etc., avec lesquels même Husserl avait essayé de construire ou, si l'on préfère, de constituer le monde.

Eliminant une certaine formulation du problème de l'identité personnelle ou de la subjectivité qui nous rend opaques à nous-mêmes, Merleau-Ponty nous a mis en état de reposer la question portant sur la subjectivité d'une manière qui nous permettra peut-être de garder dans une compréhension moins spéculative et figurée la vérité de notre passé.

Surtout, Merleau-Ponty nous a désabusé de l'idée que la philosophie, comme reflexion sur la subjectivité, peut ou doit rivaliser avec les sciences de la nature, l'idée même que la philosophie doit ou même peut être une science, la super-science que Husserl voulait toujours, ce qui atteste qu'en dépit de tout Husserl restait prisonier du rationalisme moderne. Si la vocation de l'interrogation philosophique est de se constituer en science rigoureuse, alors le rêve est bien fini, et la philosophie a effectivement touché à sa fin. Mais l'abandon de la philosophie conçue comme science rigoureuse ne signifie pas nécessairement l'abandon de la philosophie conçue comme interrogation rigoureuse et méthodique, et en fait c'est cette vocation historique de la philosophie que Merleau-Ponty revendiquait. Il voulait non pas détruire la philosophie, mais la ramener a "ses sources vives". (RC, 142) Comme il disait vers la fin de sa vie: "Une recherche comme celle que l'on poursuit ici sur l'ontologie de la Nature voudrait maintenir au contact des êtres et dans l'exploration des régions de l'Etre la même attention au fondamental qui reste le privilège et la tâche de la philosophie." (RC, 256)

En livrant combat aux avatars conceptuels et métaphysiques du passé, Merleau-Ponty nous aurait aidé à l'avenir à mieux apprivoiser le sphinx et à répondre d'une façon un peu moins indirecte à l'énigme qu'il nous propose qui est celle de notre propre être au monde.

McMaster University

NOTES

[1] Pascal, *Pensées*, 72 (ed. Brunschvicg).

[2] Husserl, *Die Krisis der europäischen Wissenschaften und die transzendentale Phänomenologie* (La Haye: Martinus Nijhoff, 1962), p. 18.

[3] Ibid., # 7, p. 16.

[4] David Carr, *Phenomenology and the Problem of History* (Evanston, Ill.: Northwestern University Press, 1974), p. 120.

[5] Voir *Die Krisis*, # 10.

[6] Voir *Die Krisis*, # 14.

[7] Voir *Méditations cartésiennes*, # 44.

[8] *Le Visible et l'invisible* (VI) (Paris: Gallimard, 1964), p. 290—1.

[9] *La Phénoménologie de la perception* (PhP) (Paris: Gallimard, 1945), p. v.

[10] *La Structure du comportement* (SC) (Paris: Presses Universitaires de France, 1942), p. 1.

[11] Voir, G. B. Madison, *La phénoménologie de Merleau-Ponty* (Paris: Editions Klincksieck, 1973), p. 34.

[12] "Il y a le corps comme masse de composés chimiques en interaction, le corps comme dialectique du sujet social et de son groupe, et même toutes nos habitudes sont un corps impalpable pour le moi de chaque instant. Chacun de ces degrés est âme à l'égard du précédent, corps à l'égard du suivant. Le corps en général est un ensemble de chemins déjà tracés, de pouvoirs déjà constitués, le sol dialectique acquis sur lequel s'opère une mise en forme supérieure et l'âme est le sens qui s'établit alors." (SC, 227)

[13] Voir *Méditations cartésiennes*, # 55.

[14] *Sens et non-sens* (SNS) (Paris: Editions Nagel, 1963), p. 209.

[15] "Le Primat de la perception et ses conséquences philosophiques", in *Bulletin de la société française de philosophie*, vol. 49 (décembre, 1947).

[16] Ibid.

[17] Voir la Préface par Merleau-Ponty à *L'OEuvre de Freud* par le Dr. A. Hesnard (Paris: Payot, 1960), p. 8.
[18] Sartre, "Merleau-Ponty", in *Situations IV* (Paris: Gallimard, 1964), p. 266.
[19] Voir VI, 323.
[20] La resemblance de certaines remarques du dernier Merleau-Ponty à celles d'un écrivain post-moderniste comme Norman O. Brown est parfois aussi frappante que déconcertante. Comme examples, en ce qui concerne Brown, je cite les textes suivants pris au hasard dans son livre, *Love's Body*:

> Psychoanalysis can be used to uncover the principle of union, or communion, burried beneath the surface separations, the surface declarations of independence, the surface signs of private property. Psychoanalysis also discloses the pathology of the process whereby the normal sense of being a self separate from the external world was constructed. Contrary to what is taken for granted in the lunatic state called normalcy or common sense, the distinction between self and external world is not an immutable fact, but an artificial construction. It is a boundary line; like all boundaries not natural but conventional; like all boundaries, based on love and hate. (p. 142)

> The boundary line between self and external world bears no relation to reality; the distinction between ego and world is made by spitting out part of the inside, and swallowing in part of the outside. (p. 143; Brown cite par la suite Melanie Klein, à qui Merleau-Ponty faisait aussi de nombreuses reférences)

> The net-effect of the establishement of the boundary between self and external world is inside-out and outside-in; confusion. The erection of the boundary does not alter the fact that there is, in reality, no boundary. The net-effect is illusion, self-deception; the big lie. Or alienation. "Le premier mythe du dehors et du dedans: l'aliénation se fond sur ces deux termes." (p. 143—4; la citation interne est de Hyppolite)

> Symbolism is polymorphous perversity, the translation of all of our senses into one another, the interplay between the senses, the metaphor, the free translation. The separation of the senses, their mutual isolation, is sensuality, is sexual organization, is bondage to the tyranny of one partial impulse, leading to the absolute and exclusive concentration of the life of the body in the representative person. (p. 249)

> Knowledge is carnal knowledge, a copulation of subject and object, making these two one. (p. 249)

> Knowledge is carnal knowledge. A subterranean passage between mind and body underlies all analogy; no world is metaphysical without its first being physical; and the body that is the measure of all things is sexual. (p. 249)

> Get the nothingness back into words. The aim is words with nothing to them; words that point beyond themselves rather than to themselves; transparencies, empty words. Empty words, corresponding to the void in things. (p. 259)

> It cannot be put into words because it does not consist of things. Literal words always define properties. Beyond the reality-principle and reification is silence, the

flesh. Freud said, Our god Logos; but refrain from uniting with words, in order to unite with the word made flesh. (p. 265)

The antinomy between mind and body, word and deed, speech and silence, overcome. Everything is only a metaphor; there is only poetry. (p. 266)

[21] "Il faut penser la chair, non pas à partir des substances, corps et esprit, car alors elle serait l'union de contradictoires, mais, disions-nous, comme élément, emblème concret d'une manière d'être générale." (VI, 193—4)

[22] Voir mon livre, *La Phénoménologie de Merleau-Ponty*, p. 223—4.

[23] *Les Aventures de la dialectique* (AD) (Paris: Gallimard, 1955), p. 296.

[24] Pour une analyse de la philosophie politique de Merleau-Ponty, voir mon article, "Merleau-Ponty et la Contre-Tradition", in *Dialogue* (Canada), vol. XVII, no. 3 (septembre, 1978), p. 469—475.

[25] "Un Inédit de Merleau-Ponty", in *Revue de Métaphysique et de morale*, no. 4 (octobre—décembre, 1962), p. 407.

[26] G. W. F. Hegel, *La phénoménologie de l'esprit*, trad. J. Hyppolite (Paris: Aubier, Editions Montaigne, 1939), Vol. I, p. 8.

[27] Merleau-Ponty disait: "Quelque chose a fini avec Hegel. Il y a, après Hegel, un vide philosophique, ce qui ne veut pas dire que les penseurs ou les génies aient manqué, mais que Marx, Kierkegaard, Nietzsche commencent par une dénégation de la philosophie. Faut-il dire qu'avec eux on entre dans un âge de non-philosophie? Ou bien cette destruction de la philosophie en est-elle la réalisation? Ou bien en conserve-t-elle l'essentiel, et la philosophie, comme l'écrit Husserl, renaît-elle de ses cendres?" *Résumés de cours, Collège de France 1952—1960* (RC) (Paris: Gallimard, 1968), p. 141—2.

[28] *Eloge de la philosophie* (Paris: Gallimard, 1953), p. 100.

MARIA DA PENHA VILLELA-PETIT

TOWARD A GENETIC PHENOMENOLOGY OF SPACE THROUGH A CRITICAL APPROACH TO PIAGET

The Husserlian project of a genetic phenomenology has not yet received the continuation it deserved, despite the inherent interest of the questions that no doubt could have been raised, and this in opposition to transcendental phenomenology itself. Yet, according to Husserl, in it would be found "the long-familiar problems concerning the *psychological origin* of the 'idea of space,' the 'idea of time,' the 'idea of a physical thing,' and so forth . . . as *problems of intentionality*, which have their places among the problems of a universal genesis."[1] Our object here is to take a few steps in the direction of a genetic phenomenology of space[2] by way of a discussion of Piaget's genetic psychology. In this discussion we shall be led in many instances to seek the assistance of psychoanalysis, for it seems of prime importance that a phenomenology of space be concerned with the experience of space in the multiplicity of its intentional aspects, and not solely with those aspects directly related to the construction of a representation of space.

Without being able to enter in any detail into the whole of Piaget's work, which covers the vast field of all the various intellectual acquisitions, we shall attempt to characterize these in broad outline and to elicit some of Piaget's theoretical hypotheses, in an effort to appreciate the nature and the limits of the teaching that can be drawn from it for a better understanding of the genesis of spatial relation.

The first thing to notice is that genetic psychology is centered around the child's cognitive development, assigning itself as the telos of this development with regard to space and the acquisition of Euclidean geometric representation.

Conceived as the result of the interaction of the biological organism, endowed with specific capacities, with its environment, cognitive development is governed, according to Piaget, by two functional principles — assimilation and accommodation.[3] The latter are intended to serve the individual's equilibrium, that is, they permit an increasingly complex and flexible adaptation to the physical and social environments. This sort of development is not due merely to growth or the requirements of acquisitions resulting from interactions between the individual and his

A-T. Tymieniecka (ed.), Analecta Husserliana, Vol. XXI, 189—209.

environment, but is instead due to structural reorganizations, displaying the differentiations of stages.

For each stage there is a correlation between the child's behavior and the way in which his "world" is organized. What is displayed here is something on the order of intentional constitution, which only our objectivist prejudice keeps us from recognizing as such. The stages determine the schemata of behavior and thought by means of which infantile experience is integrated. Each stage is itself compelled to follow its own particular development, which extends from a short initial disequilibrium with which it begins, moving by way of the phase of structural equilibrium characterizing it, and ending in the terminal disequilibrium through which another stage comes to light (by a process of accommodation).

If the stage at which each child reaches a particular stage can vary, the order of his acquisitions is constant and follows a set series (one never goes from 1 to 3 without passing through 2).

In summary, Piaget is attempting to determine how the human subject, by means of structural reorganizations which progressively set up intellectual capacities, reaches rational thought, since these capacities are conceived of solely in terms of adaptative and cognitive aspects.

Faced with this approach, and regardless of its legitimacy, we must not lose sight of the fact that it requires a certain abstraction with regard to the interconnection of the existential intentional relations of the human subject. In this respect, it is doubtful that, in connection with the experience of space which is of interest to us here, this approach can be taken as the only psychological perspective, or even as the most enlightening one. In addition, this abstracting procedure, by which the cognitive aspect is abstracted from the "development," tends not only to give priority to certain traits of human behavior over others but also to privilege a certain cultural model of intellectual development and, consequently, a particular model of practical and representational mastery of space.

To fill in these remarks, we shall limit ourselves to the first "stages" of the child's life, for a look at these should be sufficient to reveal the effects on the understanding of the spatial relation stemming from the theoretical framework within which Piaget casts his observations. The question we are examining here is that of the "acquisition of permanent objects," which is of prime interest to a phenomenology of space since it determines the possibility of constituting spatial references. This question, moreover, provides a guideline for a fruitful confrontation between Piaget's genetic psychology and psychoanalysis concerning the point at issue here.

Now, since the problem of the acquisition of permanent objects arises only at the third stage of the sensory-motor level (the child having already passed through the first two stages), the way in which Piaget determines just what is significant in the course of these initial stages cannot be unrelated to his understanding of what follows, and so requires that we first consider these early stages.

The first stage, according to Piaget, is that in which certain reflexes are consolidated through their functional exercise; this is the case in particular of the sucking and palmar reflexes, which "will be integrated in the later intentional prehension."[4] Through exercise this assimilation, which leads to better suckling, to finding the breast when it has been released, etc., will give rise to a generalizing assimilation (sucking for its own sake or sucking other objects) and to recognitive assimilation (distinguishing the breast from other objects), but in this amounts to no more than to extending reflex activity.

During the second stage certain types of conditioning and basic "habits" are formed. It should be noted that these stages, which cannot yet be considered acquisition stages (that is, stages of progressive practical and intellectual mastery of the object), interest Piaget only from the perspective of the consolidation of reflexes and the setting up of basic habits.

As for the affective exchanges between the newborn and his mother (and the other members of his social entourage) and the "traces" or "engrams" the exchanges cannot help but leave "on" the newborn, these receive no attention and are the object of no specific theorizing. In this way, whatever regards the relational, interhuman exchanges is left aside or, better, is considered as if it were included as part of the adult's assuring the child's survival, the upkeep of his organism; and, reciprocally, as if for the child it were only a matter of exercising the reflexes which assure him of the possibility of organic development.[5]

But how can we fail to problematize the influences on corporeal experience, in particular, of the affective surrounding (or its lack) which accompanies the satisfaction of the child's needs, influences without which we would be unable to understand the genesis of an affective memory or of a prememory, or, as Husserl would say, the "horizons," intentionally delimited potentialities which accompany the actual subjective process itself.[6] The almost exclusive attention accorded to "reflexes" in the first stages is given, in Piaget, as an indication of the way the body is thought of in genetic psychology, namely, as an objectified body-organism (*Körper*) and not as a lived body (*Leib*), to employ an essential phenomenological distinction. In other words, genetic psychology objectifies the body as does the physiologist when he studies another body, without taking into account the fact that this sort of objectivation can be but the work of scientific observation but in no way that of someone who experiences his body and who lives it within a context of intersubjective relations.

In this way, genetic psychology fails to see both the body as "lived body" and what, following Husserl's fifth Cartesian meditation, we could call an emerging *Paarung*.

This is not the case for psychoanalysis. The latter thematizes this initial *Paarung* — which takes on the traits of the mother-child relation — by problematizing the relations which are spun, through satisfying the needs of the newborn, between his lived body and a part of the mother's body (the "partial object"). Psychoanalysis is thus able to take into account the importance of the affective investment of certain zones of the body proper, including the awakening of certain intellectual aptitudes. This was shown, for example, by R. A. Spitz when he stressed the tie between the investment of the buccal region and the acquisition of speech.[7] In this, psychoanalysis makes us realize the relevance of a theory of "body locations" correlated with that of object fragmentation (which can also be part of someone else's body), by revealing as more complex than one would allow at first sight both our relation to objects and the very question of our corporality, which is first seen to be given as a co-corporality.

Another point concerning these first phases, where there is evident disagreement between genetic psychology and psychoanalysis, has to do with substitutive processes. This point is of particular interest to us inasmuch as here we find a contrast between an uncritical manner of approaching the question of imagination (Piaget) and an approach that accords a large role to the imaginary in regard to the way in which a human being experiences his world. The difference takes shape, more precisely, around what Piaget has called a process of generalizing assimilation (that is, the extension of the reflex schema of sucking, where, he says, "psychoanalysts mistakenly see an early symbolic behavior").[8] Piaget is referring to the coordination that enables the thumb to be brought to the mouth to be sucked in place of the breast. "The mistake of psychoanalysts," he continues, "lies in seeing this as a "representative assimilation of the thumb and the breast," for where would this symbolic or evocative power come from well before the formation of the earliest mental images?"[9]

It seems to us, regardless of the errors that can be ascribed to certain psychoanalytic discourses lacking conceptual rigor, that this reproach stems from an ambiguity in the word "representation." Where in psychoanalysis the "thumb" or any other object is taken to represent the "breast," this means that a substitution relation can be established between them ("substitute for" being one of the senses of "represent"). For this representation/substitution to be established, there is no need for the prior formation of elaborate representations taken as mental images that would be linked through an act of analogical comparison. It is important that the substitutions take place, especially those that integrate "objects" other than the parts of one's own body, "objects-other-than-myself" to use D. W. Winnicott's expression. (Winnicott more than anyone else has been attentive to these "transitional phenomena" between auto-eroticism and the object relation.)[10] These substitutions transgress the immanence of the sphere of belonging reduced to the lived experience of one's own body, thereby opening up an intermediary area, a "between" connecting the self and the nonself, the inside and the outside, the child and the mother, where something (a pillow, a piece of cloth, etc.) is integrated into experience as it "stands for," as it "can be taken for." The origin of presymbolic processes can be related to this "potential space" as a play space rather than to the simple functional substitution of the thumb for the breast.

The other aspect of the question raised by Piaget concerns what he calls the formation of mental images, which, according to him, cannot be ascribed to the newborn, as psychoanalysis would have it.

There would appear to be two sources of misunderstanding. The first would be positive, experimental, for at the time he addressed this approach to psychoanalysis, Piaget believed in an excessive physiological immaturity of the newborn, an immaturity which recent studies in no way confirm. C. Trevarthen, in fact, states the contrary, basing his remarks on solid experiments: "in the newborn human, the basic cerebral mechanisms of oculo-motor exploration are quite sophisticated, contrary to what is generally believed. . . ."[11] The second source of misunderstanding, philosophically more critical, is the lack of distinction between what Piaget himself conceives of as "mental image" and what in psychoanalysis is understood by "phantasy." Now, if for Piaget the mental image can only stem from a reproductive imagination, founded on an objectified percept, the phantasy in psychoanalysis refers to a productive imagination, to a phantasizing activity making its productions on the basis of "fragmentary recollections of interactions with human partners"[12] and, perhaps more generally, on the basis of fragments of the global experience of the child (including human partners as well as what he experiences of the world). This phantasizing activity does not, however, imply mastery or manipulation on the part of the subject. Rather in it, living it unaware, the subject, through tension, constitutes himself.

It goes without saying that the two conceptions of imagination confronting one another here deserve a much more extended discussion, but this cannot be done within the framework of our essay. Let it suffice to say that, prisoner of an altogether quite banal and limited view of imagination, Piaget has difficulty recognizing an imaginary life as being in some way presupposed by perception objectivation itself. This misappreciation of the imaginary is tied up with the lack of attention paid to the relations built up between the child and his human "partners"; Piaget ends up neglecting entirely the influences of these relations on the constitution of the subject's "affective life." And yet the work of many artists attests to the importance of the sensorial surroundings which are being formed for the child, in connection with the presence of human partners, and in which certain privileged "lines" are traced out, susceptible of being redeployed throughout life, because of their orientation toward certain choices and, in particular, toward a certain quality of beings and spaces, a selection based on affective tendencies. This is seen in terms of certain constants in the way in which their "affective world" is organized. In this way, when a Baudelaire expresses himself on the privileged path that is for him the sense of smell — "If you could know all that I see! All that I feel! All that I hear in your hair! My soul voyages on its perfume as the soul of other men on music"[13] — it is difficult not to hear in these words the reverberation of affective traces which were first inscribed in the "night" of the first months of life and which form, one might say, the horizon of his later experiences.

Of course, these experiences of adult life as well as the interpretation associated with them might not have been taken into account by genetic psychology by reason of the methodological limits it imposes on itself. But this is not so with regard to the aspects of the life of the child, where the involvement with other humans (mother, etc.) is only too evident, and yet these aspects are not sufficiently integrated into Piaget's theory of development. How can we fail to see this as an undisputable sign of prejudice which makes the relation to the physical object and to its correlative space *the* privileged relation, so that the relation to others can find a place only by being reduced to what Piaget calls the "social environment"? Piaget is content with recognizing the social environment as an external, although indispensable, factor in intellectual acquisitions. Conceived of in this way, and to the extent that the social environment includes the relation to others, it can function only as a screening concept. For it prevents us from thinking that the relation to others is not external but, on the contrary, intrinsic to the constitution of human subjectivity, marking its historicity through and through, that is, all the spatializing and temporalizing as well as the communicative aspects of this constitution.

This reduction is carried out at the very level of what is to be understood for Piaget, namely, intellectual development; this dissociation between the cognitive and the affective makes the latter appear as a factor introducing perturbations within normal development. But should one not recognize instead, or else run the risk of a disastrous split, that affectivity is less an accident that disturbs intellectual development, even if this is so in pathological cases, than it is the ground upon which what is termed normal intellectual development is constituted along one of the essential axes of existence?

And is not all interest in science and in deductive thought — not to speak of the arts — rooted in this ground, which does not mean that rationality is reduced to this but instead that the desire for knowledge draws its *strength* here?[14]

This should be clearly understood: these considerations are not aimed at invalidating the results of genetic psychology, but at indicating some of the limits of its theoretical horizon which appear to us to leave out, among other things, the creative aspects of our relation to the world. And these creative aspects themselves involve the recognition of the inextricable bond between the cognitive and the affective.

It is time to return to the question of the conquest of permanent objects, in the early stages, that is before the child has acquired the use of language. This allows us to point up Piaget's contribution, without abstaining from commenting on some of Piaget's interpretations, for we are convinced along with Merleau-Ponty that "the contents of Piaget's works are more persuasive than his interpretations of them" and "that there is often a rather wide margin between what the child expresses and what is understood by Piaget."[15]

One of the unquestionable merits of genetic psychology is to have overcome "subjectivistic sensualism," which attributed the constitution of the object and of spatial relations to "sensations" alone, that is, to the passive receptivity of external "data." On the contrary, this approach shows that the constitution of the object and of spatial relations depends on the subject's own activity, rejoining here a phenomenology for which feeling and moving are intrinsically related.[16]

Piaget's acknowledgment of the role of motor activity indicates that the object and the spatial surrounding take "shape," that is, are constituted, only because the subject goes toward them through his gestures, his body. This is a body, then, that grasps and moves as much as it looks and listens, or better yet, that coordinates vision and hearing to prehension, to movement.

At this stage in the development of the individual corresponding to Piaget's third stage, we witness the emergence of the hand as the "grasping" organ. Prehension is thus the condition for the formation of a world of permanent objects, because through it the changing "sensorial images" of the early stages give way little by little to a world of solid objects, resulting from the coordination of visual experience with the motor experience, the latter concerning the invariance and resistance of things.

In fact, recent experiments have shown that even before prehension, the resistance offered by the environment begins to be experienced by the infant through his body, especially his members, whenever he encounters an obstacle which prevents his accomplishing a movement.[17] Thus, very early on the newborn experiences through his body both kinesthesia and the resistances he encounters; the world thereby shows itself to possess a certain solidity even before the more specified actions of manipulation. Moreover, to this manipulating activity corresponds the activity-passivity of touch which seems to precede it so that we could say that the world is tangible and palpable before being manipulable.

Another aspect to consider concerning the coordination of perception and prehension as the condition for acquiring permanent objects is that of "sensorial images." For Piaget this expression designates the newborn's sensible world as an infra-objective world, and, as such, destined to disappear. As he writes, "the initial universe is a world without objects consisting only of moving and inconsistent 'images' which appear and then are completely reabsorbed, either definitively or to reappear in a modified or analogous form."[18]

These images, indicating the absence of a world of localizable fixed things — a world arising only with the development of movement — would therefore seem to form the intermittent film of a psychic life, prisoner of its own immanence, incapable of positing a world of things, outside of and separate from itself. Now, if at a certain level this is indisputable, it remains that from a phenomenological point of view an important lacuna appears to subsist in Piaget's understanding of the infant's burgeoning experience of the world. For what is not perceived in this subordination of all experience to the gradual conquest of the object and of objective space, is the opening to the sensible world that is taking place through these sensorial images. Therefore, to mention only what concerns the visual, at issue is an initiation to the visibility of the world during the early stages when the infant is but a "field of experiences" and has not yet emerged as a subject.

In other words, the infant already possesses, in these sensorial images, his first grasp on the visible that surrounds him, calls upon him, even fascinates him. He has "visions," we are tempted to say, which resemble those painters sometimes try to respond to. In this way, Cezanne wanted to find "the expression of the confused sensations we find at birth"; and this brought Henri Maldiney to say:

> These confused primordial sensations through which we communicate with the world before any objectivity intervenes, are very quickly clarified and corrected by the necessities of practical life which has to be supported by well-defined objects, distinct from one another, from which we have extirpated all the pathos which originally bound us to the world. From these sensations we have excluded the how in order to retain only the what. This color, this light on which our gaze rests is now but an indifferent quality that enables us to identify an object or an hour of the day. It is no longer a way of living with the world. But for Cezanne, it was so. . . .[19]

Now painting allows us to see what "makes [things] visible," that is, the play of light and shadow, colored modulations, the expansion, separation, and overlap of things or their "lines of force." Painting thus bears witness to all that visually determines space and depth, but which is neglected and relegated to the category of "secondary qualities" in a psychology dominated by a concern for only the geometrical determination of space.

Now these so-called "secondary" qualities are, in fact, primary, and it is through them and in them that a world can appear in the visible order and in the sensible order as such. And this is what is at stake in the sensorial images whose infra-objective character Piaget alone retains.

This omission indicates that the phenomenal dimension of space is not really understood; for, indeed, space is not given independently of light, of its modulations which cause constriction or dilation and, consequently, it is not given independently of variations in color, of which Merleau-Ponty said, "it creates itself out of itself identities, differences, texture, materiality, something." We can therefore expect to find in these "sensorial images" as intentional experience of light and the play of light as well as of its affective and rhythmic interweave, which the infant experiences from the early months.

In addition, the visual schemata which are formed and which permit the subsequent recognition of things, that is, of their "appearance," on the basis of a few significant features or profiles, possess an essential characteristic, namely, that the features out of which they are formed can in some sense be isolated, and their appearance can enable the imagination to make comparisons of things that are quite different. It is due to this structuration of vision,[20] and to the emotional reverberation caused by the way in which the visible enters into our body, that the world can be given in its essential and metaphorical character.

Of what does this metaphorical character consist if not the open substitutive chains where one thing can visually take the place of another or awake its echo in the subject, on the basis of some common "quality" (color, etc.) or configuration?

This metaphorical character is clearly displayed only when it is brought to the level of linguistic or graphic expression. But, at the same time, linguistic metaphor is significant only because it refers back to this potentiality that seems to permeate the experience we have of the world. Or, as Paul Ricoeur has remarked, "language itself, through the way in which it makes one meaning pass into another, evokes 'something of the metaphorical character of the world itself' which [the poem] proclaims."[21]

Understanding this "capacity" of the visible implies that vision is not reduced to a simple function in the construction of an objective reality, following a certain geometrical and physical description of the world, which is itself constructed. On the contrary, it is important to restore to vision its "fundamental power of manifesting, of showing more than itself," as Merleau-Ponty wrote, adding: "And since we are told that a little ink suffices to make us see forests and tempests, [vision] must have its own imaginary."[22] This means that human vision has always already filled space beyond what is said to be provided "empirically" or "objectively," and this is so by virtue of the dimensions of meaning implied in the human being's insertion in space.

These remarks, by which we have attempted to display the implications of the infant's initiation to the visibility of the world — but which are hidden rather than indicated by the expression "sensorial images" — have not led us away from the question concerning the acquisition of objects. Instead, they were intended to help us perceive the problem at issue here, which is to know whether the acquisition of the object, realized in its identity across perceptual-motor syntheses, eliminates purely and simply what was announced in the so-called "sensorial images."

Our response, although decisive, is obviously not simple. For, on the one hand, the conquest of permanence, establishing the belief in a stable world (and, correlatively, enabling the subject to constitute himself, his own identity) implies going beyond a world or a preworld experienced as a changing succession of appearances. But, on the other hand, the conquest of the object as the pole of a practical and theoretical life should not suppress the question concerning the phenomenal appearance of things. Now this seems to be what occurs in Piaget's conception of the movement from one stage to another — in other words, what he means by "genesis" — in line with the schema of linear temporality. In doing so, it is as if of the world and its things, he were to leave us only the "in-itself" of classical ontology, that is, the "thing" purified of all the intensity of its appearing, which would be taken as secondary and negligible once the possibility of its objective mastery and determination is established.

Having stated these reservations we can now move on to the pertinent aspects of what Piaget brings to light concerning the acquisition of permanent objects. The observed facts are, briefly, as follows: while the coordination between prehension and vision is developing, the infant begins (during the fourth stage) to look for what has been removed from his field of perception. This indicates that the thing that has "disappeared" begins to represent practically for him something possessing a certain permanence. At the start the object's permanence still depends on the child's grasping movement.

Let us look at one of Piaget's examples: "Obs. 15 — Lucienne, at 0.6 (0), is alone in her crib and, watching what she does, grasps the cloth covering the sides. She pulls the folds toward herself, but each time lets them go. She then brings her closed hand before her eyes and carefully opens it. She attentively gazes at her fingers and begins again. She repeats this more than ten times."[23]

Here appears a sort of tactile permanence that the child attributes to the object he has just grasped. The object's permanence, Piaget says, "remains relative to the subject's action."[24] The subject, who cannot yet distinguish the object's existence from his own sensori-motor experience, has the feeling that by his acts he can bring about the presence of the object. Little by little the experimenter is led to remark that an object that is shown to the infant and then hidden behind a pillow or a screen does not "disappear" for him, since he will look for it, proof that the infant attributes a permanence to it independent of its absence from the current perceptual field. Thus it is to the extent that the "I can," that is, the child's possibilities of action are extended — soon not only through the development of manipulating actions but also through the development of postural modes and of walking — that a stable and differentiated world can be constituted for him.

Now it is important to notice that the acquisition of permanent objects, following the structure presented in Piaget's observations, is connected to an anticipation of the reappearance of the object which has disappeared. But can this anticipation of the object through the active search for it be reduced, as Piaget would have it, to a simpler process of accommodation?

How does it happen then that when the child's actions are thwarted, he counts on "the effectiveness of his passion and his anger," to use Piaget's own terms? In other words, has not the schema of appearance followed by disappearance and then reappearance after the child's "active intervention" already played a role in the relation to others (to the mother) even before it is involved in the acquisition of permanence relating to physical objects? Hence the recourse to anger, which is more than the expression of a failure in the psychomotor operations of accommodation, but is the expressive means employed very early by the child to make his mother — the source of resolving his tensions — reappear.

Thus in the emotional behavior characterizing anger "which confirms his powerlessness to identify himself with his body as the focal point of directed gestures,"[26] the child is addressing the "Other," even if this occurs as he is involved with things (onto which, moreover, he not infrequently projects the attributes of personal "otherness," as does the adult in the course of certain emotional experiences).

It therefore seems that there exists in the very schema of permanence — regardless of what Piaget thinks — a certain precedence of the relation to the "other" over the relation to the object, so that the acquisition of the permanent object is from the outset cast in the "mold" of that already formed through the relation to another, itself possessing a certain "stability." (This stability refers her to the repeatable character of the appearance, that is, to a possible return after disappearance.)

Do we not also find confirmation in the fact that physiognomical recognition, or at least a sensitivity to physiognomy, to which the child can reply by smiling or by turning away, is one of the earliest sensitivities to form in the infant? Moreover, the subject is frequently led precisely to these "physiognomical" readings of external reality at the time of strong emotional experiences which change for him, as everyday language aptly describes it, the very "face" of things.[27] If we follow the indications contained in these experiences, we are made to see that the perceptive and operative stability of external reality is not itself independent of the subject's situation in relation to otherness.

Of course, something decisive takes place for the child when he acquires the permanent object as the pole of his actions, through which begins the active mastery of his ambient space. For, these actions directed toward the object attest to an increase in the child's possibilities, as soon he will also be able to move around by himself and to broaden the "exploration" of his "environment." This increase in the possibilities of acting on the environing world, this expansion in the sphere of his own world, should also be considered in connection with the constitution and the progressive differentiation of his corporeal schema.[28]

Finally, this is one of the conditions for what within a psychoanalytic framework could be called a progressive dis-inclusion of the child's space in relation to the maternal space, if we understand at the same time that, reciprocally and circularly, this dis-inclusion is itself a condition for a satisfactory conclusion to the active appropriation of space. In this, the increase in the child's possibilities of action seems as much the necessary (though not sufficient) condition as the indication of the investment of an "intermediary space,"[29] which would in some sense "mediate" the access to "external reality" at the same time that it extends the opening of a "living space" for the subject.

One can understand that the difficulties this investment may encounter along its path (even, perhaps, due to the lack of an interval[30] capable of giving range to a progressive dis-inclusion) are conveyed by certain problems or certain lacunae in the way in which the child's active familiarization with his environment occurs, and in the very way in which his own "body image" is being constituted, reciprocally.[31]

From what we have just said, it follows that Piaget can be reproached not with having dealt with the correlation between "action" and "external reality" — Freud himself established this correlation, as is recalled in J. Guillaumin's recent article[32] — but with having neglected certain aspects of our communication with the world before or beyond our practical and logical mastery of it, in particular, the "psychic experiences" which form the other side of our access to "external reality."

This is what we now propose to briefly point out by making explicit a certain number of Piaget's positions concerning our experience of space.

It appears that during the entire discussion of the permanent object — which is closely connected to the notion of space because it is through this part of acquisition that, as Piaget shows, localization and displacement schemata are formed and these finally structure the spatiotemporal organization of a practical universe — space is always thought of in terms of its practical aspects. But this restriction, by which space is understood only in relation to the constitution of a practical universe whose interiorized schemata will serve as a basis for metric operations and for Euclidean geometry, demands that we consider it as such, that is, as a "restriction" by attempting to show the aspects that are left aside.

Indeed, by referring almost exclusively to the mastery of space, Piaget forgets that space is not only a field of practical actions and intellectual operations of a being who must, at the end of a de-centering process, manage to represent himself as an object among others in a space that is one and homogeneous. This "homogeneity" to which the

Euclidean representation has accustomed us – and which is the result of an objectifying abstraction through which the subject objectifies himself by ceasing to inhabit his body in order to project it as an object among others[33] – should be thought of as separated out from a more radical heterogeneity: that of the space of the life-world (*Lebenswelt*) as it is given in its differentiated "places." Inhabited space is therefore not a homogeneous space but, even after the acquisition of Euclidean representation, it remains a space where various and heterogeneous places are linked together through their meanings for the subject inhabiting them.

For the child, this differentiation into places which constitutes space will be connected, as is the case later for the adult, to all that is tied to permission, prohibition, or the difficulties regarding access to these places. Thus, from infancy, the access to spaces is not determined for the child solely in relation to his psychomotor capacities but also in relation to the significations and the admonitions coming from the child's human entourage. Through this a relation is gradually established between the modes of access (or the prohibition) to spaces and a sociocultural order to which the child's conduct refers. The knowledge of space therefore passes by way of admonitions and so by way of the language addressed by adults to the child. Adults help to articulate for the child not just a small differentiated world (that of the various places in his house, etc.), but also to relate different places with different people: Mommy and Daddy's room, Daddy's chair, and so on.

It is in this way that the child becomes accustomed at once to the experience of his "place" in the family and to the social and cultural aspects of spaces and to the roles that accompany them. Language thus plays a determining role from the outset, even when the child is still at a so-called "preverbal" stage. But Piaget is unable to acknowledge this fully insofar as he brings in language only at the time of its "acquisition," at a given moment in development and insofar as he conceives of it only within an instrumental framework, as the instrument of communication for representations which are held to be constituted outside language. Language, however, is constitutive of experience in a different sense, even in the child who does not yet speak, but who is already "listening." How can we fail to recognize that all human space is a space of speech, permeated with discourse, which organizes it in accordance with certain meanings and values, and that it is in this space invested with "sense" that the child makes his way?

In summation, thematizing the child's knowledge of space in terms that are too exclusively bound up with psychomotor development tends to bypass not only the affective aspects but also the sociocultural aspects which are always involved in the child's familiarization with his ambient space, aspects without which there would be no true human access to "external reality," which itself is not just a manipulable world, the place of physical displacement and the ground of objective representations, but instead is always first of all a sociohistorical and culturally "marked" world in which one must both "situate oneself" and "orient oneself."

Moreover, it is only on the basis of a space conceived of as "heterogeneous," as it is displayed in inhabited space (and even the space of the landscape is itself never homogeneous) that one can understand how it can provide an "extension" and a "topography" to what, not without a certain substantialist naiveté, psychoanalysis calls "psychic reality." For it is indeed through spatial (or rather, spatiotemporal) "figures" and "vectors" that the human being's anguishes, conflicts, secrets, aspirations, and joys take on shape, as do the "lacunae" of his history and the obstacles offered to him by "reality," or even the imaginary schemata that underlie his creations, his thoughts.

In the following brief and partial remarks, we shall try to catch a glimpse of what is meant here by the "figures" and "vectors" that shape and orient the spatialization inherent in human existence. Let us look at the narrative form we call the fairy tale, which has recently become the object of attention of certain psychoanalysts[34] who conceive of it as a cultural form of narration, helping children (and entire peoples) in a specific way to structure their experience psychologically, to the extent that these stories exteriorize by giving shape to "intrapsychic" and "interpersonal" tensions.

Is it not evident that in a certain kind of fairy tale the hero manages to find a way out only by crossing dangerous spaces — thick forests, lakes, or grottoes inhabited by monsters — in the course of "travels" where the way is barred by obstacles to overcome and tests to undergo? It is therefore because it can be divided up into differentiated places, on the level of affectivity and of meaning, that in the fairly tale and in other narratives space can have a structuring and expressive function in "psychic life" and in human existence as such.

On the other hand, spatial figuration can represent a certain threat to the subject's psychic life. This occurs when the figure becomes autonomous, taking possession of the field of consciousness, becoming an obsessive theme. In this way, a patient of the psychiatrist Roland Kuhn expresses his schizophrenic situation of being locked inside himself by means of the boundaries surrounding the city in his drawing, adding the following inscriptions: "*Existenzcentrum – Eigenbetrieb – Geschlossene Form* . . . (that is, "Existence center – own exploitation – Closed form . . .").[35] By this, that is, by his concern for "boundaries," he expresses his "closed" existence, his internal world, shut in with no openings, no emergence toward the outside, as his words, which seem not to articulate a situation but to be sent back to him, to "face" him.

On the contrary, for creative individuals (and communities) spatial figurations and vectors give structure and configuration to existence when they enable one to better withstand difficulties and give a sense to life events. We have only to think of the figures of "exile," "crossing the desert," and so on. If we pursue this reflection we are also led to see that an exclusive attention to practical mastery and to operations involved in the conquest of objective representation hides other essential dimensions of our relation to space. Let us underscore here, in conclusion, the slight regard paid by Piaget to the sheltering dimension of space, which is concerned with "containing," with enveloping protection. Here, spaces are less to be mastered practically, than they are spaces where a living place finds its resources, nourishes and develops itself (we wish to maintain the reflexive character of these verbs, their openness and their inherent polysemy). It is through this sheltering dimension that we can measure the give-and-take and the fundamental interconnection, which we have tried to bring out, between protecting spaces and the originary relationship to others in its unconscious inscripton.

We shall once again have recourse to a literary testimony to indicate something about this give-and-take, this interconnection. This recourse to literature, in particular to "poetic images," is all the more understandable as what we hope to glimpse is on the order of a poetics of space like that envisaged by G. Bachelard. And does not D. W. Winnicott himself write, "If there is some truth in what I am saying, it appears that poets have already spoken of this" (*Fear of Breakdown*).

Let us look at a few analyses of S. T. Coleridge's work. The English poet (1772–1834), whose life knew so many painful emotional exiles beginning with that from his mother and his own home,[36] often echoes in his poetic images either the "icy cold" of his years at the orphanage (see, for example, "Frost at Midnight") or the "nostalgic desire of early childhood and of the state of communion with the maternal world."[37] Thus, addressing the loved but absent woman, Coleridge will invoke her with terms such as "bosom . . . cradle" or "lap . . . pillow."[38] Now, in this constellation of "images" we see that the part's of the mother's body — "breast," "lap" — which signify her metonymically, are placed alongside "cradle" and "pillow." Not only do these form part of the child's immediate environment, but they act as a substitute for the maternal body: the cradle being that which welcomes the child when he is not in the arms of her who carried him and who still carries him intermittently (the "holding"), the pillow being what allows him to put up with the first separations from the mother, if it becomes the object of a "transitional utilization," as Winnicott uses this term.

It is therefore significant that in Coleridge the appeal to the loved one calls upon the enveloping, security-enhancing dimension of space, insofar as the latter is primordially inseparable from an affective maternal presence. Thus something resounds in this appeal that is reminiscent of the first space in which the child found himself co-implicated with his mother and where their own "spheres" (the spheres of belonging in Husserlian terms) were not yet sufficiently differentiated. Co-corporality, or an almost "fusional" space, as it is called by certain psychoanalysts — and which Henry Moore has given plastic presence in some of his "Mother and Child" sculptures where an almost single body appears to encompass both.

Finally, if in our conclusion we have endeavored to evoke works of poetry and plastic arts, it is because they require an understanding of space which takes into account its sheltering dimension whose importance has been neglected by Piaget's genetic psychology. Indeed, it is only by taking into account this sheltering dimension, as well as the dimension of mastery and of their interrelated games — without obliterating the intersubjective and symbolic implications of the one and the other — that a phenomenological approach can hope to clarify the experience of space.

Centre National de la Recherche Scientifique, Paris

NOTES

[1] Husserl, *Cartesian Meditations*, trans. Dorian Cairns (The Hague: Nijhoff, 1973), p. 76.

[2] As a result of its expressed intention, this work is situated on the level of a "phenomenological psychology."

[3] These two concepts, which come from genetic psychology, are to be understood in the following way: "assimilation" is a mental activity of the child by means of which an external situation is perceived or processed in such a way as to allow it to be incl' ded within an existing schema; "accommodation" is a mental activity which serves to change an initial schema so as to adapt to a new situation.

[4] J. Piaget, *La psychologie de l'enfant* (Paris: P.U.F., 1966), p. 10.

[5] Because of its insertion from the outset within a network of interhuman relations charged with meaningful and symbolic dimensions, there is never with respect to the infant a "pure physiology" nor, as a result, anything that is purely "adaptive."

[6] Husserl, *Cartesian Meditations*, p. 82.

[7] See R. A. Spitz, *Le non et le oui — La genèse de la communication humaine*, trans. A. M. Rocheblave-Spenle (Paris: P.U.F., 1962), especially chapter 8: "Un cas de non-investissement de la zone orale." Spitz's thesis has received support in the case of Monica W., who in the first few months of life was unable to receive any food through the mouth and who, at 23 months, had not yet acquired any form of language (in any case, oral language or that employing gestures of the head). "This," Spitz continues, "agrees with our hypothesis concerning the critical role of the oral zone and anaclitic object-relations in the acquisition of human semantic communication. In Monica's case, when the nutrition process was moved from the mouth to the abdominal fistula, neither the mouth nor the head were specifically involved in relations satisfying a need. As a result, she had not even acquired head movements intended for semantic communication, all the more reason for her to be unable to use her mouth for verbalization."

[8] J. Piaget and B. Inhelder, *La phychologie de l'enfant* (Paris: P.U.F., 1964), p. 10.

[9] Ibid.

[10] See in this regard D. W. Winnicott, in *Jeu et réalité — L'espace potentiel*, trans. C. Monod and J. B. Pontalis (Paris: Gallimard, 1975), especially chapter 1, "Objets transitionnels et phénomènes transitionnels."

[11] See C. Trevarthen, "L'action dans l'espace et la perception de l'espace: Mécanismes cérébraux de base," in *De l'espace corporel à l'espace écologique* (Paris: P.U.F., 1974).

[12] Cf. Cobliner (W. Godfrey), "L'école genevoise de psychologie génétique et la psychanalyse: analogies et dissemblances," published in appendix to Spitz's *De la naissance à la parole* (Paris: P.U.F., 1968), p. 255.

[13] Cf. Baudelaire, "Un hémisphère dans une chevelure" in *Petits poèmes en prose* (Paris: Pléiade), p. 252.

[14] As Bachelard stated in his own fashion in *La psychanalyse du Feu* (Paris: Gallimard, 1949), p. 44: "We can study only what we first dreamed about."

[15] Maurice Merleau-Ponty, *Bulletin de Psychologie*, no. 236 (1964): 177.

[16] Cf., for example, Erwin Strauss, *Vom Sinn der Sinne.*

[17] See E. Vurpillot, "Les débuts de la construction de l'espace chez l'enfant," in *De l'espace corporel à l'espace écologique* (Paris: P.U.F., 1974), p. 108.

[18] Piaget and Inhelder, *La psychologie de l'enfant*, p. 15.

[19] H. Maldiney, *Regard parole espace* (Lausanne: l'Age d'Homme, 1973), p. 16.

[20] Let me emphasize that even from the perspective of experimental psychology, there is something lacking in Piaget concerning a psychology of vision and a psychology of perception in general.

[21] Paul Ricoeur, in *La Métaphore Vive* (Paris: Seuil, 1975), p. 316, on the subject of P. Wheelwright's *Metaphor and Reality* (Bloomington: Indiana University Press, 1968).

[22] M. Merleau-Ponty, *L'oeil et l'esprit* (Paris: Gallimard, 1976).

[23] Cf. Piaget, *La construction du réel chez l'enfant* (Neuchâtel: Delachaux, 1963), p. 25.

[24] Ibid., p. 23.

[25] Cf. Piaget, *La construction du réel chez l'enfant*, p. 79: "the child counts only on the repetition of his accommodation movements to realize his desire, and in the event of failure, on the effectiveness of his passion and his anger."

[26] Pierre Kaufmann in *L'expérience émotionnelle de l'espace* (1967), p. 60. Further, continuing his analysis of anger, Kaufmann writes, "But if anger is the anger of the subject, if it arises out of *an unheeded call to others, where identity is founded* . . ." (p. 62; my italics).

[27] E. H. Gombrich, based on psychological data, points out in his *Meditations on a Hobby Horse of the Roots of Artistic Form* (London: Phaidon Press, 1971): "Our whole perceptual apparatus is somehow hypersensitized in this direction of physiognomy vision and the merest hint suffices for us to create an expressive physiognomy that 'looks' at us with surprising intensity. In a heightened state of emotion, in the dark, or in a feverish spell, the looseness of this trigger may assume pathological forms" (p. 6).

[28] See in this regard Liliane Lurçat's recent work, *L'enfant et l'espace* (*le rôle du corps*) (Paris: P.U.F., 1976). The author writes, "the corporeal schema, far from being a primitive canvas on which later acquisitions would be added, is the result of connections which are established between postural space and ambient space. These connections are made through the activity of the subject, including the activities involved in pegging locations" (pp. 31–32).

[29] I am referring here to the notion of "transitional" and "potential" in D. W. Winnicott.

[30] For the discussion of this notion of "interval," see Pierre Fedida, "L'objeu — objet, jeu et enfance/l'espace psychothérapeutique," *Psychanalyse à l'université* 1, no. 3 (1976).

[31] This notion of "body image" contributes to that of corporeal schema (which designates the interiorized counterpart to the activities of orientation, lateralization, and even of practical spatialization in general) an imaginary component, whose somatic impact can nonetheless be quite real. It corresponds to the habitation, the imaginary and "symbolic" obsession of the lived body experience, understanding that this experience is that of a "sexual" body.

With regard to the tie between body image and spatial experience in psychosis, as well as the "symbolizing function of 'the body image,'" see the works of Gisela Pankow, *L'homme et sa psychose* (Paris: Aubier-Montaigne, 1973) and *Structure familiale et psychose* (Paris: Aubier-Montaigne, 1977).

[32] See J. Guillaumin, "Psychanalyse, épreuve de la 'réalité psychique,'" *Nouvelle Revue de Psychanalyse*, no. 12 (Autumn 1975), p. 183: "The point that seems decisive to me is that the instrument of the experience of reality has always been for Freud the motor character of relations, and, in general, action."

[33] Cf. Piaget and Inhelder, *La psychologie de l'enfant*, p. 15.

[34] Bruno Bettelheim, *Psychanalyse des contes de fées* (Paris: Laffont, 1976).

[35] Roland Kuhn, "Daseinanalytische Studie über die Bedeutung von Grenzen im Wahn," in *Monatschift für die Psychiatrie und Neurologie* (1951), quoted by H. Maldiney in *Regard Parole Espace*.
[36] The poet evokes this once in the following terms: "I dreamt of my sweet birth-space" ("Frost at Midnight," vv. 27—28). Let me recall that along with this "memory" the figure of the poet's father should also be associated, and this nuances (and corrects) the expression, cited below, regarding "communion with the maternal world."
[37] On Coleridge, see Christian La Cassagnère, introduction to *Coleridge-Poèmes* (Paris: Aubier-Flammarion, 1975).
[38] See ibid., p. 78. The poems concerned are "The hour when we shall meet again" and "A day dream."

ARION L. KELKEL

TEMPS ET FINITUDE CHEZ HUSSERL

> Jedes einzelne Erlebnis kann, wie anfangen, so enden und damit seine Dauer abschliessen . . . Aber der Erlebnisstrom kann nicht anfangen und enden.
>
> E. Husserl, *Ideen zu einer reinen Phänomenologie und phänomenologischen Philosophie*, I.

> Das *Zeitliche* gilt doch gemeinhin als das Endliche.
>
> M. Heidegger, *Kant und das Problem der Metaphysik.*

On dit parfois que la façon dont on comprend le temps dépend de l'homme qu'on est ou de la philosophie qu'on a. Peut-être serait-il tout aussi légitime d'inverser cette proposition sans qu'elle en soit moins vraie. Quoi qu'il en soit, nombre de philosophes inclinent à considérer le temps comme la marque éminente de la finitude humaine. S'il a la certitude ou l'illusion de pouvoir disposer de l'espace, qu'il s'applique à organiser à sa guise, l'homme, en revanche, ressent obscurément que le temps n'est pas vraiment en son pouvoir, qu'il est plutôt la marque de son "impouvoir", en un mot, il éprouve une radicale impuissance devant le temps qui passe, qui lui échappe et lui coule pour ainsi dire entre les doigts telles des gouttes d'eau qu'il aurait voulu recueillir et tenir dans sa main: il fait l'expérience pénible qu'il ne dispose pas en toute souveraineté du temps qu'il nomme pourtant "son temps", qu'il mesure comme il mesure l'espace, il sait qu'il n'en est ni l'auteur ni le maître. A l'approche de la mort qu'il redoute ou qu'il fuit dans le divertissement, il n'ignore point que le temps lui est compté même si, auparavant, il croyait toujours encore avoir le temps pour mieux faire, pour s'amender ou jouir des biens de ce monde, pour accomplir les exploits qu'il n'a pas eu l'audace d'entreprendre jusque là. Il appréhende avec angoisse la proximité de la vieillesse et il se lamente devant l'inexorable fuite du temps, sa fragilité et son caractère éphémère. Si l'homme se voit ainsi déterminé par ce qui le limite et le conditionne, rien d'étonnant à ce que les philosophes, depuis Kant à tout le moins sinon déjà depuis Aristote, aient cru pouvoir définir la subjectivité humaine

A-T. Tymieniecka (ed.), Analecta Husserliana, Vol. XXI, 211–234.

en son essence par la temporalité: le temps est le sens de la vie ou de l'existence, affirment-ils, poète comme philosophe.[1]

La pensée contemporaine illustrée par Husserl et Heidegger nous permettra de mieux cerner le problème que nous nous proposons de méditer: le temps humain ne peut-il se penser que sous le signe de la finitude? N'est-il rien de plus, comme l'allègue Heidegger, que l'extension de l'existence de l'homme pris entre ses deux limites: entre sa *fin* qui se dessine en avant de lui, et son *origine* qu'il devine en arrière de lui, entre la mort toujours menaçante et la naissance toujours déjà survenue, entre l'être qu'il a à être et l'être qu'il n'est plus, entre l'être du présent et le non-être du présent passé ou du présent à venir? Ou bien l'image authentique du temps est-elle celle qui nous le présente comme la succession ininterrompue de Maintenant, qui n'admet ni commencement ni fin? Le temps est-il ce jaillissement incessant d'un Présent vivant riche de tout son horizon infini de passé et d'avenir, que semble suggérer la description husserlienne?

En d'autres termes, l'opposition entre les deux conceptions phénoménologiques du temps peut-elle se résumer par l'opposition entre la dominance chez Heidegger d'un temps irrémédiablement fini, chez Husserl d'un temps infini proche de l'éternité? La thèse est-elle soutenable que Husserl rejoignant par là la "conception vulgaire" et commune du temps définirait ultimement celui-ci par l'infinité et l'objectivité, tandis que Heidegger privilégiant la subjectivité vécue et l'existentialité du temps penserait la temporalité originelle et authentique exclusivement sous la figure de la finitude de l'être destiné à mourir? Peut-être les divergences se révéleront-elles plus complexes. Car si tous deux, Husserl comme Heidegger, interrogent sur *l'essence* du temps qu'ils ramènent à la question de *l'origine* du temps, il reste que pour l'un il s'agit, après réduction du temps empirique objectif, d'analyser la "conscience intime du temps",[2] chez l'autre, c'est une analyse du temps vécu existentiel, celui de la quotidienneté la plus concrète, qui est appelée à exhiber les indices révélateurs, même s'il le dénonce comme un nivellement et une dégradation du temps originel.[3] Quoi qu'il en soit, l'opposition entre les deux protagonistes de la phénoménologie du temps est sans doute plus doctrinale que méthodologique et à tout le moins atteste combien la théorie du temps chez l'un comme chez l'autre engage profondément leur pensée phénoménologique et ontologique.

Il n'est guère besoin de le rappeler: selon Heidegger "le sens du

Dasein est la temporalité" et tout le projet de *Sein und Zeit* fut bien de penser l'être dans l'horizon du temps.[4] Pour Husserl, "l'énigme de la conscience du temps" couvre toute une dimension problématique et fondamentale de la recherche phénoménologique qu'il avoue avoir partiellement passée sous silence dans l'exposé des *Ideen*, de crainte des confusions qu'elle risque d'entraîner faute d'une suffisante clarté dans cette sphère de "la source originaire d'un absolu définitif et véritable".[5]

Quant à notre propos, notons simplement qu'il s'agit d'éclairer la conception husserlienne du temps par un aspect qui mettra plus décisivement en valeur l'option métaphysique qu'elle engage, par contraste avec certaine interprétation ontologique du temps et du sens de la temporalité qu'on peut déceler dans l'analyse heideggerienne. Il nous paraît en particulier intéressant et instructif d'examiner leur respective *accentuation* des différentes dimensions du temps à partir d'un point de départ qui leur est commun (du moins partiellement): l'un et l'autre, en effet, ont en mémoire, et y font référence, explicitement ou implicitement, la conception du temps de saint Augustin, qui demeure pour eux comme peut-être encore pour nous à cet égard un penseur exemplaire.[6] Le problème qu'il convient de méditer est le suivant: comment Husserl a-t-il été conduit, à partir de sa méditation de l'analyse significative des apories de l'expérience du temps par saint Augustin, avec lequel il constate la difficulté quasi insurmontable du problème et l'embarras qu'éprouve le philosophe dès qu'il interroge: "qu'est-ce donc que le temps?",[7] et qu'il veut "rendre raison" (*Rechenschaft geben*) de la conscience que nous avons du temps pour conceptualiser ce qui est du vécu en quelque sorte inobjectivable et irreprésentable,[8] et d'autre part, en réfléchissant sur la manière dont l'auteur des *Confessions* décrit les diverses dimensions temporelles comme situées, s'organisant et s'articulant autour de la dimension primordiale du présent qui les fonde et les sous-tend, à une indéniable accentuation et un véritable *privilège du* "*Présent vivant*" tandis que Heidegger, manifestement tributaire de sa conception ontologico-anthropologique de la finitude de l'existence humaine ("Plus originelle que l'homme, écrit Heidegger, est en lui la finitude du *Dasein*"), accorde la *primauté à l'extase du futur* lors même qu'il s'applique à penser le temps dans l'unité de ses extases éclatées, bref dans l'unité du procès même de la temporalisation.[9]

D'où la première question qu'il importe de démêler: la prééminence ontologique (quoique ce ne soit pas là exactement le langage de

Husserl, du moins dans les *Leçons sur la conscience intime du temps*) conférée au *Présent vivant*, au présent dans la vivante et frémissante densité et opacité de sa durée dans le flux du temps, n'interdit-elle pas toute conception "finitiste" du temps? Le moment présent qui n'est pas un Maintenant décharné, limite ponctuelle et quasi abstraite entre le passé qui n'est plus et le futur qui n'est pas encore, paraît chez Husserl un Maintenant "épais" (*grobe Jetzt*) et gros de toutes les protentions et rétentions qui remplissent la conscience du temps si bien que ses descriptions tendent à faire apparaître une temporalité infiniment *ouverte*, même si le flux du temps est fait de continuité et de discontinuité,[10] par opposition à la manière dont, selon Heidegger, le temps est un temps *clos*, comme refermé sur soi et sur sa propre finitude du fait que l'existant humain, en vertu du "souci" qui le caractérise fondamentalement, déploie son existence en avant de soi, vers son futur en récupérant son origine et sa *Gewesenheit*, en vertu surtout du fait qu'il est, dans son mode d'être authentique, marqué par l'être-pour-la-fin, c'est-à-dire par l'être-pour-la-mort dans lequel il est jeté dès sa naissance tandis que la préoccupation préférentielle pour le présent est pour ainsi dire signe de déchéance (*Verfallen*) de l'existant et le propre de la conception vulgaire du temps.[11]

En d'autres termes encore, la dominance du Maintenant présent, du *Präsenzfeld* chez Husserl ne se rattache-t-elle pas tout à la fois au concept de conscience ou d'évidence et de vérité?[12] Ne traduit-elle pas aussi son attachement à ce qu'il est convenu d'appeler "la métaphysique de la présence" qui finit par figer la phénoménologie en une "phénoménologie de la perception" et dès lors une philosophie de l'*ici* et du *maintenant*, du propre et du proche, de la *Selbstgegebenheit*, plutôt qu'une philosophie de l'absence, du lointain et de l'altérité, ou une métaphysique de l'être qui s'absente dans sa présence même et se retire dans le non-présent, bref une métaphysique où l'homme se reconnaît comme "lieu-tenant du néant"?[13]

S'arrêter à ces schémas d'interprétation par trop généraux voire dogmatiques pour qu'ils puissent convenir à la pensée de Husserl serait sans doute faire preuve de précipitation ou de prévention. Il y a bien chez Husserl une continuité du Maintenant et du non-Maintenant, du présent et du non-présent par le truchement justement des rétentions et des protentions qui contribuent également à la constitution du vécu présent, il y a une co-implication de l'intuition donatrice de la chose même et de la non-intuition de l'anticipation et de la présomption.

Comment contester la solidarité vivante, efficiente de tous les moments temporels? Le passé qu'est-il sinon l'ombre portée de l'avenir? ne vit-il pas entièrement de la vie du présent, laquelle pointe vers le futur qui à son tour tend continuellement à redonner au présent sa présence? [14]

Il n'en demeure pas moins que Husserl ne cesse de réaffirmer *l'unité* de la conscience vivante actuelle qui englobe présent, passé et futur, mais où le passé retenu hante la conscience sans plus l'envahir à moins qu'elle n'y prête son concours à travers des actes de remémoration ou de réflexion au point qu'on notera que le passé a une façon à lui d'être et d'exister qu'on serait tenté de décrire en disant de lui qu'il est le présent qui s'absente en tombant au-dessous de la ligne de l'actualité. Et il n'en est pas moins vrai que la "métaphysique du Présent vivant" qui régit les analyses husserliennes retient à la fois l'image d'un temps dominé par le Maintenant qui est jaillissement incessant et continuel engouffrement en soi-même du vécu présent,[15] et l'image du temps de part en part décrit comme flux du vécu qui s'écoule sans fin. Même si Husserl reconnaît que chaque vécu individuel peut avoir un commencement et une fin, "le flux du vécu lui-même ne peut commencer ni finir", il est succession ininterrompue de "Maintenant".[16] *L'infinité* de l'horizon de la temporalité paraît de la sorte clairement énoncée: le temps ne connaît ni commencement ni fin et ne saurait être l'objet d'une expérience de totalisation achevée alors que Heidegger laisse entrevoir la possibilité pour le *Dasein* d'un "pouvoir-être-tout" de l'existence à travers l'unité extatique de la temporalité s'anticipant dans la décision résolue (*Entschlossenheit*). Il est vrai que Husserl par ailleurs expliquait déjà dans le *Zeitbewusstsein* que la représentation du temps infini, contrairement à ce que suggérera en un sens Heidegger, n'est pas un *vécu* immédiat de la conscience temporelle, mais en vérité le "produit de la représentation", une construction conceptuelle tout comme l'est la série infinie des nombres entiers ou l'idée géométrique de l'espace infini.[17]

Si Heidegger peut nous aider à découvrir les caractères spécifiques de la finitude temporelle, il n'y a guère, reconnaissons-le, dans les textes majeurs de Husserl sur le temps et la temporalité, d'indications précises concernant ce problème. Il nous paraît judicieux, afin d'éclairer les rapports existant entre temps et finitude chez Husserl, de les faire ressortir comme par contraste et sur le fond de la conception heideggerienne du temps. Inversant le mouvement de l'analyse, nous partirons de ce que l'auteur de *Sein und Zeit* nomme "le concept vulgaire du temps", qui est au fond celui que nous adoptons tous, naïvement et spontanément dans notre vie, celui que nous concevons à partir de notre expérience quotidienne du temps: pure succession de Maintenant, flux ininterrompu et continu d'événements, qui en droit ne connaît ni commencement ni fin puisque après chaque Maintenant un autre Maintenant s'annonce et survient et avant chaque Maintenant présent un autre a logiquement et nécessairement dû le précéder. Tout Maintenant actuel ne provient-il pas d'un Maintenant antérieur et n'est-il pas destiné à passer en un Maintenant nouveau? De quelque façon qu'on s'y prenne, à aucun moment on ne saurait, en toute logique, être en droit d'arrêter l'écoulement du temps, la suite infinie des Maintenant qui surviennent et passent.[18]

D'où la conscience commune tire deux conclusions que les philosophes se sont empressés de prendre à leur compte pour les conceptualiser: d'abord, le temps est *continu* et *infini*, à l'image du fleuve qui s'écoule sans fin d'amont en aval qu'un spectateur anonyme regarde du haut d'un pont, ou comparable à la ligne droite sans commencement ni limite qu'une main invisible trace; mais c'est aussi l'image irrésistible de la continuité sans hiatus d'un temps qui toujours continue et passe, qui ne se laisse point suspendre ni arrêter quoi qu'on fasse pour en retenir la substance, pour le figer ou le fixer.

En second lieu, et c'est sur ce point que l'analyse de Heidegger vise peut-être implicitement les descriptions de son aîné: le concept vulgaire du temps — dont on se demandera si Husserl n'en reste pas tributaire à son insu — n'est qu'un *concept dérivé* et dégradé d'un temps au fond inauthentique qui consacre la domination et la primauté d'une des trois dimensions du temps (Heidegger, pour sa part, les nomme "extases" puisque ce sont là les trois modalités d'être par quoi le *Dasein* s'échappe de soi et s'installe dans le "hors de soi"): celle du Présent pensé sous la figure du "Maintenant" qui par sa fonction primordiale fait perdre aux autres dimensions leur originalité propre. Car, puisque le temps est "pure succession de Maintenant", suite continue et ininterrompue d'instants qui passent, le passé dès lors n'est plus qu'un Maintenant qui a cessé d'être, qui est justement passé dans le non-Maintenant, et l'avenir un Maintenant qui n'est pas encore mais est sur le point d'ad-venir.[19]

Telle est, selon Heidegger, la conception vulgaire et commune du temps marqué par son *infinité* et la *dominance du Présent*: non qu'elle soit une conception fausse ou condamnable en soi, le qualificatif de "vulgaire" ne doit pas nous abuser, en dépit des apparences il ne recèle aucun mépris pour la conscience commune, simplement la constatation que ce concept correspond à un mode d'exister inauthentique, dominé par l'exclusive préoccupation du *Dasein* pour les étants du Monde qui le sollicitent, en proie qu'il est au divertissement et à la dispersion parmi les étants sous la main, auxquels il est présent et qui remplissent entièrement de leur présence pesante et fascinante son propre présent, qui pour ainsi dire le tiennent dans leur "main-tenance".

Mais ce temps inauthentique et comme impropre à la nature profonde du *Dasein*, il l'est aussi parce qu'il est le temps de *l'oubli* et de la *fuite*: c'est une manière illusoire sans doute mais quotidienne et banale, constante de se détourner de la fin et de l'angoisse de la mort, dont la menace tout au plus pointe, de façon voilée, dans le sentiment que nous avons de la fuite inexorable du temps: ce qui n'est qu'un "savoir fugitif de la mort", explique Heidegger, un reflet de la méconnaissance flagrante qui est la nôtre de la "futurité" authentique de l'existence et de la finitude qui est la vraie marque de la temporalité humaine.[20]

Une autre caractéristique que la conception commune naïve a coutume de conférer au temps traduit elle aussi sa tendance tout à la fois à la dissimulation et au nivellement du temps authentique: c'est l'idée (ou la certitude) de *l'irréversibilité* du temps, idée qui du reste contredit l'illusoire velléité de l'existant humain dans sa quotidienneté de "maîtriser le temps", d'en disposer à sa guise puisque la réserve des Maintenant paraît inépuisable: nous croyons toujours encore "avoir le temps", pouvoir revenir sur ce qui fut. De fait, à première vue, on ne comprend pas ce qui interdirait de parcourir en quelque sorte à rebours la série infinie des instants s'écoulant inlassablement et dont on sait qu'elle se figure par une ligne droite que rien ne nous empêche de tracer de droite à gauche ou de gauche à droite. Dans la logique du temps infini, rien en effet ne s'y oppose si ce n'est justement le fait que le temps inauthentique procède par dérivation et nivellement de la temporalité authentique qui, quant à elle, est primordialement déterminée par l'extase du futur, donc impérativement orientée.

Contrairement à ce que croit dans sa naïveté la conscience commune, le Maintenant n'est pas "gros" d'un avenir qui n'est pas encore et serait appelé à ad-venir, c'est bien plutôt "le présent qui surgit du futur dans l'unité extatique originaire du procès de temporalisation authentique".[21] C'est parce que l'existant quotidien, dans sa préoccupation, ne prend pas garde à la finitude de la temporalité, qu'il fuit la mort qu'il n'ose regarder en face, détournant son regard du terme de son "être-au-monde", qu'il méconnaît *l'instance future* authentique du temps et donc la temporalité originelle. Se détournant de l'être-pour-la-fin, on croit qu'on a toujours encore le temps pour faire ceci ou cela, qu'on peut donc parfois le gaspiller et le perdre sans risque grave puisque le temps — ou la vie — continue . . . et nous offre toujours une nouvelle chance à saisir.[22]

Rappelons que l'éminence du futur dans l'analyse heideggerienne, dont il n'y a pas lieu ici de reproduire toutes les subtilités, tient à l'essence du *Dasein* défini par la structure d'être d'un étant pour lequel il y va toujours de son être même, d'un étant toujours en-avant-de-soi par le souci qu'il a de ce qu'il a à être, et principalement son pouvoir-être éminent par quoi il peut à tout moment re-venir à soi. En un mot, l'avenir originel n'est pas un Maintenant qui n'est pas encore réel, mais attendu et qui le sera, mais c'est plutôt la "venue à soi" (Zu-kunft) de l'existant par quoi celui-ci revient à soi-même et son pouvoir-être le plus propre. Or, c'est l'anticipation de la mort dans la décision résolue qui permet au *Dasein* de venir authentiquement à soi comme ce tout qu'il est en tant que "jeté dans la mort". Car, ne disons pas que l'existence a un terme, au sens où elle cesserait d'être un jour, mais disons que le *Dasein* existe comme fini ou comme dès sa naissance voué à la finitude de la mort. Le futur authentique, le seul qui en définitive compte, est celui précisément qui se dévoile comme l'anticipation (*Vor-laufen*) de la mort. C'est bien la mort qui clôture définitivement le pouvoir-être de l'existant tout en consacrant le moment de sa définitive "venue à soi" (*Zu-kunft*).[23]

Nous voilà parvenus au point aigu où la confrontation entre Heidegger et Husserl, entre une ontologie de la finitude temporelle qui privilégie le futur et la clôture du temps, et une phénoménologie d'un temps infiniment ouvert et dominé par le moment présent semble aboutir à une irréductible divergence. Qu'en est-il exactement? qu'en est-il notamment de la critique implicite que Heidegger paraît adresser à son maître Husserl à travers l'analyse qu'il fait de la conception vulgaire du temps? Y a-t-il réellement dans la théorie husserlienne un privilège exclusif du Maintenant présent? L'absence d'analyse poussée de la dimension du futur dans le *Zeitbewusstsein*[24] est-elle significative, traduit-elle la méconnaissance husserlienne de l'extase du futur et dès lors de la finitude du temps authentique? Ou bien, ne peut-on déceler chez Husserl à son tour, comme on l'a suggéré, un primat du futur face à la domination du présent pourtant difficile à nier?[25]

Husserl comme Heidegger, rappelons-le, considère la temporalité comme la structure essentielle du vécu et de la conscience et s'efforcera, sa vie durant, de décrire ce qu'il nomme lui-même "l'énigme de la conscience du temps".[26] Ce qui explique que depuis les *Leçons sur la conscience intime du temps* de 1904—05 — dont on n'oubliera pas qu'il en confia la publication en 1928 à Heidegger — jusque dans les années trente, il soit périodiquement et inlassablement revenu sur la problématique tant chaque fois il eut le sentiment, comme il le confesse dès le début, de se perdre et de s'enliser dans "les difficultés, les contradictions, les labyrinthes les plus étranges".[27] Qu'il y ait indéniablement évolution dans l'approche husserlienne, au fur et à mesure que la phénoménologie se développera d'une psychologie descriptive à une phénoménologie transcendantale voire une ontologie du Monde de la vie où la temporalité s'élargit en historicité, nul n'en disconviendra. Mais notre propos ici n'est pas de retracer ce cheminement de la méditation husserlienne sur le temps.

Notons simplement que dès 1905 le temps est bien décrit comme "la forme unitaire de tous les vécus", la conscience apparaissant comme "le flux unitaire du vécu".[28] Le temps appartient par essence au vécu comme tel ou encore, comme Husserl le précisera quelques années plus tard, "tout vécu réel est un vécu qui dure" et "s'ordonne à un continuum sans fin de durées", autrement dit encore, tout vécu "appartient à un unique flux du vécu (*Erlebnisstrom*) qui s'écoule sans fin".[29] Il semble donc bien que, d'entrée de jeu, Husserl se soit rallié à la thèse que Heidegger impute à l'interprétation ordinaire et commune du temps, laquelle retient essentiellement l'image de l'infinité du flux temporel. Car sans doute chaque vécu, par exemple une émotion brusquement ressentie, commence et finit "à un moment donné" dans le temps, à un point déterminé du flux, mais le flux du vécu lui-même "ne peut commencer ni finir", assure Husserl en conformité avec toute une tradition philosophique.[30]

Or, il est à peine besoin de le souligner, la totalité du flux avec l'enchaînement ininterrompu des vécus individuels (individualisables par le pouvoir qu'a la conscience de porter son attention sur l'un des points du flux pour l'identifier en quelque sorte à lui-même) n'est jamais donnée ni susceptible de l'être, aucune conscience n'ayant le pouvoir de l'embrasser comme d'un seul regard.[31] Et pourtant nous avons conscience de "l'absence de limite du développement", de la progression continue des vécus, qui nous fait passer d'un vécu à de nouveaux vécus, chaque vécu en appelant en quelque manière à un autre, aucun vécu ne pouvant se présenter comme le terme de la progression ou comme indépendant de l'enchaînement dans lequel il est pris.

L'idée — car c'est bien d'une idée régulative au sens kantien du mot qu'il s'agit[32] — de *l'infinité* en même temps que de *l'unité* du flux temporel se complète par celle tout aussi régulative que tout vécu qui a le sens d'un vécu actuel, d'un "Maintenant", s'insère dans une continuité et donc se relie à un Maintenant nouveau toujours se renouvelant. De surcroît, pris en lui-même et comme abstraitement le Maintenant actuel apparaît inévitablement comme une *limite* ponctuelle. Or comment s'en tenir à pareille conception qui frise l'absurdité et reviendrait à abolir tout simplement l'idée même de temporalité dans la mesure où elle consiste à considérer le Maintenant comme une limite absolue, un *Nullpunkt* et dès lors comme une limite pour ainsi dire insaisissable tel un néant s'intercalant entre deux autres néants: entre le passé qui n'est plus et l'avenir qui n'est pas encore? En fait, Husserl le notera, le Maintenant actuel n'est en aucune façon détaché des Maintenant ou des vécus qui lui succèdent ou qui l'ont précédé, il s'unit à eux dans une continuité d'enchaînement nécessaire. En outre, les vécus ou les Maintenant se succèdent dans la continuité ininterrompue selon un *ordre* donné, l'ordre de l'avant et de l'après, qui est *irréversible.* Il y a transition continue des uns aux autres: tout nouveau vécu étant nécessairement précédé "dans le temps" par un vécu antérieur et suivi d'un autre vécu, d'un Maintenant qui lui est postérieur.[33]

En d'autres termes, si le flux temporel est bien décrit comme une succession de Maintenant, il n'en reste pas moins que le Maintenant lui-même se présente d'emblée avec un *triple horizon* que Husserl tend à mettre en évidence: un horizon d'antériorité, de postériorité et de simultanéité, chaque moment d'actualité du vécu comportant ce que Husserl, dès le *Zeitbewusstsein*, nomme, d'une part, *rétention* et, d'autre part, *protention.*[34] On connaît les grandes lignes de sa description. Bornons-nous à rappeler que le point de départ est tout de même le Maintenant présent pris comme point d'origine, comme "*Quellpunkt*" que constitue du reste l'impression originelle (*Urimpression*), qui est définie comme "le commencement absolu, la source originaire (*Urquelle*) d'où tout procède et tout jaillit".[35] Toute rétention, Husserl prend soin de le souligner, est nécessairement précédée d'une impression originaire à laquelle elle correspond et dont elle est comme la réplique.[36]

Ainsi donc le Maintenant ne subsiste point à la manière d'une limite ponctuelle idéale, ou comme un élément atomique, il traîne avec lui et derrière lui, lorsqu'il passe en un nouveau Maintenant, sa propre ombre que Husserl représente sous l'image insolite de la "queue de comète". Le son entendu présentement continue de résonner à mes oreilles même une fois suivi d'autres sons de la mélodie, sans quoi il n'y aurait point de mélodie. Telle est ce que Husserl nomme la rétention: "Avec le surgissement d'une donnée originaire (*Urdatum*), d'une phase nouvelle, la précédente n'est pas perdue, mais "gardée en tête" (*im Begriff behalten*)", ce qui veut dire précisément "retenue" sans que la conscience s'évade de la phase présente, bien au contraire, et, ce qui est tout aussi significatif, sans qu'elle cesse pour autant d'etre orientée vers la phase à venir dans la protention.[37] En somme, c'est la conscience originaire qui passe elle-même dans une modification rétentionnelle d'elle-même. Dans le vécu actuel, dans le Maintenant est contenu un horizon rétentionnel, j'ai conscience d'un moment qui "vient-juste-de-passer", d'une phase du flux à peine disparue, d'un Maintenant passé. Chaque Maintenant apparaît ainsi accompagné d'un horizon d'antériorité qui ne s'efface pas mais en quelque manière se remplit continuellement de nouvelles phases rétentionnelles au fur et à mesure que se succèdent les Maintenants affectés dans ma conscience du temps du coefficient d'actualité.[38]

Ce qui vaut pour la phase d'antériorité qui accompagne la phase actuelle vaut aussi, naturellement, explique Husserl, mais avec un "indice inversé" (*umgekehrten Vorzeichen*), pour l'horizon de postériorité: chaque Maintenant s'ouvrant pour ainsi dire sur les Maintenant qui viennent vers lui, qui se hâtent à sa rencontre et en lesquels se convertira le Maintenant actuel dans son actualité même: telles sont ce que Husserl nomme les *protentions.* En fait, comparées avec les analyses qu'il consacre aux rétentions, les pages décrivant la conscience ou l'intentionnalité protentionnelle sont peu importantes sinon insignifiantes. Constatation que rejoint une autre remarque de Husserl qui ne manque pas de surprendre: à savoir que la protention immédiate est l'exacte contrepartie de la rétention immédiate, ou plus précisément, comme il le formule dans les *Leçons sur la conscience intime du temps*: "L'intuition de l'attente est une intuition de *souvenir retourné* car dans cette dernière les intentions qui visent le maintenant ne "précèdent" pas le processus, mais le suivent".[39] Elles font simplement signe "dans la direction opposée". Ce qui suggère l'interprétation que le futur se détermine par principe à partir du passé.

L'attente se remplit toujours par la présentation du présent, et lorsque le moment attendu survient, le futur se mue en présent tandis que le Maintenant présent est devenu relativement passé.[40] Tout se passe comme si le futur ne pouvait se constituer que dans l'horizon prétracé par le passé.[41]

Néanmoins, quoi qu'il en soit de la triple "horizontalité" du présent, le flux du temps est bien primordialement axé sur le Maintenant présent même si la conscience tient encore "sous sa prise" et comme en main (*im Begriff behalten*) ce qui est du passé immédiat et que le Maintenant actuel ne cesse de se convertir en Maintenant retenu et de sombrer de plus en plus dans le passé; s'il sombre de la sorte, il le fait toujours en traînant après lui telle une queue de comète ses rétentions qui s'emboîtent les unes dans les autres de même qu'il projette en avant de lui ses protentions qui à leur tour s'enchaînent avec le Maintenant présent tout en étant préfigurées par les souvenirs du passé. Le flux du temps se constitue ainsi non pas à la manière d'un collier de perles qu'on enfile les unes après les autres, mais au contraire, au lieu de se juxtaposer dans l'ordre de leur apparition, les différentes phases temporelles semblent se superposer (*überschieben*) continuellement si bien que le résultat en est un véritable "emboîtement" des phases antérieures, qui précèdent le Maintenant actuel, avec celui-ci et avec les rétentions de rétentions.[42]

Seulement, ce processus n'est pas sans limite, il ne peut se poursuivre à l'infini car le champ de conscience ou *Präsenzfeld* est lui-même limité par le pouvoir qu'a la conscience attentionnelle de garder encore comme à portée de la main ce qui est entré dans le processus rétentionnel et tend à sombrer dans un passé toujours plus éloigné. L'horizon rétentionnel est *restreint*, il y a un point de fuite qui se confond avec l'imperceptible si bien que Husserl, assurément, ne rejoint pas Bergson pour dire qu'en droit tout mon passé à chaque instant est virtuellement présent ou me suit dans la continuité du flux, même s'il considère que l'horizon du passé est infiniment ouvert: tout mon passé non-immédiat ou lointain requiert pour resurgir un acte spécifique de la conscience, une remémoration ou re-présentation (*Ver-gegenwärtigung*) explicite qui peut bien s'accompagner du sentiment que "je peux" en toute liberté et à tout moment me ressouvenir et cela "aussi souvent que je le veux" (43).

Nous voilà sur le point de retrouver notre problème initial: qu'en est-il de la finitude temporelle chez Husserl? Tient-elle ou non à la dominance du Maintenant présent sur les autres dimensions temporelles et à la limitation de l'horizon rétentionnel? Et d'abord est-on réellement en droit d'affirmer qu'il y a une telle prédominance ou primauté du Présent? Observons que dans l'analyse husserlienne le temps est loin d'être décrit comme une série monotone ou pure succession de Maintenant représentable par une abstraite ligne droite puisque chaque nouveau Maintenant n'efface pas totalement ceux qui l'ont immédiatement précédé, mais, bien au contraire, ces derniers sont retenus sous le regard de la conscience présente. Et même s'il faut se résigner à ce qu'au cours du procès de temporalisation de la conscience les anciens Maintenant peu à peu sombrent toujours plus loin dans le passé et se vident pour ainsi dire de leur actualité, il n'en demeure pas moins vrai que le Maintenant est gros des Maintenant passés immédiats et des Maintenant à venir immédiats au point qu'il constitue justement non une limite ponctuelle, mais bien plutôt un "champ de présent" ou de présence (*Präsenzfeld*) recelant tout à la fois impression originaire, protention et un certain nombre de rétentions voire de rétentions de rétentions.

Privilège, par conséquent, du "*Präsenzfeld*" où la conscience coïncide entièrement avec elle-même, mais où sont retenues aussi constamment dans une co-présence les phases du passé et du futur immédiats, et ce dans l'unité d'un même flux de conscience.[44]

Il semble donc bien que chez Husserl la prééminence parmi les trois modalités temporelles en revienne au Maintenant présent qui sans cesse se constitue en "champ de présence" avec son horizon fluant de rétentions et de protentions, de passé et de futur partiellement "sous la main" et quasiment insaisissables, dont chaque moment renvoie nécessairement à l'instant présent de l'impression originaire constamment jaillissant et rejaillissant. Sans doute ce Présent privilégié n'est pas le Maintenant compris comme l'instant-limite sans épaisseur et sans vie, ce pur Maintenant dont Husserl dit qu'il est une "limite idéale", quelque chose d'entièrement abstrait" qui ne peut rien être en lui-même" et qui, en définitive, n'est guère "différent *toto caelo* du non-Maintenant".[45] Un tel Maintenant à la limite du non-être et comme privé de tout contenu de sens ne se révèle comme Maintenant qu'en cessant d'être ce qu'il est, limite abstraite entre le passé et le futur et en s'unissant continuellement aux Maintenants précédents qui à leur tour n'apparaissent comme tels que dans ce mouvement de transition continue de l'impression originaire à la rétention et la protention, chacun des moments ne prenant sens que par et pour les deux autres.[46] Le Maintenant "épais" que constitue le champ de présence est plein de ces phases qui constamment le mettent en question et le menacent, et par quoi il se trouve engagé dans un mouvement de *transcendance*, pour parler avec Heidegger, tout en étant la seule source de jaillissement de la conscience du temps et donc de la conscience tout court.

Comment refuser la primauté à ce Présent là, à ce Maintenant épais d'où procède toute temporalisation même s'il se constitue avec son horizon rétentionnel et protentionnel? Toutes les dimensions du temps sont vues à partir de ce Maintenant présent que Husserl qualifiera plus tard de *Présent vivant* justement parce qu'il ne cesse de renaître de ses cendres. Interprétation à tout le moins que paraissent confirmer les textes postérieurs, et notamment les célèbres inédits sur le temps répertoriés sous le sigle *Manuscrits du Groupe C.* Husserl y décrit la présence du Présent comme "*strömend-lebendige Gegenwart*" qui constitue une "unité de coexistence", mais dans le mode du changement fluctuant. Il le désigne sous le nom de "*urtümliche Gegenwart*", de Présent originaire parce qu'il est le moment où s'origine le procès même de la temporalisation et d'où jaillit la conscience même du temps.[47]

On le notera, la manière dont Husserl décrit dans les textes des années trente la double (ou triple) "horizontalité" du Présent vivant n'est pas foncièrement différente des descriptions des *Leçons* de 1905 ou 1917. La seule nuance qu'on relèvera éventuellement concerne l'accentuation de la bi-polarité et aussi la constatation que rétention et protention ont ceci en commun qu'elles ne sont ni une "présentation" (*Gegenwärtigung*) ni une "re-présentation" (*Ver-gegenwärtigung*), mais en quelque manière une "dé-présentation" (*Ent-gegenwärtigung*). E. Fink avait jadis clairement mis en lumière cette caractéristique de la rétention et de la protention dans son essai sur *Re-présentation et Image* de 1930.[48] La rétention est pour ainsi dire la manière dont le Présent s'absente en fusionnant avec son propre évanouissement, et dès lors ressemble à un mode de recouvrement de soi par quoi la conscience, dans son Maintenant, échappe à soi-même (peut-être évoquera-t-on là encore le concept heideggerien d'*extase* qui décrit le mouvement de "*Ent-rückung*"). La protention de même n'est pas le Maintenant à venir, elle est le Présent lui-même en tant qu'il n'est pas encore mais déjà là, "en avance sur soi": c'est dans le mouvement protentionnel d'anticipation de soi que "le Présent va à la rencontre du futur afin de l'accueillir pour ainsi dire les bras ouverts".[49]

Sommes-nous en droit d'en conclure, comme certains le suggèrent, qu'il y a bien finalement, chez Husserl comme chez Heidegger, un incontestable *primat du futur*? Suffit-il d'alléguer que la protention est comme le moment actif, productif de la synthèse temporelle pour que le Maintenant soit réduit à n'être rien d'autre, en un sens, qu'une continuelle approche de soi à partir de la protention? Ajoutera-t-on que, par ailleurs, la conscience de part en part marquée par l'intentionnalité inclut nécessairement un mouvement de dépassement de soi (de transcendance, comme l'interprète Heidegger) et en appelle constamment au remplissement de ses intentions vides, de ses "*Vormeinungen*" qui sont comme des anticipations de remplissements à venir? Quant à la rétention, elle se bornerait somme toute à "retenir" tandis que la protention ayant littéralement le sens d'un "pro-jet" de possibilités jouirait ipso facto d'un privilège sur l'actualité du Présent.

Nous voici parvenus au stade de l'analyse où se réintroduit dans notre problématique la question de la finitude chez Husserl. Ne tient-elle pas justement à la manière dont il ne se lasse pas de décrire la radicale inadéquation de l'expérience actuelle du Monde, expérience toujours présomptive et infiniment ouverte en appelant à la confirmation et la vérification des évidences présentes et passées? On connaît la thèse husserlienne du "*leibhaft Selbstgegeben*" du perçu qui, en dépit de sa valeur d'indice de vérité, n'est pourtant jamais absolument "donné en personne" (*selbst gegeben*) puisque rien n'y est jamais purement ni adéquatement donné une fois pour toutes, le perçu se composant toujours d'un horizon d'anticipations et de co-intentions non remplies, de simultanéités et de présomptions, au point que le noyau du perçu pourrait bien, dans l'infinité des anticipations, n'être rien de plus que "l'entrecroisement des horizons, plus exactement l'infinité des entrecroisements possibles".[50] Le présent réel se réduirait-il donc à n'être que "l'infinité des possibles", la réalisation incomplète de tous les possibles impliqués dans l'horizon d'anticipation infiniment ouvert? Sans doute pas s'il est vrai par ailleurs, comme nous l'avons observé, que le Maintenant vivant est gros aussi de rétentions de vécus immédiatement antérieurs qu'il traîne après lui telle une "queue de comète".

Quoi qu'il en soit, on est enclin ici à traduire l'idée de l'infinité de l'horizon d'anticipation ou d'antériorité du perçu en termes justement de finitude. Husserl ne se lasse pas de le répéter: la perception transcendante se définit par son imperfection, lisons sa radicale inadéquation, ce qui signifie ultimement l'inachèvement essentiel de la synthèse temporelle dont elle est le corrélat. Déjà les analyses portant dans les *Recherches logiques* sur la corrélation intention/remplissement permettraient d'illustrer le privilège qui revient de la sorte au futur. La découverte du concept même d'*intentionnalité* et de sa structure foncièrement *téléologique* semble le confirmer. La conscience intentionnelle qu'est-elle donc sinon une anticipation du remplissement à venir par une intuition donatrice de l'objet visé? L'intention est bien visée de l'objet, mais non encore possession de l'objet "en personne" (*leibhafte Selbsthabe*): elle ne le sera que dès l'instant que par une anticipation la conscience vide se remplit d'une intuition et qu'une sorte de coïncidence s'est établie entre la visée intentionnelle et l'intuition remplissante.[51]

Dans le procès même de la perception d'un objet situé dans l'espace, qui s'effectue par "esquisses ou profils" (*Abschattungen*), comme l'expliquent les *Ideen* I, les faces données de l'objet suscitent en même temps des intentions relatives à des aspects inexplorés, qui pourront toutefois se remplir progressivement au cours de nouvelles séries de perceptions partielles. Si bien que l'appréhension de la chose elle-même "dans son être-donné en personne" (*leibhaften Selbstgegebenheit*) est le résultat infiniment différé d'un processus de synthèse d'identification, disait Husserl dès les *Recherches logiques*, entre les intentions non-remplies et les intuitions venant les remplir au fur et à mesure.[52]

Toute conscience perceptive est de la sorte affectée d'une structure téléologique d'anticipation ou de pré-somption (*Vor-meinung*) qui définit l'intentionnalité de la conscience que Husserl nommera plus tard "intentionnalité d'horizon" (*Horizontintentionalität*), laquelle est caractéristique de toute espèce d'expérience du Monde objectif. Elle ne se comprend, reconnaissons-le, qu'à la condition qu'on admette qu'il y a une sorte de prééminence du futur sur le présent en vertu de laquelle la réalisation de l'idéal de la validité définitive (*Endgültigkeit*) de l'expérience, somme toute l'idéal de l'évidence adéquate, de la coïncidence entre l'intention proprement dite de l'objet et sa donation "en personne", est littéralement reléguée dans un futur indéterminé: tout processus perceptif étant régi par un *télos* qui n'est autre chose que l'indice de l'anticipation d'un futur état de remplissement parfait de l'intention.[53]

La thèse bien connue et jamais abandonnée par Husserl selon laquelle la chose perçue est toujours tout ensemble du visible et de l'invisible, pour paraphraser Merleau-Ponty, et que si elle est bien saisie directe de la chose même, “elle n’en est pas moins aussi anticipation”, non pas par accident mais *par essence*, ne saurait se comprendre hors de la structure temporelle de la conscience. L’expérience est un mixte, fait tout à la fois de “véritable donation de la chose elle-même” (*Selbstgebung*) dans un présent actuel et de pré-intentions et de co-intentions vides, mais en appelant à leur remplissement. En tout état de cause, l’objet ou le Monde dans l’expérience que j’en ai n’est jamais que le pôle identique et idéal des perceptions concordantes possibles et réelles, l’évidence adéquate ou la vérité n’étant jamais que *provisoire* et *relative*, la confirmation définitive n’étant qu’un espoir, sans doute motivé, un *idée* située à l’infini dont la réalisation dépend somme toute de la non-rupture du système harmonieux de l’expérience, expliquera Husserl dans *Erste Philosophie*.[54] Merleau-Ponty fidèle en ce sens à Husserl renchérira en précisant à son tour que “la croyance à la chose et au Monde ne peut signifier que la présomption d’une synthèse (temporelle) achevée”.[55]

La *finitude* de l’expérience, selon Husserl, qu’est-elle dès lors sinon *l’idée de l’infinité* d’expériences concordantes possibles triomphant de toutes les discordances qui menacent, de toutes les ruptures éventuelles du flux harmonieux de l’expérience du Monde, intervenant sur fond de “certitude de croyance” (*Glaubensgewissheit*). L’idée de Monde vrai, de Monde “en soi” n’est aux yeux de Husserl que la traduction d’un idéal d’approximation saisi dans l’anticipation même de la chose et s’inscrit comme corrélat dans une conception de “l’infinité” posée *idealiter* et opposée à l’incontournable “finitude” de l’expérience réelle, condition pourtant de ma certitude de pouvoir toujours aller plus loin dans l’exploration du Monde, de ma conviction de l’inachèvement de mon flux d’expérience, que “je peux” à tout moment élargir mon expérience de la chose et du Monde se donnant dans l’horizon d’un “et ainsi de suite” infiniment ouvert. La finitude de notre être réside ainsi dans le pouvoir d’infinitisation (*Verendlichung*) qui nous est conféré, elle tient à notre pouvoir de saisir l’infini dans le fini de l’expérience.[56]

Pour résumer nos analyses et pour faire retour à la problématique du temps, nous serons peut-être tentés de soutenir qu’il n’y a point chez Husserl, au bout du compte, de *primat réel du Présent*, du moins au sens où l’entend Heidegger comme ce Maintenant qui appartient à la série indéfinie des Maintenant qui se succèdent et se chassent pour ainsi dire l’un l’autre, chacun faisant oublier celui qui le précéda, incapable aussi d’anticiper celui qui le suivra sinon en se le représentant comme un présent non encore advenu mais déjà prétracé dans le passé. En fait, il n'en est rien: non seulement Husserl n'a pas la naïveté d'accorder un quelconque privilège au Maintenant présent qui ferait fonction d'unique source d'évidence, mais encore, conclura-t-on en allant plus loin dans l'interprétation, il semble même admettre avec Heidegger que le futur jouit d'une certaine prérogative dans la mesure où il contribue décisivement à la constitution du temps de la conscience et par suite de l'expérience tout entière. Le futur n'est-il pas cette dimension ouverte dans laquelle tout s'origine y compris la conscience intentionnelle? En déduira-t-on tout uniment que le temps se temporalise (*sich zeitigt*) à partir du futur, et ce parce que nulle autre explication ne permettrait de comprendre l'irréversibilité du temps et le fait que son mouvement a un sens et un ordre, c'est-à-dire qu'il est toujours orienté selon l'avant et l'après?

Certes le futur tel que le décrit Husserl — à la différence de Heidegger — est une dimension *ouverte*, infiniment ouverte même comme l'est la conscience intentionnelle qui en appelle en quelque sorte au plein de l'intuition pour s'accomplir. Il convient peut-être après tout d'interpréter la protention comme si elle n'était pour la conscience présente que *promesse* ou *espoir* de remplissement, d'enrichissement continu du vide des intentions anticipatrices. Pour Heidegger aussi, on le sait, l'existence est fondamentalement "pro-jet", non pas être mais "pouvoir-être" de ses possibles, fût-ce de l'impossible possibilité de la mort, et donc soumise à l'instance du futur. Pour Husserl, de manière analogue, c'est le futur qui est promesse de donation de la vérité et de la présence du Monde; c'est par le futur que le passé prend figure et cesse d'être vécu passivement pour être remémoré activement, c'est-à-dire connu comme tel. C'est par l'horizon d'anticipation ou de protention que trace l'avenir que la conscience cesse d'être engluée dans son présent éphemère et dans l'en-soi compact du passé pour recouvrer une vraie vie. L'avenir serait-il donc ce qui fonde tout ensemble passé et présent en unifiant sous sa férule la vie de la conscience? Serait-il vraiment, comme on est tenté de le soupconner, la source cachée d'où procèdent présent et passé pour autant justement que la conscience, de part en part intentionnalité, est ce mouvement de tension s'ouvrant sur la chaîne des plénitudes de conscience possibles, dont parle déjà l'essai sur le temps de 1905, quoique timidement, qui l'arrache à l'actualité réalisée pour lui révéler à chaque fois son "inactualité", son incomplétude et son inachèvement?[57] L'interprétation résolument téléologique des textes postérieurs de Husserl le confirmera à nouveau en attestant que la vie du Moi est toujours vie dans l'anticipation et la précursion en direction de ce qui est à-venir, non au sens simplement où le futur serait ce que je serai plus tard et que je m'attends à être, mais au sens où je *suis* dès à présent engagé dans le futur que j'ai à être.[58] Je suis ainsi continuellement "en avance sur moi-même", je ne suis jamais au terme absolu de mes anticipations de l'état de remplissement de mes intentions, je suis toujours dans un mouvement d'approche infini de moi-même.

Il s'ensuit qu'il nous faudra rectifier l'interprétation proposée initialement. L'horizon d'avenir, selon Husserl, est *sans fin.* Il est toutefois toujours l'horizon du futur vu du présent lequel est cette vie renaissant sans cesse à elle-même. La conscience n'est jamais accomplie, jamais en coïncidence absolue avec elle-même, jamais en possession totale de la chose elle-même qu'elle vise. Le futur, de surcroît, est ce par quoi le passé lui-même se vérifie et se rectifie: seulement cette vérification n'est jamais achevée une fois pour toutes, elle demeure l'idéal à poursuivre et dont la réalisation est indéfiniment différée. Telle est, répétons-le, la *finitude* selon Husserl: le paradoxe qui fait qu'elle est *l'infinité* (*Unendlichkeit*) ouverte de l'horizon temporel dans lequel s'inscrit la vie de la conscience.

Et pourtant ne convient-il pas là encore de faire preuve de prudence? Contrairement à l'interprétation suggérée quant à la primauté du futur, comment la dominance du Présent vivant pourrait-elle être réellement mise en question? Car le Moi décrit par Husserl est à jamais impuissant à quitter sa condition d'être qui est celle d'un être "présent au Monde" et "présent à soi". Le Présent vivant est bien cette dimension d'où surgissent et où convergent toutes les directions temporelles: vers le passé comme vers l'avenir, et même vers le Présent lui-même, fût-il vrai par ailleurs que le Présent vivant est animé et comme travaillé par une double tension qui le pousse, d'un côté vers le passé par le biais de l'intentionnalité rétentionnelle, laquelle vise un Maintenant tout juste écoulé qui au moment où il était Maintenant présent visait à travers la protention le Maintenant actuel;[59] de l'autre côté, la conscience se porte en avant de soi à travers ses protentions, mais celles-ci, à leur tour, visent ultimement le Présent à venir lors même qu'elles réagissent sur les anciens Maintenant en tant qu'elles constituent des intentions d'attente prétracées dans le souvenir.[60] Tous les vécus de conscience, Husserl l'affirme avec vigueur, se caractérisent toujours par leur "orientation changeante vis-à-vis du Maintenant vivant" qui ne cesse d'être centre et pôle vers lequel tous les vécus convergent (ou d'où ils divergent?): le Maintenant présent constamment se renouvelant.[61]

Là encore les inédits du groupe C viennent confirmer l'interprétation et affirmer la prééminence du Présent: "Le Présent est la réalité absolue, y lit-on, il est la réalité authentique par excellence en tant qu'il est temporalisation originaire (*urzeitigende*)".[62] Aussi n'est-il pas besoin de se demander pourquoi seul le Présent mérite d'être qualifié de "vivant". Déjà les *Leçons sur le temps* de 1905 avaient suggéré l'idée que c'est la capacité du "Maintenant présent" (*Jetzige Jetzt*) d'assurer sa propre "survie", sans quoi il n'y aurait pas de continuité temporelle ni de flux du temps, qui confère au Présent sa prééminence: il survit encore dans les rétentions et s'anime déjà dans les protentions qui l'annoncent. Et enfin c'est bien le Présent qui figure comme "source du temps" s'il est vrai, comme Husserl le formule dès 1905, que "l'impression originaire est la source originaire de toute conscience et de tout être", qu'elle est "le point-source vivant d'où l'être jaillit (*lebendige Quellpunkt des Seins*)": c'est de la vie du Présent vivant que jaillit et vit le flux du temps tout entier.[63] Merleau-Ponty dit-il autre chose lors-qu'il souligne que "le présent (au sens large, avec ses horizons de passé et d'avenir originaires) a cependant un privilège parce qu'il est la zone où l'être et la conscience coïncident"?[64]

Evoquons ici un dernier problème tout aussi ardu, qui à son tour est susceptible de jeter quelque lumière sur l'originalité de l'approche husserlienne: il s'agit du très vieux problème de la corrélation de l'espace et du temps, ou, si l'on préfère, celui de leur détermination réciproque souvent discutée par les philosophes. Sans entrer dans le fond du débat, rappelons simplement que Husserl, dans les *Leçons* de 1905, le rattache à la question du temps pour autant qu'il est tout ensemble succession et simultanéité. Il ne saurait y avoir de conscience temporelle sans cette double structure de la simultanéité — laquelle a l'air de renvoyer à l'espace — et de la succession qui semble être du temps pur. Or, Husserl s'applique à le montrer, simultanéité et succession dérivent d'une certaine manière l'une de l'autre, renvoient l'une à l'autre mutuellement et se constituent corrélativement l'une avec l'autre. Car, au fond, on peut déceler deux types de simultanéité ou de co-existence: l'un est la co-existence, l'être-ensemble (*Zusammen*) du perçu et du percevoir, qui constitue la simultanéité proprement dite et en quelque sorte l'inter-dépendance du temps immanent et du temps objectif, l'autre est l'être-ensemble de l'impression présente et de l'impression juste passée, qui constitue la continuité de l'écoulement des vécus et donc de la succession temporelle. Le temps est ainsi, selon Husserl, toujours synthèse de continuité et de discontinuité, d'unité et de multiplicité, de succession et de simultanéité car il est fait de l'émergence et de l'immersion continues des moments temporels.[65]

Contrairement à Heidegger pour qui l'espace est du temps déchu, un présent qui s'absorbe en lui-même en oubliant qu'il vit du futur et du passé,[66] Husserl se garde bien de déduire l'espace du temps; chacun a sa place et son rôle propre: si le temps est succession, il ne peut l'être que parce qu'il est aussi permanence et simultanéité, l'espace qui paraît indépendant du temps ne recouvre en fait sa structure propre de simultanéité que sous la forme de la con-temporanéité des consciences l'appréhendant et s'appréhendant mutuellement d'un même mouvement.[67]

Il est temps à présent de conclure ces réflexions au cours desquelles nous avons essayé de mettre face à face quelques-uns des éléments significatifs à nos yeux de la métaphysique du temps pensé sous le signe de la finitude, positivement ou négativement, à laquelle sacrifient en dernière instance et Heidegger et Husserl. L'opposition constatée entre les deux penseurs pourtant également soucieux d'éclairer la question de l'essence du temps nous a servi de guide: l'un nous a paru privilégier essentiellement une vision "*infinitiste*" du temps, ce qui, paradoxalement, tendait à rattacher la finitude humaine à l'idée de l'essentiel inachèvement du flux du temps et devait d'emblée frapper toute expérience du Monde du sceau de l'inadéquation et de l'imperfection et affecter le temps du coefficient de l'imprévisible; l'autre, pour qui la mort était la limite absolue de la temporalité et donc le signe de l'irréductible finitude de l'homme qui dans le souci qu'il a de soi et de son authenticité ne peut pas ne pas se porter de préférence "en avant de soi" vers son futur — le futur ultime étant ce terme extrême de l'existence à partir duquel il se voit définitivement renvoyé à soi et à l'impossible possibilité de son être, qui le détermine dès l'origine pose le primat absolu du futur ou de l'être-pour-la-mort saisi dans l'instant (*Augenblick*) décisif de la résolution qui permet à l'existant humain de s'appréhender dans son être-tout. N'est-ce pas simplifier à l'extrême l'opposition entre Husserl et Heidegger que de suggérer que chez le premier la vie l'emporte en tout état de cause tandis que chez le second la mort paraît avoir le dernier mot? Sans doute la pensée de l'un comme de l'autre est-elle plus complexe et plus subtile. Seulement le primat ontologique conféré à l'être-pour-la-mort ne consacre-t-il pas, en dernière analyse, l'arrêt du temps en annulant pour ainsi dire le mouvement même de la temporalisation, par conséquent, en mettant fin à la prédominance du futur et en repoussant l'existant dans l'inauthenticité de la déchéance auprès des étants du Monde et dès lors dans les bras du présent qui n'est que la présence pesante des choses de sa préoccupation quotidienne?

L'opposition que nous venons ainsi d'apercevoir entre deux théories majeures du temps ne se résout-elle pas, en définitive, en l'opposition fondamentale de deux philosophies centrées différemment: l'une sur une métaphysique de la subjectivité et de la conscience connaissante, dont le mode d'être éminent est la *re-présentation* (*Ver-gegenwärtigung*) qui entraîne inévitablement une métaphysique de la *présence* et du *Présent*, l'autre sur l'existence et sur l'être, qui n'échappe pas — quoiqu'elle veuille — à la tentation d'une forme d'anthropologisation, du moins à l'époque de *Sein und Zeit* où l'exister défini par la temporalité qui est signe de l'impuissance et de l'échec et donc de la finitude de l'homme, l'emporte encore sur l'être même si déjà s'y dessine, à travers la prépondérance de l'être-pour-la-fin, l'ombre d'une pensée qui réserve le dernier mot à la totalité close où risquent de s'engloutir et l'être et le temps?

Université de Grenoble

NOTES

[1] Cf. P. Claudel: "Le temps est le sens de la vie". *Art Poétique* et M. Heidegger: "Der Sinn des Daseins ist die Zeitlichkeit". *Sein und Zeit* (cité plus loin sous le sigle: SZ), p. 331.

[2] E. Husserl, *Vorlesungen zur Phänomenologie des inneren Zeitbewusstseins*, éd. par M. Heidegger, Halle, 1928, trad. fr. par H. Dussort sous le titre *Leçons pour une phénoménologie de la conscience intime du temps*, préface de G. Granel, Paris, 1964. Nous citerons plus loin: *Zbw.*

[3] SZ, §65, p. 331 sq.

[4] SZ, passim, surtout p. 331 sq.

[5] *Ideen I*, p. 162—3, tr. fr. p. 274—5.

[6] Zbw. Introduction, p. 368, tr. fr. p. 3 et P. Ricoeur, *Temps et récit*, I, Avant-propos et ch. I (consacré à un commentaire du livre XI des *Confessions*). Husserl reconnaît à saint Augustin le mérite d'avoir été le premier à ressentir les immenses difficultés que

recèle le problème de la conscience du temps; il constate aussi qu'en la matière l'époque moderne n'a rien produit de mieux. Quant à Heidegger, s'il se réfère plus volontiers à Platon et surtout à Aristote et à Kant, il n'oublie pas saint Augustin. Cf. *Grundprobleme der Phänomenologie*, in *Gesamtausgabe*, tome 24, p. 325.

[7] La phénoménologie du temps, comme saint Augustin, inaugure sa réflexion sur le temps sous l'égide d'une question ontologique. Cf. E. Husserl, Zbw. p. 368, tr. fr. p. 3—4.

[8] "Dès que nous cherchons ... à rendre raison de la conscience du temps, à mettre correctement en rapport le temps objectif et la conscience subjective du temps ... nous nous perdons ans les difficultés, les contradictions, les labyrinthes les plus étranges". Id. ibid.

[9] *Kant und das Problem der Metaphysik* (cité plus loin KPM), p. 207, tr. fr. p. 285, et aussi SZ, p. 325 sq.

[10] Zbw., p. 399 et 424, tr. fr. p. 56 et 89.

[11] SZ, §81, surtout p. 423 sq.

[12] Cf. M. Merleau-Ponty, *Phénoménologie de la perception*, p. 476 et E. Husserl, Zbw., p. 391, tr. fr. p. 46. Husserl y parle du "originäre Zeitfeld".

[13] Cf. J. Derrida, *La voix et le phénomène*, p. 69 sq. et M. Heidegger, *Was ist Metaphysik?*, p. 34, tr. fr. in *Questions* I, p. 66.

[14] Pour cette interprétation, cf. par ex. G. Granel, *Le sens du Temps et de la Perception chez E. Husserl*, p. 70 sq.

[15] Id. ibid., p. 84.

[16] E. Husserl, *Ideen I*, p. 163, tr. fr. p. 275.

[17] Sur la "*Ganzheitsstruktur*" du *Dasein* selon Heidegger, cf. SZ, §61, p. 301 sq. Sur le temps infini chez Husserl, cf. Zbw., §4, p. 377, tr. fr. p. 23.

[18] Cf. SZ, §81, p. 420 sq., surtout p. 421 et aussi p. 330—1.

[19] SZ, p. 423.

[20] SZ, p. 426.

[21] SZ, p. 427.

[22] SZ, p. 424.

[23] Pour toute cette analyse, cf. SZ, §65 et 66, p. 325 sq.

[24] Les *Leçons* de 1905 ne consacrent qu'un seul paragraphe à la notion capitale pourtant de la "protention".

[25] G. Granel, op. cit., p. 94.

[26] E. Husserl, *Ideen I*, p. 163, tr. fr. p. 275.

[27] Zbw., Introduction, p. 368, tr. fr. p. 4.

[28] *Ideen I*, §81 à 83, surtout p. 161 sq., tr. fr. p. 272 sq.

[29] Id. ibid., p. 163, tr. fr. p. 275.

[30] Id. ibid.

[31] Id. ibid., §83, p. 166, tr. fr. p. 276 sq.

[32] Id. ibid.

[33] Id. ibid., p. 164, tr. fr. p. 276 sq.

[34] Id. ibid., §82 et Zbw., passim.

[35] Zbw. p. 451, tr. fr. p. 131.

[36] Zbw., §13, p. 271 sq., tr. fr. p. 159 sq.

[37] Zbw. Suppl. IX, surtout p. 471—2, tr. fr. p. 159.

[38] Id. ibid.

[39] *Ideen I*, p. 147, tr. fr. p. 248 et Zbw. §26, p. 413 sq.
[40] Zbw. p. 414, tr. fr. p. 76.
[41] Cf. *Erfahrung u. Urteil*, p. 34 et *Cartesianische Meditationen*, p. 141.
[42] *Ideen I*, p. 164, tr. fr. p. 277.
[43] Zbw. §18, p. 402, tr. fr. p. 59.
[44] Ibid., §38.
[45] Zbw. §16, p. 424 sq., tr. fr. p. 56 sq.
[46] Id. ibid.
[47] Manuscrits C III, p. 24, cité par G. Brand, *Welt, Ich und Zeit*, p. 76.
[48] Cf. *Studien z. Phänomenologie* 1930—39, p. 23 sq.
[49] Manuscrits C 17 I, p. 44, cité par G. Brand, op. cit., p. 87.
[50] E. Husserl, *Erste Philosophie* II, p. 45 sq. et G. Granel, op. cit. p. 192.
[51] E. Husserl, *Recherches logiques* III (cité plus loin RL), VIème Recherche, ch. III et V, et aussi *Ideen I*, §24, p. 43 sq. tr. fr. p. 78, et *Formale und transzendentale Logik*, §§59 et 60 et 107.
[52] RL, III, §16 et 24, et *Ideen I*, p. 74 et 83, tr. fr. p. 132 et 144.
[53] Cf. FTL, Appendice II et §107.
[54] Op. cit., tome II, p. 45 sq. et 383, tr. fr. p. 62 sq., cf. aussi p. 392.
[55] Op. cit., p. 380.
[56] E. Husserl, *Erste Philosophie* II, p. 386—7. On y lit: "A priori ist Welt nur erfahrbar von "endlichen Wesen", ihre Endlichkeit ist es, dass sie im wesensmässig begrenzten Jetzt Welt erfahren, Unendliches im Endlichen. Das Unendliche ist a priori nur präsumierbar, a priori nur seiend als eine im Stil endlicher einstimmiger Erfahrung intentional beschlossene Idee, beschlossen im offenen Und-so-weiter".
[57] Zbw. §24 et 25.
[58] Manuscr. C 2 III.
[59] Zbw. Suppl. III.
[60] Zbw. §24 et 25, p. 412 sq., tr. fr. p. 74 sq.
[61] Zbw. p. 412, tr. fr. p. 74.
[62] Manuscr. C 17 I, p. 42.
[63] Zbw. §31, p. 423 sq., tr. fr. p. 88 sq.
[64] Op. cit., p. 484—5.
[65] Zbw. §38 et surtout p. 432, tr. fr. p. 102 sq.
[66] SZ, p. 369 et 417 sq.
[67] Zbw. §38 et Suppl. VII, surtout p. 432 sq., tr. fr. p. 103 sq.

FRANÇOISE DASTUR

MAN AND HISTORY

Is it possible to understand the historical dimension as a whole without referring to an authority other than the anthropological one? This question may seem meaningless at first glance, so evident does it seem that Karl Marx was right when saying that men *make* history.[1] Are we not, in modern times, all agreed with his fundamental thesis which constitutes man as *the* historical being by definition? But what does the historicity of man mean in the end? Does it mean the decline of the Absolutes of all kinds and the domination of the most unbridled relativism — or the rise of a new alliance between truth and history, being and time? When history begins to enter philosophy, when it becomes not only an "object" of philosophical reflection but also the milieu of its unfolding, it is precisely in order to put an end to the permanent scandal of the plurality of philosophies compared to the unicity of truth. Immersed as we are in the anthropological interpretation too often given of Hegelianism (of which the most astounding example is the interpretation of the *Phänomenologie des Geistes* by Alexandre Kojève)[2] we have forgotten that the history of which Hegel speaks is not so much the history made by man or suffered by him than it is the history of *spirit*, that is of an eternal entity. In the Hegelian sense, there is a history *because* there is the Absolute and not *in spite* of it or in the space created by its absence. The view supported by Hegel is not the idea of a *vérité à faire*, of a truth which has to be realized, but is on the contrary a truth which requires time to unfold and to expose itself otherwise it would remain pure virtuality. In this paradoxical sense, which is also the real Hegelian depth of thought, that which is eternal is precisely what requires in itself the historical dimension, because it is not only substance but also subject, not only in itself but also for itself, and therefore needs human incarnation and mediation. We consequently cannot speak of anthropocentrism in the case of Hegel because in this philosophy man is not the being which makes history but the being through which history is realized, infinite reason employing finite human passions for the sake of its own triumph. The growth of anthropocentrism during the 19th century has eclipsed

A-T. Tymieniecka (ed.), Analecta Husserliana, Vol. XXI, 235—246.

this conception of history. The origin of the prevalence of historicism, and of the relativism which is implied by it, is the breaking of the link between truth and history. However the philosophies of historicity which have developed in the relativistic climate which marked the collapse of Hegelianism announce a new confrontation with Hegel and present a new perspective on the relation of truth and history which is not to be confused with a sheer anthropocentrism.

Let us take first Dilthey's case, probably the most difficult to defend because he is the most likely to have fallen into psychologism and historicism. Nevertheless he was the first to give notability to the concept of "historicity." His friend and correspondent, Yorck von Wartenburg, "invented" the term of "Geschichtlichkeit" to qualify the genetic difference of man in comparison to the rest of nature.[3] It is remarkable that the concept of historicity originally takes its sense out of a Christian and Lutheran perspective which is not without analogy with Hegel's philosophical Christianism: For Yorck, Christ is the principle of history because He incarnates a supranatural life which is shared by all mankind, but which is not yet to be found in Greece (which ignores history, is "*ungeschichtlich*"), nor even in Jerusalem where this supranatural life is still considered to be concentrated in God alone.[4] This philosophical Christianism proceeds from Yorck's opposition to positivism and mechanism which constitute the then dominating *Weltanschauung*. Like Dilthey, he aspires to grasp reality in its liveliness (*Lebendigkeit*) and to combat all forms of metaphysics as attempts to eternize and stabilize what each representation contains of relatively intemporal material. Being as such is nothing more than a special manifestation of life,[5] and the real question is to think life itself. Dilthey, who is nearly a pantheist, does not evidently share this philosophical Christianism, but, like Yorck, he identifies the two concepts of "Geschichtlichkeit" and "Lebendigkeit" — of history and life.[6] For him too, it is not possible to go behind life in order to find there a stable entity which could support it: all reference to an intemporal transcendence comes from a transposition of the concept of substance to the world of living experience. Yorck's philosophical "monotheism" and Dilthey's immanentist pantheism are two different versions of a same philosophy of life which refuses to proceed with intemporal entities that could be separated from the process of events. But what is the origin of the prestige of the concept of life? The commentator of Hegel's initial works cannot ignore the importance of this concept in the Iena period,[7] that is, during a time when Hegel was more interested in describing human condition than devoted to pure philosophical speculation. This idea of life, which is a synonym of infinity and totality, of nonseparation of unity and multiplicity, is common to the entire German romantic way of thinking which Dilthey inherited. The explanation of the importance of this concept in this period consists in the fact that it refers to a being which is able to produce itself and its own mobility, to an autonomous being — and not to a universal and abstract one.[8] Dilthey himself speaks of "*Leben überhaupt*," of the unity of life in general that is dispersed in an infinity of living beings. It is finally the concept of life which throws light on the concept of historicity, and it becomes clear that what life and spirit have in common is precisely "infinity." Dilthey indeed has asserted the rights of finite consciousness against Hegel,[9] but without denying the *potential* infinity of spirit that is implied in the idea of the universality of comprehension (*Verstehen*). The philosophies of historicity, in the case of Hegel's, Dilthey's, and Yorck's philosophies, certainly think there is a relation between life and death, but without seeing in death and finiteness the very fundament of historicity. For them, what makes historicity possible is still an Absolute capable of overcoming its own limitations.

We find the same pattern in Husserl's philosophy. Fink was the first to show the similarity of the Husserlian and Hegelian way of thinking, as far as the subject of history is concerned.[10] We can in fact find a philosophy of history in the *Krisis*, and its most simplified version is to be found in the Beilage XXVI[11] where three different levels of historicity are differentiated, in an apparent similarity with the Hegelian ternary scheme.[12] Of course, there are some Husserlian commentators who think that in reality Husserl did not succeed at all to think history in itself.[13] Undoubtedly the article of 1911, *Die Philosophie als strenge Wissenschaft*, was written to contradict the Diltheyan historicism in order to affirm the a-historicity of logical laws and logical truth. Nevertheless, transcendental phenomenology must be completed by a genetic phenomenology[14] which reveals that all reality is constituted by the absolute subjectivity. Thus the genetic relativity of scientific truth is demonstrated, in spite of the fact that the *sense* of this truth remains the absolute irrelativity itself, sense and genesis being of course not at all identical. What Husserl tried to formulate in his last works was the "absolute historicity" of the life of consciousness — it was at least the program he announced as early as 1930 in a letter to Georg Misch, the author of "*Lebensphilosophie und Phänomenologie*" and the publisher of Dilthey.[15] Husserl finds again the theme of "life" in the *Lebenswelt* perspective which dominates his last philosophy, but it may be asked if he was really able, in a better manner than Dilthey, to think of life in its concretion and facticity. Because he too thinks of life as infinity.[16] The subject considers itself as finite as long as it has not achieved transcendental reduction, but as soon as it understands itself as absolute subjectivity, as the origin of all reality, its own finiteness vanishes and makes room for the infinite life of the transcendental ego.[17] It is true that this infinity has a teleological nature and that it is the correlate of the "regulative" idea of truth, an idea in the Kantian sense. There is indeed a transformation of the Hegelian scheme of the philosophy of history: the way out of the alienation of spirit opens on the infinity of time, on the infinite traditionality of science. In *Experience and Judgement*, Husserl explains how the eternity of truth should rather be understood as "omnitemporality," that is to say is referred to an eminently historical subjectivity which constitutes it.[18] Thus the knowledge of the Eternal is interpreted as a form of human existence. But human existence as a whole, can Husserl think it up to its roots? The few passages where Husserl mentions death seem to demonstrate that this is not the case. He speaks of the immortality of men in the sense that men have inherited the entire former evolution and will continue to contribute to the future evolution. There is then a possibility for the dead to survive in the memory of the living.[19] But as far as the absolute ego is concerned and not only man, who "dies inevitably", Husserl declares that it is "necessarily immortal", because it cannot proceed, like any other thing of the world, from a temporal nothing and it cannot return to nothing in the same temporal sense.[20] For the ultimate subject, there is neither birth nor death, but an awakening to the natural and mundane life that is as unjustified as the suspension of this natural life, that is, death, will be.[21] The ego has an out-of-time being and this is the reason it can neither be born nor die, birth and death being intratemporal events, and unmotivated and unintelligible ones. Dilthey, in the same manner, saw in death the "enigma of life," which life cannot understand by definition.[22] Husserl's absolute ego "comes down" into time as the Hegelian spirit does, and that is the reason Husserl does not yet succeed in understanding the absolute historicity of consciousness. Transcendental phenomenology and genetic phenomenology are finally incompatible: A first world constitution which is not in itself historical is required if the transcendental reduction wants to maintain its claim, otherwise the distinction between subjectivity and nature on which the *époque* is based would be deprived of all validity.[23]

If there is, as in Dilthey's thought, an attempt in the Husserlian phenomenology to make room for the finitude of human existence, it is nevertheless only with Heidegger that finitude and historicity are thought of as being essentially connected to each other, to the point that he sees in mortality the hidden ground of the historicity of existence.[24] History, not of course as the account of what happened (*Historie*) but as the process of events (*Geschichte*), *is* only if its "subject" is finite. Are we again, with Heidegger, confronted with a new kind of "historicism" and thrown back into a relativistic point of view? This was probably the reaction of Husserl who, when reading *Sein und Zeit*, saw in it a return to the anthropocentrism he fought against all his lifelong. In the following years, Husserl was able to perceive the historicity of *truth*, to show how history depends on the possibility of the infinite repetition of the idealities, and to form a new idea of what constitutes *traditionality*, in the sense that the traditionality of truth involves not only a limited group of people but all mankind.[25] He saw then in history not the mere unrolling of events in time, but the process of retention and transmission of *sense* through a mankind which is only the factual and anthropological index of subjectivity and intersubjectivity in general.[26] History can no longer be considered as being made by man as a factual being, but by the transcendental intersubjectivity itself which establishes its realm in the world through the "revolution of historicity" that was brought about by the birth of philosophy in ancient Greece.[27] Nothing could be more foreign to Husserl than the idea of historicity as being rooted in facticity.

In this respect all depends on what is exactly meant by "facticity." For Heidegger, what he calls "facticity of the existence" does not refer to the mere fact of being among other beings of the same or of a different species, in the sense of the being-at-hand (*Vorhandenheit*) of mere things. It means the "fact" that man has to become responsible for his own being[28] which is not a "substance" but a being in the modus of possibility (*Seinkönnen*): Man is a being who has to become what he is, who has to deliver to himself the possibilities he inherits. This is the reason the structure of his being is historical. Historicity does not mean here the extraneous fact of the immersion of the ego into the "nightmare"[29] of history, as if man would remain at the mercy of circumstances and events in an odyssey-like history, but the very essence of human existence that temporalizes itself as a "thrown project" (*ein geworfener Entwurf*) and which consequently has to transfer to itself the possibilities it inherits and that it chooses to "repeat."[30]

Heidegger calls "*Schicksal*" – destiny – this self-transmission of its own heritage and sees in it the kernel of history. He then concedes that historicity in his sense is nothing more than a more concrete elaboration of temporality,[31] but this concretion indeed throws a new light on the original phenomenon of "temporalization": To be originally temporal and not only intra-temporal means for the human being that its existence has a *destinal* structure, that is to say that it is characterized by an internal *self-traditionality.* Are we once again brought back to a solipsistic point of view, in the sense that the acknowledgment of the destinal structure of the self could not account for history, which in its full meaning refers to an "intersubjective" history? This could be the case if the self of which Heidegger speaks in the §64[32] could be simply identified with the traditional ego, that is with an "isolated" subject. But to be a self does not mean to be worldless, nor to be separated from others, because selfness is not the substantial fundament of personality, but a conquest over the impersonality of the "one" which is the primary level of existence. Existence is, in the first instance and for the most part (*zunächst und zumeist*), impersonal existence in the daily world of preoccupation, and this implies that being with the others (*Mitsein*) is a primary determination of being-there (*Da-sein*). Therefore the real "subject" of history is not the single individual, but a collective entity, the collection of the heirs of a determined period, that is what Dilthey described under the name of "generation", which shares a common destiny, and in which Heidegger sees the achieved historicality.[33] It is true that Heidegger does not dwell very long on this subject of the collective dimension of history, but his whole conception of what is historicity would be unintelligible if his point of view remains a solipsistic one. There is a relation of belonging between the dead and the living which is the ground of true tradition. And true tradition becomes explicit only with "repetition."[34]

What does the word of repetition (*Wiederholung*) mean here? Not the mere reproduction of a pattern of life which has been realized in the past, but the seeking (*holen*) of a possibility of existence in the past and the *reply* to it.[35] History is then not the realm of reality, but of possibility, because it is the dimension in which a dialogue between men is made possible. That is why true history has to be understood as the "return" of the possible,[36] which does not mean the return of identical events but the essential *continuity* of history. This is, may be, the postulate of the Heideggerian analysis. But it must be clearly expressed. First of all, continuity is not *given* but is only *possible* on the level of authentic existence. The question then is not to state the reality of continuity, but rather to "create" it. And secondly, this continuity presupposes — and does not conversely neutralize — death. This is the most important point: history cannot in this sense become the refuge of the deceived, of those who try to find in it a new kind of transcendence, a new chance of survival proposed to the individual. It is on the contrary the modus of existence of a being which is destined to die. The dialogue that takes place between the generations does not abolish the finitude of those participating in it; on the contrary, finitude is what gives them the possibility of questioning and replying to each other. Mortality is not for Heidegger a limitation in a negative sense, but rather a "capacity," which opens — and does not put an end — to a being in the world, because the possibility of the impossibility of existence, that it to say death[37] is the foundation of all presence in the world. The finitude of the human being is then no longer opposed to the infinity of being and thought as being finally surmountable, as was the case with Hegel and even Dilthey; to the finitude of man corresponds the finiteness of being itself, which discloses itself only through the transcendence of human existence as a capacity of being related to nothingness.[38] Being is not an Absolute that requires man to unfold and become "effective." Being in itself is neither God nor a product of man but the finite partner of man, and it requires man not only to become "self-conscious" in a Hegelian way, but to "be," because there is being only as long as man exists,[39] that is, in an eminently *temporal* manner.

It is not easy to understand how a finite and temporal being can differ from a being produced by man, himself a finite and temporal entity, but this is the culmination of the Heideggerian thought during the period following the publication of *Sein and Zeit* in 1927. What makes the understanding of this close relationship between being and time even more difficult is the fact that the third section of the book, under the title "*Zeit und Sein*," was never published, because, as Heidegger explained later on, it was not possible to express the conversion of the exposition of the temporality of existence (sections I and II) into the explicitation of the temporal "nature" of being itself with the assistance of the metaphysical language in which, however, this conversion has to be accomplished.[40] The reversal in question (*Die Kehre*) should not be understood as the dialectical inversion of the negativity of finitude into the positivity of a being which wants unconditionally its own effectivity,[41] but rather as the transmutation of the opposed values of eternity and temporality by the light of a completely "assumed" finitude. Yet man has not only to assume his own finitude but above all the finiteness of being itself. Here the real "turning point" takes place: in the capacity of thinking the finitude and historicity of man as the *counterpart* of the finiteness and historicity of being itself, that is the thinking of the advent of being and existence on the background of a more essential and more fundamental imprisonment of being into the night of nothingness. On this basis, Heidegger will speak later on of a history of being (*Seinsgeschichte*) and not only of man, of a dispensation of being (*Seinsgeschick*) and not only of the destinal structure of human existence, in the sense that being itself destines man to become the locus of being (*Da-sein*) and not man who through his own transcendence creates being.[42] History becomes then a dimension of being and not only and not primarily of man; it is in a very different way from the Hegelian history of the Absolute that the anthropological authority upon history is denied. It is not in order to dispossess man of his power to "make history" that the authority is given to being but to insist on the belonging together (*Zusammengehören*) of man and being in a dimension that the late Heidegger will call: "*Ereignis*" — appropriating event.[43]

What does the relation between truth and history become then? Hegel showed the temporal dimension of truth which, because it is essentially a process, is only to be found at the end, as result.[44] Husserl developed the idea of a truth which, like the logical idealities, can generate history, in the sense that once emitted it can infinitely be repeated and he saw in the omnitemporality of truth a modus of temporality.[45] Heidegger went further in the same direction: Truth is relative to man. He says about truth what he already said about being: that there is truth so far as and so long as man exists.[46] Does this mean the "subjectivity" of truth, that is to say relativism again? Not at all if we are able to think of man on another basis than the solipsistic subject. Relativity does not mean here in a Protagorean manner that the individual is the measure of what is true and false, but that man can have a relationship with being itself and it is this relation that Heidegger calls "truth" in the sense of the disclosure (*Entdecken*) of what *is*.[47] Man has the capacity to set free what is *as* it is and not to get lost in the private world of its own representations. Man is initially as well in truth as in fallacy.[48] That is the reason, in opposition to Husserl, Heidegger does not need to presuppose an "ideal subject," different from the empirical ego, as the only possible basis of truth and science. Such an ideal subject is a mere construction, a phantasmagoria in regard to the only existing "subject": man in his facticity. In the same manner, it is necessary to revoke the idea of an eternal truth which is part of the legacy of Christian theology, not of philosophy.[49]

Facticity of human existence and not the infinity of a transcendental and ideal subject is the actual *a priori* of philosophy and science, because for Heidegger existence means being originally in truth as well as in fallacy and not, as for the entire metaphysical tradition, *a being originally in delusion only* — according to the archetypal image of the human condition that is given in the Platonic allegory of the Cave. Therefore it is not necessary to escape from our mortal condition in order to attain to the eternal truth, it is not required to neutralize history in order to lay the foundation of science, nor to find in a transcendental history the basis of the empirical one — in an Husserlian way. Facticity of the human existence, that is to say the fact that we do not choose to enter existence, historicity of the human existence, that is to say assumption — repetition in the Heideggerian way of speaking- of facticity, are the sources of truth, and they can also be, in their unauthentic modality, the sources of fallacy. Mortality, which is the original principle of individuation — because death only, as unconditional possibility, isolates the individual in itself[50] constitutes the basis of truth as well as of history. The acknowledgment of the human finitude does not mean the impossibility of a history of truth, but on the contrary the possibility of a new alliance of truth and history. For history is neither the actual nightmare of the people of the Cave, nor the irresistible process of rationalization of the world; it is the *free* perpetuation of human existence through a *free* process of traditionalization.

University of Paris I

NOTES

[1] For example, in "Die Deutsche Ideologie."

[2] Alexandre Kojève, *Introduction à la lecture de Hegel* (Paris: Gallimard, 1947).

[3] Cf. Leonhard von Renthe-Fink, *Geschichtlichkeit*, Zur terminologischer und begrifflicher Ursprung bei Hegel, Haym, Dilthey und Yorck (Göttingen: 1964).

[4] P. Yorck von Wartenburg, *Bewusstseinsstellungen und Geschichte* (Tübingen: Niemayer, 1956), pp. 45ff.

[5] "Sein ist ein Derivat des Lebens, eine partikulare Lebensmanifestation," letter 127 from Yorck to Dilthey in *Briefwechsel zwischen W. Dilthey und dem Grafen Paul Yorck von Wartenburg 1877—1897* (Halle: Niemayer, 1923), p. 203.

[6] "Leben ist seinem Stoff nach eins mit der Geschichte." Dilthey, *Gesammelte Schriften* (Stuttgart: Teubner, 1927), Band VII, p. 256.

[7] Cf. J. Hyppolite, "Vie et conscience de la vie dans la philosophie hégélienne de Iéna," in *Revue de Métaphysique et de Morale*, 1938.

[8] "Leben — das meint das *sich aus sich* erzeugende und in seiner Bewegung *sich in sich* haltende Sein," M. Heidegger, *Gesamtausgabe* (Frankfurt am Main: Klostermann, 1980, Band 32), p. 207.

[9] Cf. H. G. Gadamer, *Wahrheit und Methode*, Mohr (P. Siebeck) (Tübingen: 1975), pp. 218ff.

[10] Speaking of the Husserlian teleology, E. Fink writes: "Dieses geschichts philosophische Schema ist nicht neu. Es ist eine 'Phänomenologie des Geistes' mit den analytischen Mitteln einer intentionalen Bewusstseinsexegese . . ." in *Nähe und Distanz* (Freiburg: Alber, 1976), p. 170.

[11] *Die Krisis* (Den Haag: Nijhoff, 1962), pp. 502—503. This appendix is not reproduced in the English edition.

[12] Cf. A. Pazanin, *Wissenschaft und Geschichte in der Phänomenologie E. Husserls* (The Hague: Nijhoff, 1972), pp. 163ff.

[13] For example, Walter Biemel in "Les phases décisives dans le développement de la philosophie de Husserl" in *Husserl*, Cahiers de Royaumont (Paris: 1959), p. 58: "Les essais de Husserl pour saisir thématiquement l'histoire peuvent être considérés comme des échecs."

[14] Cf. principally *Cartesian Meditations*, §§37 and 38.

[15] In this letter to Misch dated 16.11.1930, Husserl gives a retrospective interpretation of his philosophical development from logical and real ontological problems to the foundation of the theory of transcendental subjectivity and intersubjectivity. He writes: "Denn mit der 'tr. Reduction' war meiner Ueberzeugung nach die letztlich wirkliche und concrete Subjektivität in der ganzen Fülle ihres Seins u. Lebens gewonnen, in ihr das universale leistende u. nicht bloss theoretisch leistende Leben: die abs. Subj. in ihrer Historizität — Subjektivität, Wissenschaft, Welt, Kultur, ethisch-relig. Streben etc. . . . — alles — in einem neuen Noem. u. Sinn. Das seit 10 Jahren vorbereite, nun wirklich werdende Buch, als ganz system. Aufbau wird hoff. vollste Klarheit schaffen . . ." This letter is cited by A. Diemer in *Husserl* (Meisenheim an Glan: Anton Hain, 1956), pp. 393—394.

[16] Cf. for example, *Formale und transzendentale Logik* (Halle: Niemayer, 1929), where Husserl speaks of "die Unendichkeiten des Lebens."

[17] Cf. once more E. Fink in "Welt und Geschichte", *Nähe und Distanz* (Freiburg: 1976), Alber, p. 171: "... das absolute Subjekt, solange es seiner selbst noch nicht gewiss geworden ist, hält sich für 'endlich', hält sich für ein begrenztes Geschöpf unter anderen Naturgeschöpfen. Wenn es aber seiner schöpferischen Leistung innegeworden ist in der Reduktion, streift es die Endlichkeit ab wie ein schäbiges Kleid."
[18] *Experience and Judgement* (Evanston: Northwestern University Press, 1973), §64c: "The timelessness of objectivities of the understanding ... proves to be a privileged form of temporality."
[19] Cf. the manuscripts cited by R. Toulemont, in *L'essence de la société selon Husserl* (Paris: P.U.F., 1962), p. 270.
[20] Ibid.
[21] Cf. A. Aguirre, *Genetische Phänomenologie und Reduktion* (The Hague: Nijhoff, 1970), p. 164.
[22] Dilthey, *Gesammelte Schriften* (Stuttgart: Teubner, 1927), Band III p. 241 and Band VIII, p. 80.
[23] Cf. E. Tugendhat, *Der Wahrheitsbegriff bei Husserl und Heidegger* (Berlin: W. de Gruyter, 1967), p. 254.
[24] Cf. *Sein und Zeit* (Tübingen: Niemeyer, 1963), p. 386: "Das eigentliche Sein zum Tode, das heisst die Endlichkeit der Zeitlichkeit, ist der verborgene Grund der Geschichtlichkeit des Daseins."
[25] Cf. *Die Krisis der europäischen Wissenschaften und die transzendentale Phänomenologie*, Husserliana Band VI, pp. 368ff.
[26] Cf. J. Derrida, *L'origine de la géométrie* (Paris: P.U.F., 1962), p. 80.
[27] *Krisis*, o.c. p. 325: "Wissenschaftliche Kultur unter Ideen der Unendlichkeit bedeutet also eine Revolutionierung in der ganzen Weise des Menschentums als kulturschaffenden. Sie bedeutet auch eine Revolutionierung der Geschichtlichkeit, die nun Geschichte des Entwerdens des endlichen Menschentums im Werden zum Menschentum unendlicher Aufgaben ist."
[28] Cf. *Sein und Zeit*, op. cit., p. 135: "Der Ausdruck Geworfenheit soll die *Faktizität* der Ueberantwortung andeuten."
[29] According to Joyce. Cf. Derrida, *L'origine de la géométrie*, op. cit., pp. 104—105.
[30] *Sein und Zeit*, op. cit., pp. 383—384.
[31] Ibid., p. 382: "So erweist sich im Grunde die Interpretation der Geschichtlichkeit des Daseins nur als eine konkretere Ausarbeitung der Zeitlichkeit."
[32] The title of the §64 is: "Sorge und Selbstheit" (Care and selfness).
[33] *Sein und Zeit*, op. cit., p. 384: "Das schicksalhafte Geschick des Daseins in und mit seiner 'Generation' macht das volle, eigentliche Geschehen des Daseins aus."
[34] Ibid., p. 385: "Die wiederholung ist die ausdrückliche Ueberlieferung."
[35] Ibid., p. 386: "Die Wiederholung *erwidert* vielmehr die Möglichkeit der dagewesenen Existenz."
[36] Ibid., p. 391: "Die eigentliche Geschichtlichkeit versteht die Geschichte als die 'Wiederkehr' des Möglichen."
[37] Ibid., p. 262: "Er (der Tod) ist die Möglichkeit der Unmöglichkeit jeglichen Verhaltens zu .. , jedes Existierens."
[38] "Sein und Nichts gehören zusammen (...) weil das Sein selbst im Wesen endlich ist und sich nur in der Transzendenz des in das Nichts hinausgehaltenen Daseins

offenbart." "Was ist Metaphysik" (1929) in *Wegmarken*, Klostermann, Frankfurt am Main, 1967, p. 17.

[39] *Sein und Zeit*, op. cit., p. 212: "Allerdings nur solange Dasein *ist*, das heisst die ontische Möglichkeit von Seinsverständnis, 'gibt es' Sein."

[40] Cf. "Brief über den Humanismus" in *Wegmarken*, op. cit., pp. 159 and 187—188.

[41] Ibid., p. 190.

[42] Ibid., p. 168: "Das Werfende im Entwerfen ist nicht der Mensch, sondern das Sein selbst, das dem Menschen in die Ek-sistenz des Da-seins als sein Wesen schickt."

[43] Cf. *Identität und Differenz* (Pfullingen: Neske, 1957), p. 32: ". . . Wenn anders das Zusammen*gehören* von Mensch und Sein in das Wesenslicht des Ereignisses gelangen soll." *Ereignis* is then understood as a capacity of appropriation (*Eignung*), according to the etymology of the word.

[44] "Es ist von dem Absoluten zu sagen, dass es wesentlich *Resultat*, dass es erst am Ende ist, was es in Wahrheit ist . . ." Hegel, *Phänomenologie des Geistes* (Hamburg: Vorrede, Meiner), p. 21.

[45] Cf. *Experience and Judgement*, op. cit., §64c.

[46] *Sein und Zeit*, op. cit., p. 226: "Wahrheit 'gibt es' nur sofern und solange Dasein ist."

[47] Ibid., p. 227.

[48] Ibid., p. 222: "Der volle existential-ontologische Sinn des Satzes: 'Dasein ist in der Wahrheit' sagt gleichurprünglich mit: 'Dasein ist in der Unwahrheit.'"

[49] Ibid., p. 229: "Die Behauptung 'ewiger Wahrheiten', ebenso wie die Vermengung der phänomenal gegründeten 'Idealität' des Daseins mit einem idealisierten absoluten Subjekt gehören zu den längst noch nicht radikal ausgetriebenen Resten von christlicher Theologie innerhalb der philosophischen Problematik."

[50] Ibid.; p. 263: "Die im Vorlaufen verstandene Unbezüglichkeit des Todes vereinzelt das Dasein auf es selbst."

PART IV

SOME FURTHER PERSPECTIVES UPON THE HUMAN CONDITION

LUDWIG GRÜNBERG

FROM PHENOMENOLOGY TO AN AXIO-CENTRIC ONTOLOGY OF THE HUMAN CONDITION

Philosophy was born from the need to provide answers to several theoretically relevant questions, namely, (1) concerning the search for truth and the value grading of the forms of existence, (2) concerning the relationship between man and the universe as a whole and between man and the values he has created, and (3) concerning man's specific way of being, which Montaigne named "the Human Condition." Values allow man to transcend his natural condition by his human condition: philosophical thinking applied to the fascinating world of values works as "one of the threads of *man's existential self-interpretation in existence*" and, at the same time, as "a thread of the Human Condition itself, without which the human being would not be *human*."[1]

Starting from these considerations, I shall try to resuscitate the attempt made by Lucian Blaga and D. D. Roşca in Romanian philosophy between the two world wars[2] to give *value* the prerogative of a central concept in philosophy, to reinterpret the ontological problems in an axiological perspective, and, implicitly, to trace the coordinates of an axio-centric ontology of the human condition. Such a project is in close agreement with the program of "philosophical reconstruction" initiated by the World Phenomenology Institute — which, by its theme "Phenomenology of Man and the Human Condition," proves to be a common ground for the "cross-cultural" philosophical dialogue.

I. THE "OPENINGS" OF PHENOMENOLOGY TOWARD AN AXIOLOGICAL OUTLOOK ON THE HUMAN CONDITION

As early as 1938, Camil Petrescu — one of the most competent Romanian exegetes of Husserl's work — reached the conclusion that phenomenology means a divorce from traditional metaphysics, but offers "openings" toward the foundation of a new ontology of the human condition, by a shift of emphasis from the transcendent to the transcendental, from the "given fact" to "meaning," from contemplation to the practical and valuation responses of man to the world.[3]

Such a consideration was consonant with Husserl's programmatic

A-T. Tymieniecka (ed.), Analecta Husserliana, Vol. XXI, 249–273.

credo: before deciding to direct my sight toward the world, "I am within the world"; this original state supposes "a practical-axiological relationship," so that "covering the world with one's sight (*überschauen*) does not mean contemplating it," but keeping it as such, "with the practical and valuation reactions it incessantly imposes upon me" ("in meinen Wertsetzungen und Handlungen gestaltete und immer neu gestaltete Welt").[4]

It is now, at last, after the publication of Husserl's ethical researches by Alois Roth and after some relevant studies and monographs on the philosopher's later manuscripts have been issued, that the practical reason and axiological prospects of Husserl's phenomenology have been properly defined. Today we are aware of the fact that Husserl's ethical preoccupations materialized in his manuscript entitled "Kritisches aus Ethik-Vorlesungen" (Sommersemester 1902), had started as early as 1891 — therefore one decade earlier than the year when *Logische Untersuchungen* was published; moreover, we now know that Husserl from the very beginning approached ethical problems by investigating their roots originating in emotional life and in value. Even if at first "analogy-drawing considerations" appeared, referring to the "parallel between theoretical and practical reasons," Husserl would not replace the outdated psychologically based reductionism by a logically based reductionism; he emphasized the fact that "in spite of the essential analogy between the logical and the (practical) axiology, the setting up of a pure form of ethics does not mean at all a duplicate of logic; rather, some fundamental dissimilarities between the two spheres come up."[5] He tried to reveal the synthetic nature of reason — both theoretical and practical — and to find an adequate context for the irreducible, specific features of value; along with it, "he also developed the idea of interdependence of ethics and value theory on social and historical factors," because "for Husserl, the context of narrowly ethical must be seen as part of the *broader axiological concerns* of a developing society."[6]

The axiological significance of Husserl's outlook can be properly understood through establishing the preeminent position held in his writings by the problems of practical reason (the original background which can account for his evolution toward understanding man as "the substratum of a teleological reason," i.e., as a valuing being) as well as through considering the multiple implications of the concept of "intentionality of consciousness" within the "ego-cogito-cogitatum" structure. A promising prospect to reinterpret Husserl's conception of the human condition thus opens, from the viewpoint of the way that brought him, by reductions (also applied to the "empirical" ego itself), to the "transcendental ego" and at last to the capital idea of *Lebenswelt*, which associates knowledge with valuation and action.

The remark was made as early as 1929, while explaining the shift from passive syntheses to active syntheses (which is also to be found in his manuscripts written at the time of *Krisis*, with direct references to the problems of axiology).[7] In fact, Husserl stated that, by means of phenomenology, "each of us and all together should search for the latest possibilities and necessities on the basis of which we respond to realities: judging, valuing, acting."[8]

The triptych of the latter notions (knowledge, valuation, action) is relevant for the manner in which Husserl implants all "responsibilities of the human life" in the "life world," conferring a surprising axiological dimension to the synthesis between *Weltverständniss* and *Selbstverständniss* (understanding of the world and understanding of the self).

Was it not Husserl himself who associated with the "life world" an intentionality of consciousness which he called "affective" or "practical" and which — as E. Levinas remarked — implies an "axio-logos"? I suppose that the return to the "life world" in Husserl's opinion does not mean a vague longing after primary, antepredicative experiences, but a practical-axiological foundation of any theoretical attempt in the *Lebenswelt*, which is — in and by itself — an "originator of values" (*Werte-setzend*). This "founding of values" does not represent a valuing directed from outside, but it is rather a condition and an ontic compound of the "life world"; it lies at the basis of the irreducible nature of the human, detectable by a methodological procedure thus exceeding the conceptual framework of a phenomenology of perception: it opens the prospect of a phenomenology of value (and action), implicitly of a new type of ontology of the human condition.

The coordinates of such an ontology of the Human Condition were outlined by A-T. Tymieniecka in her paper delivered at the 17th World Congress of Philosophy (Montreal, August 1983),[9] starting from her own conception of "man's self-individualization of life," which is at the same time "man's self-interpretation-in-existence." This new type of ontology of the human condition continues Husserl's project, but it exceeds the limits of the eidetic and transcendental approach; it brings about a shift of direction as well as a change of context and especially a change of vision.

First of all, we note a fertile shift of direction from the phenomenological "descent" to the origins to the phenomenological "ascent" along the evolution phases through which Tymieniecka elucidates the germinal "virtualities" of the specifically human "creative orchestration" of faculties promoting manifestations of the irreducible human condition.

Second, we find equally fertile change of context, by placing the ontological approach to the human condition in a "creative context," so that the cognitive, meaning-bestowing function of man's self-interpretation (which Husserl emphasized) is doubled by an operational-actional side. Supposing a dialectical understanding of the relationship man-nature and individual-society, "man's creative activity . . . , without which the human being would not be able to endow his existence with specifically human meaningfulness" becomes the privileged context that "not only renders accessible the whole expanse of the Human Condition, but also expresses man's specific means for extracting his unique being from the other types of being, things, and from an anonymous Nature."[10]

Last but not least, the new stage of phenomenology brought a promising shift of vision, appealing to another initiating intuition to open the universe of man, and all that is of concern to him. More precisely, the question is that of passing on from "the transcendental/structural vision, which assumed human consciousness as its root-source, and placed the human being at the beginning, the initiator," to "the metaphysical vision of the *human condition within the game of life*," for which man is "the vortex of the universal sense," whose beingness culminates in the creation of values.[11] Such a metaphysical (ontological) vision is at once cosmological and axiological. It leads to the idea that the axiological is the proper way to exist as a human way for others, and so it opens new prospects for the liberation of the philosophical discourse from gnoseological contemplativism, conveying a genuinely ontological ring and centering it round the problem of man as a "valuing being," who perpetually creates himself by an active response to his needs and aspirations in the world (i.e., by aiming at and achieving values!); his "conditions of possibility" are implicit in his very creative activity.

In this way, the phenomenological problems of intersubjectivity, of meaning, of history, of the "life-world," and of creativity will meet the concerns of Marxist philosophy as regards the foundation of an axiocentric ontology of the human condition starting from the dialectical relation individual-society, from praxis. The dialogue is really possible, because "Marxist philosophy and phenomenology are approaching the same kind of man, man subject to determination."[12]

The project of philosophical reconstruction initiated by the World Institute of Phenomenology is the frame of reference and a stimulus in my attempt at basing the ontological approach on the axiological dimension of the Human Condition. As a matter of fact, it is a sequel to the "openings" offered by phenomenology – starting from Husserl's first studies where the idea of practical reason was prevalent, to the "phenomenology of creative context." And, if there will be some special accents, I should like to appeal to Angela Ales Bello's words: "in this vast project of the philosophical communication there is room for immeasurable variety in original thought."[13]

II. HOMO AESTIMANS AND AXIO-CENTRIC ONTOLOGY

In my view, the creation of an ontology represents the crowning of philosophical thinking. I will spare no effort in pleading for this thesis, which was so unjustly discredited by logical positivism. A philosopher who does not want to become the author of a world, of an ontological construct, renounces his vocation; he may be anything, a subtle thinker, even one of genius, but a partisan of unfulfillment. True, an ontological outlook which proposes both an image of the world and an attitude to the world can never be a permanent thing, or a victory ruling out another attempt. Yet, it is worth trying all the more so as when it sets not to follow the pattern of reductionistic or anthropomorphic-speculative metaphysical constructs and accepts the idea shared so far by Husserl (in his *Ideen I*) of regionalizing the ontological problem range. Following Husserl's suggestion, it is possible to demonstrate the irreducible specificity of each axiological region without neglecting that all values were participating in one and the same essentiality (*Wesenheit*), where the perennial metaphysical issues ("individualization," "the one and the many," "being and becoming," "the possible and the actual," "life and modes of beingness," "differentiation through unity," etc.) have their vortex in the issues concerning the Human Condition. It is not hard to understand that the Human Condition differentiates itself within the infinity of the material world and within the unity of everything there is alive as *a-determined-way-of-being-and-becoming*. Creative action, conscious teleological institution, the search for truth and beauty, freedom and moral responsibility, the triptych "knowledge-valuation-action" constitutes determinations of "regional ontology," inherent to a manner of being, having sociohistorical source-conditions and axiological dimensions. Such an approach is convergent with "the metaphysical vision of the Human Condition within the game of life" elaborated by A-T. Tymieniecka in "the creative context." The originality of the viewpoint I am standing for[14] lies in acceptance of value as a key concept in the project for the building of a regional ontology of the Human Condition.

From the very beginning, the project mentioned gives rise to a cardinal question: why does value (and therefore the axiological approach) play a privileged part in the project of setting up this ontology? The question is so much the more justified as traditionally the idea has taken root that philosophy could be ontology only through its cognitive dimension and not through its axiological dimension as well. But this means to create an artificial hiatus between scientist ontology (reduced to the synthetic-totalizing cognition) and an anthropocentrist ethics (reduced to the coordination of projects, purposes, hierarchies of values and ideals). We invalidate such a point of view for the simple reason that philosophy itself appeared out of the need to meditate on the human condition, to understand man — as a whole — in his interaction with the environment, that is, in the dialectics of the relations (the individual in society) in interaction with the world. Consequently, it is not man that should be conceived according to the way in which we conceive philosophy, but vice-versa: it is necessary that philosophy should be conceived according to the way in which we conceive man. Man, conceived from the point of view of the praxis, is both an object and a subject at the same time, he is an expression and a creator of nature, of a human nature. He is always an open being, never completely created, but always creating himself in the process of human work, in history and through action, implicitly through the elaboration and achievement of projects which give new meanings to human existence. He is always in tension toward something, and this tension toward the outside — springing from the need to satisfy his practical necessities — is in fact the tension toward the infinite inclusion in the limited and gives rise to that "essential force" of man who directs all his energy toward his "object." This essential force is value. Value is the specifically human reaction — projective, attitudinal, preferential, ideatic — to the world. It is present in the passion for the new, for the unprecedented, in discontent with what has been achieved and in the wish to achieve more, in the dialectics necessity-tension-pain-passion-project-fulfillment-joy-new necessities-new existential problems of tension and adaptation. If man were not a being who valuates then he would always be satisfied with what he is, he would no longer be active, creative, passionate, an author of projects and ideals, he would lose, together with his will, his capacity for action, the defining attributes of his way of existing. "The world fails to give man satisfaction, hence the latter needs to change it through his action" — this is the concentrated formula conveying the idea that man is, through his very way of being (and of acting), *a being who valuates.* That is why a regional human ontology cannot be built without the combined participation of the cognitive dimension, but neither without the axiological dimension of philosophy. The generalizations about the human ontos will not rely only on the data supplied by the sciences about man (although it will rely primarily on them) but also on the data of human action and of lived experience (for man lives and creates values through the actual interaction between the concrete situation and the ethos).

Value directly or symbolically expresses the synthetic project of one man or of a collectivity, his way of choosing the shape of his polymorphous relations with others. The entire human action (therefore the entire human existence) is pervaded by an axiological dominant. Value is the axis and the "nerve" of action, it is the possible expression of a satisfactory adaptation between the ethos and the situation, between the concrete man and the concrete world in which he acts (and thus exists). Conditioned by a biochemical preparation, by combinations of chromosomes and hereditary habits, by an individual psychism and a collective psychology, by mental structures and languages, man was born out of the animal chaos thanks to work and he asserted himself through work, through practice, as the only being able not only to adapt himself to reality, but aso to adjust reality to his needs. The attributes of existence come from the antecedents, the social and natural environment, but go *toward* reality, too; that is, they objectify values through action. But this action, which is creation and self-creation, takes place in an axiological climate by means of a total involvement of all manifestations of human life in the pursuit and creation of values. Value has an ontological precedence over cultural assets, over norms, and even over cognition. Man cultivates science because he attaches to it the value of an instrument of efficiency. Even human action itself is truly human thanks to its axiological load. The couple cognition-action is necessary, but not sufficient, for grasping the ontological status of the Human Condition.

Let me try to be more precise. Man cannot be treated exclusively as *homo faber* or as *homo cogitans.* He knows and acts and thereby bestows new significations upon the objects of knowledge and upon the objects of action in his capacity as a being who valuates, *homo aestimans.* In the human realm values belong to the order of existence itself. They are inherent to human action (which they substantiate and orient). They are principles modeling the human being, who individualizes himself from life — from nature and social-life systems — by his own self-interpretation in existence. And we ask ourselves, how could it discharge this admirable function if it were not rooted in the very core of existence? With chimeras and spiritual phantoms man's self-interpretation in existence is impossible. Producing human action and being produced by it, structuring practice and being structured by it, values represent a unity of the possible through which our freedom is expressed, of the project in which cognition is carried on and of man's self-interpretation in existence which is the "motive power" of human action. That is why values are determinations of the specifically human way of being. And precisely because they pertain to the very ontological status of man, values cannot be conceived without being loved, searched, or felt. Without antithetically opposing the Human Condition to the natural condition, without removing the human from "the thrill of life," values facilitate the understanding of the qualitative specificity of a human way of being.

The coordinates of the human way of being are not limited to a mere *in-der-Welt-sein*, because man's individualization-in-becoming is suspended upon a set of constructive coordinates of the Human Condition: "the otherness," "the life-world," and "the other self in the social circuit of the life-world."[15] Neither are these coordinates reducible to a cognitive relation between ego and the world (which renders scientism and the imperialist pretentions of the *homo technocraticus* devoid of any meaning). The transition from the concrete-sensible horizon of nature to the horizon of values, the transition from nature to culture, implies an ontological mutation which can be accounted for by an axiocentric outlook, apt to reveal a new coordinate of the Human Conditon. Values specify man in a distinct region of the universe, so that tracing back his biological filiation remains only a prelude to the understanding of the human. As a morphological form, man belongs to the animal kingdom, representing the highest stage of its development. As an existential modality, man distinguishes himself qualitatively from any other living creature, with labor providing a key to the ontological leap from nature to culture, from a dumb and horizontal existence to a meaningful and vertically hierarchized scale of values. The analysis of the correlation between the elements inherent in the working process, such as necessity, transcendency, project, and decision, allows us to draw the conclusion that work implies a human activity patterned by values, but which at the same time also patterns new values, which is not only a specific human way of transforming the environment, but concurrently a way of an ontic forming of man. Therefore, as Marx point out in his *Economic-Philosophical Manuscripts*,[16] for a better understanding of society, one must neither mix up the human with nature, nor separate the social man from nature; finally, one must not place the individual beyond the general (the anthropological), considering society as an abstraction exterior to the individual. Man is the product of his own activity, since he integrates into his environment in the process and by the action of transforming this environment, thus adapting it to the needs of man, to a human project. Work outlines the specificity of man's existence and life as a being able to elaborate and to achieve projects through which the knowledge of the objective data of the structure are congruous with the elaboration of an ideal pattern, with a preeminently valuating optics, capable of ensuring the correlation of the results of the action with the envisaged purpose. Based on practice, placed at the confluence of nature with history and of causality with finality, the human project (imaginative or conceptual) confers on the sociocultural phenomena a character of finality that is axiological and superorganic (consequently, transcendent in relation to nature). This project always has its roots in the objective data of the structure, but presupposes, on the one hand, an interaction between men (individuals or groups) through the meditation of the "society as a whole," and on the other hand, involves an act of revaluation that throws a bridge between the sphere of possible and that of desirable, permits the human action to surpass always its immediate and direct aim. The horizontal plane of abstractions appears always framed or enveloped in the vertical plane of values, that is, in a plane in which preference, choice (consequently renunciation), even choice among preferences (therefore sacrifice) are implied. The exercise of this choice, as socially conditioned and unconditioned from an absolute point of view, is necessary and free, is essential for the ontologic status of man. This choice is permanently for the ontologic status of man. This choice is permanently possible as long as man is alive; "death alone changes life into destiny" (Malraux).

Freedom is not reducible to the cognitive moment ("the understanding of necessity"). It should be considered to be, at the same time, the specifically human capacity for values. Even if, understanding necessity — within the historical limits of the praxis — man, through the medium of knowledge, resorts to explaining and optimizing the means which he has at his disposal — with a view to increasing the efficiency of his transforming activity without promoting certain moral ideals — without an axiological intentionality man's freedom of choice would be reduced to spontaneity. Without a purpose, without criteria of value discernment, without an axiological intentionality, the choices would become equivalent and, even if they did not die away, they would cease to express human freedom. In a way, an animal also chooses among concrete realities, but it cannot be free because it chooses among given (not possible) terms, passively giving itself up to the vital impulses, while man chooses by referring his aims to the ideal values which he wants to materialize. The spontaneity/creativity groundwork for "man's self-individualization-in-existence" proposed by A-T. Tymieniecka allows the understanding of the complex process by which we pass from proto-valuation to value judgment: out of sheer, blind aspiration finality becames conscious, the strain of the effort experienced becomes will, and the intentional object of desire is granted value.

Freedom does not imply any kind of choice, but *sui-generis* choices, which are axiological options. It is precisely because it relies on the cognition of necessity and because it consciously pursues the actualization of values that free human action appears to us as an itinerary between two adhesions: a passive adhesion to the stimuli of empirical reality of the spontaneous impulses of the self which gives itself up to its conditioning, and an active adhesion to the consciously assumed ideals which have become guides for creative action based on cognition combined with the will consciously to promote, to achieve values.[17]

Having reached this point of our argumentation, we have gone past the compulsory moment of understanding necessity. This moment is correlated with that of the choice resulting from an axiological option, in order to prepare the introduction into the "equation" of freedom of a new, decisive moment: the moment of creative action. Man is a valorizing being, able to ensure and to entertain a creative activity in continual tension toward the new, thanks to the always open references of the action to the values pursued.

In the framework of an axio-centric ontology, freedom is an infinite action springing from the capacity, historically and socially acquired by man within the life-world, to project several possible acts and to engage in options among several ways of actions in relation to the values he wants to create and which give human significance to his projects. To say that a certain human action determined by a situation is tantamount to saying the situation was valued by man's self-interpretation-in-existence in such a way in order to meet the exigencies of the situation and to attain the finality. Thus, the human attitude appears to us full of the individuals' moral responsibility for the good or bad consequences of the axiological options. Man is a valorizing being, able to ensure and to entertain a creative activity in continual tension toward the new, thanks to the always open references of the action to the values pursued.

This is precisely the reason why man does not lose hope because of the reality's refusal to grant privileges to science and to its theoretical relation with man. In virtue of his ontological status, not only does man know, but also knows that he knows and that he cannot know everything. In virtue of the uniqueness and finitude of human life, man is not only a being capable of recording and knowing, but he can reply to the challenges of reality through valuation — by decision, which becomes decisive for his destiny. Man reveals himself — not in his existentialistic "atomization" but as articulated in the game of life and in an historical context of social relations — as a center of valuation and decision. The individual cannot be dissociated from the social, because the criteria of selection are not at the beck and call of subjectivist whims but are historically and socially determined.[18] He projects himself and his life-world not from his own center alone, but, on the contrary, in existential reference to the otherness of his condition. Our own approach is, again, convergent with A-T. Tymieniecka's "approach to the human being which focuses upon the vital bounds with the other rather than upon his singularized being" which "seems to show that man does not unfold his individuality from his subjective selfhood alone."[19] On the contrary, his very individualization into an autonomous subject takes place within the organic, social, and spiritual world-process. Society is the specific form (never axiologically neutral!) in the frame of which, on the ground of historical practice, individuals with their own behavior, roles, and finalist actions, preserve or change aims, norms, and values. That is why we should not overlook the fact that any kind of intercourse with our fellow men is always a human action, inevitably involving trial and error, meditated decision-making and exercise of will, good and evil, beautiful and ugly. Man is always "a valuing-being-in-a-social-context," for whom the interhuman relations represent an inalienable ontic determination, even if he is obliged to rediscover and to reinvent in some way these relations through each new action and each new deed he initiates.

We are consequently placed in the theoretical framework of an ontology which throws into relief the dialectics action-communication and brings forth the relation between individuals not only as a simple means of carrying out a pragmatic aim, but as a preeminently social function and as an aim *per se.*

Approached in such an axiological perspective, culture can rediscover a dialectical, integrative meaning, because it envisages the axiosphere, that which pertains to the value-realm. If the diversity of cultures is a source of polyphonic harmony (and not a potential source of conflicts), if the historical development of cultures engenders common cultural assets for humanity and a common responsibility for its preservation and continuation, this proves that any cultural creation contains something "irreducible": value, as a specific human way of self-interpretation, self-edification, and self-fulfillment.

III. IRREDUCIBILITY OF THE EXISTENTIAL HUMAN STATUS: CREATIVITY AND VALUE

Proceeding from the above, we consider that Marx's well-known formula "Man's roots are in himself"[20] may be interpreted also as an assertion of the irreducibility of man's ontological status. Man is not only "a thinking reed" (Pascal), but a social being who valuates, a being that creates and creates himself through culture, excelling the natural state through the Human Condition.

Man's creative activity is — in Tymieniecka's view — the vehicle without which the human being would not be able to endow his existence with a specifically human meaningfulness. The building up of man's manner of being is the outcome of an ontological mutation that characterizes the passing on from nature to culture, from iterative adaptation to value creativity, through what we may call a *coupure ontologique*, paraphrasing G. Bachelard. The essence of this ontological mutation consists in the following fact: while other beings integrate in the given circumstances of existence according to their prevailingly inborn needs, man, as a valorizing being (*homo aestimans*), establishes, through creativity, a fundamental specificity of a new modality of relating to his environment. "Thus the issue of creativity appears as the focal point of interest within contextual phenomenology. By reflecting upon the creative process, on the one hand, and the creative product, on the other, philosophy is able to grasp and articulate both the fundamental specificity of human nature and the cosmic ground from which it is derived."[21] If we are to tackle the problem axiologically, the creative process appears as the *sui-generis* work process through which we actually introduce an object into the value-realm, thereby giving a new fulcrum to one of the human needs and aspirations to value. And the creative product is always a work, i.e., a concrete and relatively durable reality endowed with a value. The core of creativity remains value in both cases.

In spite of some frequent prejudices, creativity is not peculiar to any human activity that produces something novel, but only of the activities generating a qualitative novelty. Among works there are some not only novel, but also unique. Their special quality does not derive from their position in a homogeneous series, but from the fact that they belong to no series at all, that they have not been repeated and are irrepeatable. Such are the works of science, of philosophy, of art. Their novelty is originality. Therefore, there is a quantitative novelty, a qualitative novelty and a qualitative-original one. Only the latter two types mentioned fall within the framework of creativity.

At this point of our argument, another distinction appears to be necessary. Though original, the works of science and philosophy, with their respective specific features, have in common the fact that they are not attached to the matter they are expressed through otherwise than by soluble links. The same scientific and philosophical theory, the same description of facts or the same reasoning can be expressed by different words, in another order or by other expositive means. The works of art alone are indissolubly linked to the shape of their material aspect, so that any change in their form brings about a change in their meaning and value. The originality of a work of art therefore expresses the creativity specific to a peculiar relation between the artist's conception and the material body which bears it. We define the nature of this relation by stating that the originality of the work of art is immutable. We can equally assert that the artistic originality is also unlimitedly symbolic, for, beyond their substitute of words, sounds, lines, shapes, volumes, colors, etc., works of art contain a meaning which cannot be subordinated to a concept or a concatenation of concepts having a univocal sense; it is a meaning richer than that, apt to permanently transcend the concept and call for an ever-unconfined participation in the establishment of the meaning of the work by those who contemplate it.

The previous distinction allows us to maintain that creativity is present *in actu* in any human attempt that materializes in a work. That is the reaon why "the creative action is the *prototype* of action" and the work of art is the highest expression of "the originality of the creative object" and of "the invention of the originating creative work."[22]

The definition of the work of art — as the unitary and multiple product, endowed with a value, achieved through final causality, made by a moral creator out of a material, thus constituting a new qualitative object, originally immutable and unlimitedly symbolic — implies the consideration of a number of features whose conveyance marks the evolution from nature to the value domain of technics and, finally, of art. So, when the work of art is deprived of some feature or other specific to it, then it falls back from art to technics and to nature. One could thus eventually find out the articulations of the boundaries between creative and noncreative activities, and one could also reveal a typology of creativity.

The first two characteristics listed, i.e., unitary and multiple product, do not yet touch the field of works. Nature knows unitary and multiple products of integration as well. Obviously, such products may have a value, sometimes, but that only happens through the work of man, through their being turned to account by a moral agent who confers on them the character of works, able to comply with his needs that are determined historically and socially.

The third characteristic — namely, value — brings us to the field of works proper. But, for the time being, we are in the zone of technical works. Any technical work is a material product of integration, valuable, bearing a finality, due to a creator, being a new object of reality. But does the type of creativity here described imply the seventh feature too, i.e., the constitution of a qualitatively new object? Can we claim original novelty for a technical work? The answer is not quite simple. Certainly, the first technical work in a homogeneous series, the first airplane, the first telephone receiver, the first wireless set and radio station were original works. All the other technical works of the same kind, which repeated the initial pattern — with variations and improvements — were but a quantitative novelty. That is so because the first technical works in a series, the so-called inventions, are akin to the scientific works, by the quality of their originality. The inventor is not only a technical expert, but a scientist, a scholar too, who is acquainted with principles and laws. The other makers of works, who repeat the original invention, are just technicians. By the feature of originality, we enter the domain of creativity supposed by the work of science and, *mutatis mutandis*, of philosophy. And by the immutable originality and unlimited symbolism, we enter the specific area of the creation of works of art.

In this perspective, happiness is the supreme form of aesthetic creativity, aiming at the construction of the main work of a man's life: his own life as a man. We shall thus try to give a theoretical relevance to the inspired aphorisms of Stendhal. ("Beauty is a promise of happiness") and N. I. Pirogov ("Happiness is an art").

In fact, the search for happiness through value creation and granting of meanings is happiness itself. However subtle might appear to be those alluring philosophical attempts at exclusively defining happiness as a state and at systematizing a number of its constant characteristics, they all seem to have one common shortcoming; they disregard the fact that happiness has one important feature: it is eternally pursued and never fully achieved. Aristotle, Spinoza, and Marx take the notion of happiness to point to an activity, a dynamic process of self-achievement, infinitely open at an axiological level. From such a point of view, happiness is not a kind of terminus, a comfortable state of self-contentment, with all wishes having been fulfilled (as in a hedonistic view) or totally suppressed (as in the Stoics' conception). It is active life, oriented toward values; it is creativity aimed at self-achievement and self-improvement. He who does not let himself be overwhelmed by inauspicious circumstances and uses the favorable ones as best he can, he who masters every event of life from within; he who leads an activity of self-fulfillment, keeps updating his ideal of happiness, and feels the spurring satisfaction of new lines of evolution, of new prospects.

Man's striving after happiness is implied in his peculiar manner of existence and becoming, as a *homo aestimans* (a valorizing being). If he were not a valorizing being, he would be always satisfied with whatever there is, he would no longer be active or ceative, he would lose both his desire for improvement and his capacity of giving a sense to the world.

alongside with his availabilities in struggling for the achievement of the ideal of happiness. Do we not find here durable and inciting satisfaction of the kind experienced by the creator of a work above all works of art and so simply named "a man's life"?

Happiness appears to be the intense labor of building oneself as the most sublime human work one is able to attain. It means exerting creativity through the human construction of a work in which values are structured and harmonized under the badge of beauty. The assertion should not be surprising. If art is life transfigured into a relatively autonomous world of values, why should it not be possible to mold life by means of human creation, just like art. So as a man of art is a man among men when creating a work out of life, any man who turns his life into a work is an artist among artists. And a creator among creators.

Such an axio-centric ontology of the Human Condition acknowledges servitude in the face of bodily or moral pain, of implacable phenomena such as merciless diseases or death, keeping alive the awareness of the ephemeral and relative, of the finiteness and uniqueness of life, of imminent perishing, not as a source of sterile anxiety but as an indispensable support for man in order to live and create. For it is this very awareness of finiteness, simultaneously understood in terms of destiny (essence) and in terms of history that generates tremendous human energies. Thanks to it, the immediate does not satisfy the existential valencies of *homo aestimans*, dominated by the desire to comprehend the infinite in the finite, by the thirst for the absolute, by a kind of nostalgia for paradise. By transcending the immediate, man becomes the creator of a supernature (yet, not supernatural), of a culture, as a constituting — and irreversibly constituted — dimension of his specific way of existence, and as a result of living in the axiological horizon. The very obstacles against which man comes in this process, expressing the resistance of certain values, are of particular importance in the assertion of value, as René Le Senne has pointed out. Man's aspiration to surpass himself (without which no progress is possible) is not so much determined by what he has obtained but by what he tries to obtain; not only by what he has, but also by what he lacks. In the human order any object obtained which cannot find resources for dynamization tends to lose its value, so that the feeling of incompleteness, of dissatisfaction (if it is not morbid and if it is grafted on an axiological creed) is warrant for spiritual health. For it is a symptom of man becoming aware of one the coordinates of his ontic modality, of the fact that for a being who valuates, the human order is not a datum, but a process that should be understood in its historicity. It is precisely because man is a being characterized by praxis that for him nothing has already been played. And yet, nothing is being played without him, without that plus which his creativeness introduces through the subtle dialectics of interiorization and exteriorization and in restoring through his actions more than that totality received through his bio-psychosocial conditioning. To do good does not mean to comply with certain pre-existing rules, with an existence already given and concluded, but to attach a new meaning to existence, to add something which did not exist before, to participate in the human history proper, whose script is not written in advance and which essentially consists in the creation of man by himself.

From this perspective, the meaning of life already exists, because the historical initiatives of the previous generations, crystallized in the institutions produced by moral codes which bear the mark of one social structure, leave their stamp on the further development of events; but the meaning of life always remains open to debates that leave a large enough margin for human freedom (for the future is still to be created, although the starting point is represented by the inherited conditions). The traditional problem of the ultimate basis of morals is no longer posed in abstract-metaphysical terms, but in a concrete-historical creative context, being debated time and again, always open and solvable in terms of human responsibility, through choice, hope, engagement, decision, struggle. That is why the conceptual analysis of the conditions in which historical initiative must work is always a guide for action, but it can never be exhaustive, it does not dispense with the association of new significations, with options, with the decision which assumes responsibilities. Even in the most difficult situations, for man there is a solution, a hope. Sometimes, life does not submit to certain rigid rational canons, and it may create a context of unexpected, unpredictable circumstances that may ennoble a vile man, or change a coward into a hero. The axio-centric perspective may introduce into the Human Condition not only the project, success, joy, but also disillusionment, failure, anxiety, yet never despair. For man nothing has already been decided and nothing is decided without him. As Blaga used to say with subtle irony, an animal, when facing a break between itself and its surrounding conditions has but one chance of survival: to become a man.

Ontology built from an axiological viewpoint is the opposite of a philosophy that offers a human would deprived of an orienting center, deprived of any significance or appeal able to set this world in motion. A dialogue between somebody who does and somebody who does not admit an axio-centric perspective in the understanding of the Human Condition would be something like the dialogue between Caligula and Cherea in Albert Camus's well-known play:

Cherea: I think there are some deeds which are better than the others.
Caligula: I think they are all equivalent.

To the contemplation of existence as a show that is synchronically performed, our axiologic perspective opposes the committing of oneself to existence in a vertically ordered, historically constituted hierarchy in which the horizontal level of abstractions reveals itself as framed up by the vertical plane of values.

Such on ontology differs from nonhuman ontology, to which we often resort in philosophical literature in order to explain the objective basis of the Human Condition; it cannot be built exclusively on science, even less so on the natural sciences. Of course, reflection upon man is impossible today if the circumstances man is placed in are not taken into account. By means of science, we have to make full use of all the devices at our disposal in order to increase the efficiency of man's transforming power. To assert today that one may do without science, to give up the effort for scientific training, or to renounce clear-sightedness and rigor is absolutely inconceivable. But the quest for truth by science does not nearly exhaust human resources and potentialities, because meaning exists only for man and is solely constituted during and through praxis and history. At present there are, and there always will be, many questions of immense interest that science is not able to answer: questions about ultimate aims, about the selection and hierarchy of values, and about the meaning of life. Science cannot tell us what is good and what is bad (as an aim, not as a means). That accounts for the obstinacy of certain contemporary philosophical orientations (such as the immanentist structuralism promoted by M. Foucault) and even of certain Marxian philosophers (such as L. Althusser) to follow obsequiously the work of science, remaining within the perimeter of the achievements obtained at a certain moment and fearing that everything which does not bear its marks is full of dangers and burdens the wings of "imaginatio creatrix," which is as necessary in philosophy as in art.

Contrarily toward the pragmatic orientation toward daily occurrences, and to the structuralist orientation toward the formal-abstract, the axio-centric perspective points to a philosophical approach that rises above the whims of the moment, without, however, seeking shelter under the glaciality of a closed structure. To states it opposes the process; to contemplation — action; to disparate elements — totalizing significations; to mechanical determinism — dialectical determinism (implying freedom); to the morals wavering between metaphysical pessimism and optimism — a committed morals of the praxis; to scientism it opposes a militant and efficient kind of humanism, concomitantly based on a scientific and philosophical cognition. This perspective renders not only unjustified but even chimeric in the perimeter of man's ontic modality, both the pathos deprived of theoretical strictness of the "philosophy of life" and that cold objectivity toward which ultraformalist rationalism aspires, proclaiming "the death of man," forbidding the human subject any right to emotional participation, to committal, to valuation, and thus turning philosophical thinking in a taxonomic instrument, making it unable to take wings.

The objective of knowledge is truth, beyond any doubt. The philosophical Olympus, however, cannot be built on the arid soil of synchronical structures, but on the summit of Golgotha that Man has climbed by experiencing everything that is human. This is precisely one of the huge creative resources of philosophy. And that is why the absolutist doctrinary claims of a "closed system" become unconceivable in an axio-centric human ontology. But without striving to attain a totalizing ontological vision, a philosopher would deny his vocation. The reason is easy to understand: the philosopher is a seeker of the absolute, even though he knows that neither his instrument nor his product can have an absolute value.

IV. RECONSIDERING A MYTH: ORPHEUS

By appealing to a myth — in our attempt to understand the axiological dimension of the human condition — undeniable advantages emerge. The language is more flexible, with more shades of difference; it is a metaphorical language. In point of fact, the myth is an extended metaphor. It is through the events narrated that the myth is a codification of practical wisdom, a metaphorical expression of the belief in those values that have transfigurated the people's vital existential needs and proffered ontological determining factors to shape the Human

Condition. Perhaps more than any other myths, the one about Orpheus possesses the virtue of comprising within its register of polymorphous significance the idea that ever since the dawn of human society the realm of values has been the necessary link between needs and wishes, between ideal and reality, between nature and culture, between knowing and acting, between the pursuit of practical efficiency and the joy of life intensely experienced. The myth of Orpheus suggests and calls to mind an idea that is fundamental for every humanistic philosophy: the sphere of values is defined as far as the ontological status of the Human Condition is concerned.

To appeal to the myth of Orpheus does not imply at all that philosophical thought sinks into mythology. We call a mythical figure not because it miraculously highlights what thought could not state, but because the mythical figure states it differently, by symbols, by double-sense expressions. We know "there is . . . perhaps no symbolic creation which is not in the final analysis rooted in the common symbolic ground of humanity." We need — in consensus with Paul Ricoeur — hermeneutics "to the interpretation of symbols, that is to say to the making explicit of the second — and often hidden — sense of these double-sense expressions."[23] We want to make rationally intelligible — through the mythical symbol — motivations, attitudes, and acts replete with moral significance essential for understanding the human condition. Is not Orpheus' the very myth that highlights the subtle overtones implicit in understanding the axiological dimension of the Human Condition?

We have made the myth of Orpheus our choice, considering that it leads us to points of significance quite different form the myth about Narcissus. We are wide apart in this respect from Herbert Marcuse, who in chapter 8 of his *Eros and Civilization*, thinks that both Orpheus and Narcissus express the same anti-Promethean ideal. Narcissus, does. The myth of Narcissus — which Marcuse looks upon as the symbol of "the pluridimensional man" — expresses an anti-Promethean ideal symmetrical to the myth about Sisyphus, a myth placed by Camus on the frontispiece of a key work in the sphere of the existential outlook on the Human Condition. Prometheus, he who rebelled against Zeus and gave mankind fire at the cost of perpetual sufferings, is the archetype-hero of the principle of efficiency, the hero who cultivates labor, the symbol of a pyramid like axiological hierarchy the summit of which consists of practico-utilitarian values. Both Narcissus and Sisyphus stand for axiological hierarchies which do not grant labor any privileged value. Narcissus rejects labor and withdraws within the bounds of inner life, while Sisyphus accepts labor as an ordeal, striving to feel illusive happiness in repetitive, ineffective activity devoid both of meaning and value. ("We have to imagine Sisyphus happy," Albert Camus says at the end of his essay.) The introvert conduct of Narcissus and standardized one of Sisyphus are the symbol of a universe with no value axis. The consequence in either case is failure. It is the failure of escapism in one case, the failure of automatism in the other; it is the failure of not considering the values that connect the plane of knowledge with that of effective human action, in both cases. That is why it is neither Narcissus nor Sisyphus but Prometheus who become the symbol of the progress of Western culture to this day.

Quite different is the situation of Orpheus. Orpheus is no anti-Promethean hero. Like Prometheus', Orpheus' effort is labor, but in its major variant: original creation. His ideal is labor without alienation, which should not be a mere means of satisfying vital necessities, but labor as a source of joy and a factor of complete fulfillment of man's creative ability. The work he does is satisfaction and play, his language is a song. Creation is the way he conceives life. Love is the style of his life. For the legendary poet and musician, to live means to communicate. The aspiration, mythically fulfilled by Orpheus, to communicate through his speech with everything-there-is-alive is not only an aspiration to achieve the unity of man and Nature, but the communication between man and man as well. Orpheus' creation tames the elements, alleviates suffering, and radiates beauty. His very failings are genuinely human; they express what A-T. Tymieniecka calls "the heroic fervor," "the thrill of life," and "the ardent joy of existence," so as to acquire strong axiological resonance.

The trajectory of Orpheus' destiny calls to mind simultaneously "the natural condition and vital background of every type of experience" and "the unchanging stability of ideal structures,"[24] an irreducible axiological experience in which all values are cultivated, in which the superior ones convey their meaning on to the whole range of values. The place assigned to love within this range of values is an implicit protest against a social order based on efficiency for the sake of efficiency, on standardization and repression. As Marx would put it, it is the rejection of "a world deprived of heart and of soul." If the world whose symbol was Prometheus meant the denial of a repressive order, Orpheus makes a stride ahead and, after denying the existing order, he discovers a new reality possessing a different axiological order, an order in which "imaginatio creatrix" turns into a hymn dedicated to life, an order in which man opposes ignorance and injustice, cruelty and death, an order in which aesthetic sensibility becomes a force opposed to the superrepression of one's instincts. The major moral values of the community are in the forefront in this axiological universe, not aesthetic values. But the hierarchy of values so harmoniously integrates all the values, as the pattern of the coordination of the system of values becomes an aesthetic pattern. We might agree with Baudelaire who said: "Là tout n'est qu'ordre et beauté."

This does not lead us to a paradisiacal outlook devoid of contradictions, tensions, sufferings. On the contrary, Orpheus goes through the great tragedy of option. Orpheus makes his choice and sacrifices. Orpheus regrets and feels pangs. He regrets the consequences, but he does not regret his having challenged the gods by turning to look at Euridice, when they are on their way out of Hades; he did so on behalf of the lofty values of freedom and love. Orpheus' pangs are filled with pathetic axiological resonances. Orpheus pangs exceed the strictly personal background of his suffering, caused by his loss of the most beloved human being; they become the pangs for the servitude of the Human Condition and enable him not to disavow his axiological creed, to convert his suffering into a source of a new ideal on behalf of superior human values transcending individual pursuits. Even while experiencing a misfortune, Orpheus appears as a figure of magnificent stature. He suffers as a victim, not as an accomplice, since he feels the discrepancy between the order of values and the order of Nature. To Orpheus' mind this discrepancy can and must be removed. Orpheus' attitude to the world is that of a Faust "avant la lettre," for whom to create is a way of living, a human being who understands that he is a genuine creator whose action, intense as it is, is capable of intensifying other people's creative activity as well. For Orpheus, happiness itself is not to be taken for granted, but something man builds by relying on a fundamental agreement between the order of the universe and an order of human values. Nowadays, the terms of this ratio are different. The ratio itself is different. But Orpheus is still the symbol of man's perennial effort to build such a ratio. When Hemingway states in his masterpiece *The Old Man and the Sea*, through his mouthpiece the old fisherman, "Man is not made for defeat," "Man can be destroyed but he can't be defeated," it as though we heard the present-day resonance of Orpheus' lyre.

For Orpheus was not defeated. It was neither the uncontrolled elements, nor the occult forces of Hades that made him lose his beloved for ever. Orpheus was fully aware that, were he to look back at Euridice, he would lose her. Nonetheless, Orpheus attempted to do something impossible. His conduct metaphorically calls to mind a profoundly human reaction. It is the same reaction Heraclitus proclaimed in oracular-form: "He who does not put his hope in the unhopeable will never find it." How many strides ahead have not been made in the history of his self-interpretation-in-existence because man was tempted to try to do what was sacredly held impossible! The realm of the impossible is being restrained with every progress of social practice, with every important gain of philosophy, science, or art. Though man's existence is finite in time and space, he tends to attain the infinite thanks to his thought and his peculiar values. It is as though something urges man to push the boundaries of what is possible ever farther away, to critically assess the data of knowledge, to aim at attaining values beyond the data of his effective knowledge. Orpheus' sublime act is evidence of an awareness which, while observing itself, has recovered — by the intentional acts of his consciousness — the objective and constraining significance of cultural values. It is the evidence of a conduct which, taking upon itself all the risks, steadily aspires to the happy union of creative action, valuation, and knowledge. Orpheus's conduct may be looked upon as symbolically foreshadowing what the phenomenological researches in the "creative context" will be able to express at the level of concepts millenniums after: the "I can" of man is no more seen in the Cartesian/Husserlian fashion as "I can: I think, I value, . . . I act," because the priorities are reversed and cognition is no longer identified with its "rational" accomplishment. "Man's 'I can' means, 'I create, I value, I cognize.'"[25]

Orpheus metaphorically recalls the exemplary truth that man is not a mere thinking creature but a valuating creature as well. The axial principle of Orpheus' life is not Cartesianism with its "cogito ergo sum" ("I think, therefore I am"), but "valeo ergo sum" ("I valuate, therefore I am"). Orpheus is the symbol of *homo aestimans.*

Thus, despite the ambiguities inherent in a mythical language, Orpheus' conduct testifies to the fact that values confer a peculiar ontological status on the Human Condition.

University of Bucharest

NOTES

[1] Anna-Teresa Tymieniecka, "Philosophy in Man's Self-interpretation-in-existence," *Phenomenology Information Bulletin* 5 (October 1981), p. 4.
[2] Lucian Blaga, "Trilogy of Values" (1939—42); D. D. Roşca "Tragic Existence" (1934).
[3] Camil Petrescu, "Edmund Husserl," in *Istoria filosofiei moderne*, vol. 3 (Bucharest: Societatea Română de Filosofie, 1938), pp. 375—427.
[4] Edmund Husserl, *Husserliana*, vol. 8, *Erste Philosophie*, pt. 2 (1959), p. 157.
[5] Alois Roth, "Husserls ethische Untersuchungen," *Phaenomenologica* 7 (1960), pp. xii—xiii.
[6] Dallas Laskey, "Husserl as a Humanistic Moralist," *Phenomenology Information Bulletin* 7 (October 1983), p. 40.
[7] I refer to the manuscripts M III, 3, 5 and K III, 2—K, III, 26; I could also mention the set of manuscripts E from the Husserl Archives of Louvain which discuss the theme of the "possible life based on practical reason" and of the "free development of personality" — two axiological problems of utmost importance nowadays.
[8] Edmund Husserl, *Formale und transzendentale Logik* (Halle, 1929), p. 5.
[9] A-T. Tymieniecka, "Metaphysics of the Human Condition and the Creative Origin of Sense"; see also "The Praise of Life: Metaphysics of Life and of the Human Condition: A Treatise 'in a nutshell,'" *Phenomenology Information Bulletin* 6 (October 1982), pp. 81—96.
[10] A-T. Tymieniecka, "The Theme: The Human Being in Action," *Analecta Husserliana*, vol. 8 (Dordrecht: Reidel Publishing Co., 1978), pp. xii—xiii. As Tymieniecka has indicated, the results of the transcendental and formal-ontological stages of phenomenology are not to be rejected, but rather reinterpreted in the third stage and placed within the wider framework of contextual phenomenology.
[11] A-T. Tymieniecka, "The Praise of Life: Metaphysics of Life and of the Human Conditon," pp. 22, 89.
[12] J. M. Broekman, "Sozialphilosophie in Phänomenologie und Marxismus," in *Phänomenologie und Marxismus*, vol. 3 (Suhrkamp, 1978), p. 13. Editor's note: In this interpretation of Tymieniecka's thought, the author speaks for himself.
[13] Angela Ales Bello, "The Phenomenology of Man and of the Human Condition," *Phenomenology Information Bulletin* 7 (October 1983), p. 18.
[14] This viewpoint is presented in my *Axiology and Human Condition* (Bucharest: Politică, 1972); see also my "Value-revaluation and the Axiological Perspective in Philosophy," *Journal of Value Inquiry*, no. 2 (1969), pp. 100—12; "Nietzsche versus Marx in Modern Culture," in *Crisis and Consciousness*, ed. R. M. Faris (Amsterdam: B. R. Grüner Publishing Co., 1977), pp. 33—54; "Rationality and the Basis of the Value Judgment," *Journal of Value Inquiry*, no. 2 (1978), pp. 126—33; "The Problem of Happiness. An Axiological Approach," *Revolutionary World*, pp. 37—39 (1980), pp. 225—38; and *Value-Fundamental Concept of the Philosophy of Culture*, in *Proceedings of the 17th World Congress of Philosophy* (Montreal, 1983).
[15] Cf. A-T. Tymieniecka, "The Creative Self and the Other in Man's Self-Interpretation," *Analecta Husserliana*, vol. 6 (1977); see also "The Praise of Life: Metaphysics of Life and of the Human Condition," *Phenomenology Information Bulletin* 6 (October 1982), pp. 93—94.

[16] Cf. Karl Marx, *Economic-Philosophical Manuscripts*, in *Scrieri din tinerețe* (Bucharest: Politicà, 1968), p. 579.

[17] In a theoretical discussion of freedom, five types of solutions have been proposed in the history of philsophical thinking, corresponding to the phases of the progressive process of getting freedom: (a) freedom as a passive yielding to the spontaneous impulses of the self; (b) the neutralization of this passive yielding through conquering the independence of motives; (c) the rise above this attitude which "shuts off" the horizon of values, through man's capacity of carrying on the voluntary action apt to model existence; (d) the replacement of the vainglory of absolute freedom in itself with the internal freedom of choice of the individual engaged in concrete situations; (e) finally, the transition from subjective (internal) freedom to its objective realization *in actu*, concomitantly oriented by the understanding of necessity and the active adhesion to the promotion of values.

[18] Husserl characterized value as the objective correlate of an intentional act of feeling (*Gemütsakt*). From this phenomenological point of view, values, as intentional objects, are stable structures of experience and can be investigated in their costitutional development, in the "horizon of community", where "the self-determination of the 'I' is a part of the self-determination of civilization itself." According to Husserl, each individual "operates the choices within man's scale of values, but the criteria are beyond himself"; "feeling is subjective while value is objective" (Aloïs Roth, *Husserls ethische Untersuchungen* [The Hague: Martinus Nijhoff 1960], F I, 28; F I 24; pp. 92, 150, 15).

[19] A-T. Tymieniecka, "The Theme: The Human Being in Action," *Analecta Husserliana*, vol. 7 (Dordrecht: D. Reidel Publishing Co., 1978), p. x.

[20] K. Marx and F. Engels, *Opere*, vol. 1 (Bucarest: E.S.P.L.P., 1967), p. 421.

[21] Veda Cobb-Stevens, "Contextual Phenomenology and the Problem of Creativity," *Analecta Husserliana*, vol. 7 (Dordrecht: D. Reidel Publishing Co., 1978), p. 173.

[22] A-T. Tymieniecka "The Prototype of Action: Ethical or Creative?," *Analecta Husserliana*, vol. 7 (Dordrecht: D. Reidel Publishing Co., 1978), pp. 183, 209.

[23] Paul Ricoeur, "On Interpretation," in *Philosophy in France Today*, ed. Alan Montefiore (Cambridge: Cambridge University Press, 1983), p. 192.

[24] A-T. Tymieniecka, *Why is There Something rather than Nothing*? (Assen: Van Gorcum & Co., 1966), p. 19.

[25] A-T. Tymieniecka, "The Theme — The Phenomenology of Man and the Human Condition," *Analecta Husserliana*, vol. 14 (Dordrecht: D. Reidel Publishing Co., 1983), pp. xvi—xvii.

YNHUI PARK

THE HUMAN CONDITION: A PERSPECTIVAL VIEW

If the ontological question of what there is, and the epistemological question of what it is to know, are central in philosophy, the question of what a human being is is no less fundamental. For answers to such questions either presuppose a certain view on what a human being is and thus depends on it, or implies it. Ontology and epistemology, on the one hand, and a philosophy of human nature, on the other, are inevitably related to each other. It is more than natural that from Plato to Sartre, speculative philosophers as well as analytical philosophers, philosophers as well as scientists, have been explicitly or implicitly preoccupied with defining the nature of man.

However, in the face of the persistence of the seemingly unsolvable, different, and even conflicting views on the issue, many philosophers have come to conclude that we cannot fathom what man is, and thus should give up the very project of defining man. Thus, Arendt, for instance, writes:

> The problem of human nature ... seems unanswerable in both its individual psychological sense and its general philosophical sense. It is highly unlikely that we, who can know, determine, and define nature or essence of all things surrounding us, which we are "not," should ever be able to do the same for ourselves ... this would be like jumping over our shadows. Moreover, nothing entitles us to assume that man has a nature or essence in the same sense as other. ... The fact that attempts to define the nature of man lead so easily into an idea which definitely strikes us as "superhuman" and therefore is identified with the divine may cast suspicion upon the very concept of "human nature."[1]

The question of what man is like is customarily understood as the question of defining man as an object of knowledge in terms of certain stable properties or essences. But the very attempt reveals that man is not a mere object of knowledge, that he is also a subject attempting to know himself as a mere object. That he is above all a knowing subject betrays the absurdity of the very attempt to define himself as a mere object. If defining what man is appears an impossible task, it is due, perhaps, not so much to the inherent impossibility of doing so as to the wrong way in which the problem is raised and approached. Instead of

A-T. Tymieniecka (ed.), Analecta Husserliana, Vol. XXI, 275–291.

looking at man as an object of knowldge, that is, as "what," we might concentrate on him as a conscious subject, as a "who." Instead of trying to define and explain him in terms of objective property or nature, we should describe, to use Tymieniecka's phrase, "the actuality of the real individual within the context of actual existence."[2] As Sartre says, we should look at "the human condition as an indivisible unity"[3] and to try "to establish human reality, human condition, being-in-the-world of man and his being-in-situation."[4]

In this present paper I shall try to sketch out in rough outline the human condition in such a way as will make ourselves more intelligible to ourselves in relation to nature and to our actual experiences. It is hoped that our description of the human condition would clarify the nature of many important controversies, more specifically the ontological and epistemological debates that have been plaguing us for so long. I shall propose that the human condition should be described in two mutually complementary perspectives, which I shall call respectively "ontological" and "semantical" in a sense to be explained.

I. THE PROBLEM AND SOME PROPOSED SOLUTIONS

The central problem of the human condition or of human reality is the problem of the conflict between two equally indubitable intuitions about man and his place in nature. According to the one, man is an indivisible whole and inseparable from nature. He is a product of nature, thus continuous with it. If nature is described in terms of matter, man is ultimately describable in terms of material, namely, bodily properties. According to the other, man is more than body, distinguished radically from nature because of his mental or spiritual properties. It is consciousness or mind that distinguishes him from nature, makes him a unique and peculiar being. The conflict between these two intuitions has repeatedly manifested itself throughout the history of philosophy. It is the conflict between the monist metaphysical vision as it is found in Taoism, Spinozism, Merleau-Pontian phenomenology, on the one hand, and the dualist metaphysical vision as it is found in Plato, Descartes, Sartre, on the other.

Seen in this manner, the problem is logically incapable of solution. We are faced with the choice between the Charybdis' whirlpool of the monistic vision and the Scylla's rock of the dualistic vision. The problem of man becomes the problem of resolving this dilemma, if there is a solution at all. There are many proposed solutions, which can be more or less conveniently grouped into two categories: analytical and phenomenological.

An analytical solution consists in reinterpreting the concepts occurring in the description of man in such a way that the conceptual conflict disappears. The concepts of body and mind are central in describing man. It is often understood that these two concepts refer to two distinctive properties of an actual man, which appear to be in conflict. One analytical solution consists in showing that in reality they refer to one single property, not two different properties. What is called "consciousness" or "mind" is in fact identical with brain-states describable in chemical or physical vocabularies. A certain strange reality supposedly referred to by mental concepts is in fact the same thing referred to by physical concepts. If we use mental concepts in describing human beings and human behavior, it is either because of practical convenience in ordinary life or because of the incompleteness of our scientific knowledge. Eventually human beings can be exhaustively and precisely described and explained in terms of chemical and physical laws. In a nutshell, mind can be reduced to body, consciousness to brain-states, man to nature in the sense of physical reality.

Against this form of reductionism Putnam points out that "the brain has non-physical properties in the sense that brain properties have functional properties which are definable in terms that do not mention the brain's physics or chemistry."[5] What then is a functional property? On Putnam's account function refers to performance of certain things which can be described in physical terms. Clearly performance of *X* is not identical with *X* itself. A computer performs certain functions. But a function is different from the computer which can be perfectly accounted for in physical language, for the same function can be performed by different computers, which are physically not identical. Mind or consciousness is a function of brain states. This function cannot be reduced to brain states.

In line with Putnam, Rorty explains the meaning of functionalism. The force of Putnam's functionalism is, according to Rorty's interpretation, "not to have discovered the nature of mental states, but to say that they don't have nature — or if you like, to say that the mental states are those functional states which are investigated by psychologists as opposed to those investigated by physiologists."[6] He goes on to say that "mental states are simply those whose descriptions occur in a *given* vocabulary used by psychologists, the one that contains, e.g., both "belief" and "pain."[7]

There are some difficulties in both Putnam and Rorty. Although it is logically self-evident that function of *X* is clearly different from *X*, and although mental states are functions of brain states, mind or consciousness cannot be understood only in terms of function. Function is not necessarily mental. There are different functions, and only some kind of function can be said to be mental. There is no way of attributing mental states to a computer or to an automobile even though they perform certain distinctive functions. What is needed for Putnamian functionalism is to draw a distinction between two different kinds of function, between physical function and mental function. The problem of making a distinction between mind and body, between the mental and the physical, and of relating them remains intact. As it turns out, Putnamian functionalism, if successful at all, is in reality not a rejection of reductionism, a form of physical reductionism in disguise. At best it begs the question.

If the problem of Putnam's functionalism is to make intelligible the difference between the function of, say, a computer and the function of living human being, the problem of Rorty's solution in terms of different descriptions in the light of certain perspective which depends on particular pragmatic needs is to account for such needs found uniquely in human beings. Rorty explains that man cannot be describable only in physical vocabularies because he has moral and political life in addition to chemical and biological function, and desires to give to himself "self-images which are worthy of our species,"[8] "the poetic species" having "the ability to change its own behavior."[9] But to invoke such a desire, such an ability, is to admit a certain property not accountable for in physical terms, and to accept, in spite of his argument against it, Searle's point that the problem of mind and body is not simply a problem of matter of description, for as he writes, "the mental-nonmental distinction cannot be just in the eye of the beholder but it must be intrinsic to the systems. Otherwise it would be up to any beholder to treat people as nonmental and, for example, a hurricane as mental if you like."[10] Our problem is not so much as to how we, as "poetic species," choose different descriptions of things or events, as to how the moral or political dimensions of this species itself should in the first place be accounted for. In Rorty's own account, insofar as we are in part moral and political, the descriptions of ourselves require mental concepts in addition to physical vocabularies in describing the moral and political dimensions of human beings.

Identifying both persons and physical bodies as two irreducible basic particulars, Strawson refuses to reduce a person to a body. Persons and bodies are not the same, for although both mental and physical attributes can be ascribed to persons, only physical attributes can be ascribed to physical bodies.[11] Thus Strawson suggests that as a person, a human being cannot be understood only in terms of physical properties, but must be looked at as a unique single indivisible whole. However, insofar as both mental and physical attributes could be ascribed at the same time to a person, the problem is to explain how these radically different kinds of attributions can be related. This is precisely the problem of the relation between mind and body. In essence our initial problem of accounting for what man is remains as it is.

Shifting our focus from conceptual to experiential perspective, we move from analytical solutions to phenomenological answers. Instead of looking at ourselves through conceptual grids such as "mind," "body," "consciousness," and "nature," phenomenology proposes to describe and thus understand ourselves as reflective beings as they are presented to our consciousness before conceptualization. It intends to provide us with an unmediated virginal vision of human being. Despite the difference of emphasis in their respective philosophical reflections, Heidegger, Sartre, and Merleau-Ponty, among many others, have tried to give phenomenological descriptions of human reality in its indivisible unity. The Heideggerian notion of Dasein ("human existence"), the Sartrean conception of man in terms of human reality, and the Merleau-Pontyian category of "flesh" are all intended to stress the unity of man within himself. "The compound expression 'being-in-the-world,'" Heidegger writes with reference to Dasein, "indicates in the way we have coined it, that it stands for a unitary phenomenon."[12] As quoted at the beginning of this paper, Sartre stresses the necessity of looking at "the human condition as an indivisible unity." Merleau-Ponty claims that "unlike what classical idealism thought, the relationship of subject and object is no longer that cognitive relationship in which object always appears as constructed by the subject, but is a relationship of being, through which, to use a paradox, the subject is his body, his world, his situation, and in a certain sense, enters into interaction with it."[13]

There is a methodological problem in phenomenological descriptions of man. It concerns the phenomenological fundamental assumption that we can have preconceptual, unmediated knowledge of the world and ourselves. However, as we all know, this presupposition is questionable. Let us leave this question aside, and look at different phenomenological accounts of the human being. What kind of being is this unitary phenomenon referred to as Dasein, as human reality, as flesh? In order for them to be more than empty words, they must be described — this is, the way in which what is called mind and body, transcendence and immanence within a man are related to each other must be elucidated.

Heidegger refers to Dasein as, among other things, "shepherd of Being" and truth as "aletheia." As "shepherd of Being," Dasein is transcendence, cognitive consciousness irreducible to its physical reality; as "aletheia," truth is presence or revelation rather than constitution. As "shepherd of Being," Dasein is in a sense constitutive of the world and truth. As "aletheia," truth is not simply constituted by Dasein, but rather constitutes, as it were, Dasein's constitution. Interpreted in this manner, Dasein is not transcendental in the sense of the Husserlian phenomenological transcendentalism, nor is it material in the scientific sense of the term. It is Dasein which determines consciousness as well as object, mind as well as body, not vice versa. The latter concepts are derivative of, and contingent upon, Dasein, and thus do not constitute it. Mind and body, transcendence and immanence should be understood in terms of Dasein, not the other way around.

If on the one hand the notion of "shepherd of Being" refers to the transcendental dimension of man, or to consciousness, and if on the other hand the concept of "aletheia" emphasizes the natural dimension of the human being, or body, and finally if these two dimensions are somehow related to each other in an indivisible reality, which is Dasein, how could this relationship be made describable? Introducing such concepts as "pre-understanding," "pre-ontological knowledge," and "pre-categorial relationship," Heidegger tries to make the relationship intelligible in making it at once less intellectual or ideal and less physical or natural. However, no matter how hard one tries, the connection between mind and body, transcendence and immanence, ideality and materiality is conceptually unintelligible, cannot be mixed, therefore cannot be explained. For in order to understand such concepts as "pre-understanding," "pre-ontological knowledge," "precategorical relationship,” and ultimately “care” or *Sorge* as the unifying principle, we have to rely again on such concepts as mind and body, transcendence and immanence, and on the distinction between them.

It is in this context that the shameless radicalization of transcendentality of consciousness in Sartre can be understood. For Sartre human reality, equivalent to Dasein, is defined through and through in terms of consciousness transcending its object, be it nature or itself, without compromise. Consciousness is *real*, that is, exists. But its mode of existence is “nothingne: it’ in the sense that is not part of the natural world, does not occupy space, and in the sense that it denies constantly its being part of an object of any sort in relation to another consciousness. Sartre calls consciousness ‘the being-for-itself’ as distinguished from the object of consciousness, which he calls “the being-in-itself.” But the being-for-itself is not possible without having its counterpart: the being-in-itself. For insofar as the being-for-itself, that is, consciousness, is necessarily consciousness of something, that is, the being-in-itself, the one is not possible without the other. They are mutually dependent. For this reason Sartre claims that his philosophy is not another idealism, even though he is clearly not materialist. “The error of idealism is to put mind first,” Sartre writes: “The error of materialism and of all forms of naturalism is to make out of man a natural being.”[14]

However exciting it may appear at first sight, the Sartrean ontology of human reality has at least two major difficulties. First, even if he claims that his ontology transcends idealism, he is one of the most radical idealists. Insofar as human reality is designated in terms of consciousness, that is, the being-for-itself, which is in turn described as “nothingness,” what constitutes the human being is nothing but ideality or transcendence. Once human reality is thus defined, the problem of the relationship between consciousness and body within a concrete living individual pops up immediately, for clearly a concrete human being cannot be separable from his body. Human reality as consciousness is never a phenomenological fact. It is only an abstraction of it. If we cannot encounter a human being without consciousness, we cannot encounter a consciousness without body. If consciousness is radically separated from body, the Sartrean ontology of human reality must account for the relationship between them, which is not separated in a concrete living individual. In the second place, although Sartre claims that his ontology of human reality provides us with a man as a unitary, indivisible totality, in reality he fails to do so as long as he is not successful in making intelligible the relationship between consciousness and body within a concrete human being. The unity he finds is only psychological in the sense that whatever a human being does is ultimately related to his initial existential project. But if a concrete human reality cannot be accounted for purely in terms of consciousness, the Sartrean unity is not a unity of an individual, but only of consciousness.

It is against the background of Sartrean idealism that Merleau-Ponty's philosophy of body can be better understood. "It is by combination of words that I *form* the transcendental attitude, that I *constitute* the constitutive consciousness,"[15] Merleau-Ponty writes. "It is through body that I perceive things."[16] Thus the empirical immanence of human being and his body are restored without rejecting the transcendental or ideal dimension of human being, for, as he says, "we are condemned to meaning."[17] In other words, the relation of man to nature is not simply natural, determinable in causal terms, but semantical or intentional. With him we are much closer to Heidegger than to Sartre. From the perspective of Merleau-Ponty's phenomenology, a concrete human being cannot be definable either in terms of pure ideality or in terms of pure materiality, that is, either as consciousness or as body. A concrete human being can be described only as "flesh" in which mind and body, the visible and the invisible, the seer and the seen, the knower and the known, subject and object are seen as various forms of abstraction, and are criss-crossed in chiasm into an indivisible unity. To use a metaphor, body is conscious and consciousness is embodied. Such concepts as "conceptual-less knowledge" and "wild being," along with such notions as "brute being" and "incarnated cogito," abound particularly in the later writings of Merleau-Ponty and indicate his unitary vision of man as well as nature transcending both idealism and materialism as well as all forms of dualism.

Granted that ideality and materiality, or mind and biological properties, are integrated in our body, in our "flesh," the problem is to make this integration intelligible. Granted that man and nature are unified in the indivisible "wild Being," the problem is to account for this unity. Just as for Heidegger, for Merleau-Ponty it is not enough to proclaim, however loudly, his monistic and unitary vision of man and reality at large. As long as he cannot avoid applying mental concepts to the description of his indivisible man, he must clarify the relationship between what is referred to by such concepts and what is referred to by nonmental concepts. Surely Merleau-Ponty, just like Heidegger, tries to avoid such mental and physical concepts by replacing them with such new concepts as "wild Being," "body proper," "flesh," and "incarnated cogito." However, these new categories do not elucidate the relationship between mind and body, between our transcendence beyond and our immersion in nature. However, inadequate they may appear, the concepts of mind and body cannot be dropped altogether when we describe the human being, for all ontological concepts are ultimately reducible to either the concept of mind or the concept of matter. These two concepts appear to be the ontological archi-concepts. If this is the case, Merleau-Ponty must clarify the relationship between what is referred to by nonmental concepts and what is referred to by mental concepts. This problem brings us back to the problem of the relationship between mind and body, consciousness and object, transcendence and immanence, freedom and determinism, meaning and reality.

If neither analytical nor phenomenological solutions of the traditional problem of the relationship between mind and body, between the idealistic view and the naturalistic view of man, between the monistic vision of reality and the dualistic vision of reality has succeeded in giving us a satisfactory account of human reality, it may be that the problem is not solvable by nature, or it may be that this difficulty is due to the way in which the problem has been set. Perhaps it is now time to look at the problem from a radically new point of view, to see and understand human reality, that is, ourselves, in a totally new light. What is needed is a more careful examination of, in Schrag's words, "the originary matrix of lived experience."[18]

II. ONTOLOGICAL CONTINUITY AND SEMANTICAL DISCONTINUITY

Schrag makes a distinction between a metaphysical sense of origin and an epistemological sense of it. In the first sense, we are concerned with "the source and status of metaphysical entities, both real and ideal."[19] In the second sense, "origin" refers to "the conditions of psychological acts, or the prior logical conditions of categorical operation."[20] He then urges the necessity for searching out a "more vital sense of origin than that which has been delivered to us through metaphysical and epistemological construction."[21] What Schrag is suggesting is that traditionally human reality has been looked at from two different perspectives, and he wishes "to propose a suspension of this notion in the hope of disclosing a sense of origin which is "more originative vis-à-vis the question of the philosophy and science of man."[22] Metaphysically we try to look at a man as an object of knowledge, and epistemologically we attempt to look at him as a subject of knowledge. In other words, metaphysically we are concerned with reality to be known, whereas epistemologically we are concerned with consciousness of this reality. Schrag is suggesting that we should find a perspective which fuses into a single perspective the metaphysical and epistemological viewpoints.

Although I believe that Schrag's distinction between the two perspectives is essential to understanding the fundamental structure of the human condition, I would call one perspective "ontological" instead of calling it "metaphysical," and the other "semantical" instead of "epistemological." For, on the one hand, the term "ontological" is more neutral than the term "metaphysical," which is often associated with speculative philosophy, and, on the other hand, the word "epistemological" is much narrower in its application than the word "semantical," which covers not only the cognitive dimension of consciousness but all dimensions of meaning. By the ontological perspective I understand simply the perspective of reality, and by the semantical perspective I mean simply the perspective of meaning. To put it a little differently, they refer respectively to reality-space and to meaning-space.

I am arguing that a human being has necessarily at once the ontological dimension and the semantical dimension, and that he can be made intelligible only in terms of the above two simultaneous perspectives. Ontologically a man is natural, definable in terms of his chemical and physiological properties; semantically he is ideal, intelligible only in terms of consciousness and meaning. Just as a disembodied mind or consciousness has no phenomenological justification, it is impossible to conceive of a concrete human being without consciousness. Just as a body devoid of mind cannot be considered as a human being, it is impossible to meet mind without body. This is, in a concrete human being the ontological materiality and the semantical ideality, mind and body, the knower and the known, meaning and reality are unified in an indivisible whole in the way in which, according to a Buddhist view, Nirvana and Samsara refers to the same single whole.

We are tempted to stand on a comprehensive perspective from which both the ontological perspective and the semantical perspective could be fused into a single perspective. Schrag's search for "a more vital sense of origin" appears as a search for such a comprehensive perspective.

However tempting it may be, such a search for a comprehensive perspective which would encompass the ontological perspective and the semantical perspective is as deceptive and futile as the goal of the being-for-itself trying to be at one and at the same time the being-for-itself and the being-in-iteslf in Sartre's philosophy. In an attempt to free itself from its anguish and responsibility, the being-for-itself wants to be the being-in-itself, a being which is self-sufficient. But the trouble with the being-for-itself is that it wants to be the being-in-itself without losing its original status, which is the being-for-itself. In other words, he wants to be at the same time the being-for-itself and the being-in-itself, as Sartre calls it "perfect being" or God. However, within the philosophy of Sartre the being-for-itself and the being-in-itself are mutually exclusive. Likewise, whatever perspective on ourselves we try to take, we are already either on the ontological perspective or on the semantical perspective, not on the two perspectives at once.

This is to admit that the distinction between the two perspectives is ultimate. If we look at ourselves from the ontological perspective, we are already involved in the presupposition of the distinction between the ontological dimension and the semantical dimension, for to look from the ontological perspective is still a looking, a cognizing, and thus involved in the semantical dimension. If we look at ourselves from the semantical perspective, we are already admitting the ontological dimension of ourselves, for to assert something is to presuppose the object of this assertion. Our fate is analogous to the fate of Midas. Just as Midas could not touch anything without transforming it into gold, we can neither describe, talk, nor think without already admitting the ontological dimension and the semantical dimension. To take another analogy, we are like the Japanese Dharma-figure. The figure depicting the enlightened Buddha is made in such a way that no matter how this figure is placed, it stands always in a same position smiling and showing his state of enlightenment. Like this Dharma, we are standing always at two perspectives: ontological and semantical.

To admit the impossibility of reducing the one perspective to the other, or of fusing the two perspective into a single one would appear to admit the unsolvable conflict between mind and body, reality and meaning, transcendence and immanence. But is this so? Is this conflict really real? Upon a closer examination we realize that there is no real conflict. Contrary to appearance, the ontological perspective and the semantical perspective are not mutually exclusive of each. They constitute rather two different dimensions of the human being, which are complementary to each other. This is, mind and body do not refer respectively to two different and separable entities or properties. They refer simply to different perspectives or two separate dimensions of the same thing. An actual human being is not made of mind and body, but one single indivisible entity to be looked at from both the ontological and semantical perspectives. Just like the color and the weight of an object do not refer to different components of the object, but simply to two different aspects of it, the ontological perspective and the semantical perspective, body and mind, reality and meaning do not refer to two different constituents of a concrete human being, but only to two different dimensions of it. The problem of the relationship between mind and body, which is the problem of understanding the human condition, arises and appears unsolvable only because "mind" and "body" have been understood as referring not to two aspects or dimensions of a man as an indivisible unity, but to two mutually exclusive properties. And yet, they are two different sides of the same coin.

Ontologically mind is *really* continuous with body, man with nature; semantically mind is *logically* discontinuous with body, man with nature. Ontologically the semantical dimension is continuous with the ontological dimension; semantically the semantical dimension is discontinuous with the ontological dimension. Ontologically we are continuous with nature, but semantically we are discontinuous with it. Ontologically we are body, object, chemical properties, but semantically we are mind, subject, mental properties. Semantically we are conscious, transcendental, intentional, and free, but ontologically we are nonconscious, empirical, material, and determined. Ontologically continuous with nature, we are determinable, explainable in a kind of causal terms just like any other objects in the universe, but semantically discontinuous with nature, we are free, comprehensible only in terms of reason entirely unlike any other objects in the world. The ontological perspective and the semantical perspective are not in real conflict but complementary to each other in the same way in which the concept of wave and the concept of particle are complementary in quantum mechanics. If the concept of wave and the concept of particle are equally essential in describing quantum mechanics, the ontological perspective and the semantical perspective are equally essential in describing human reality. These two perspectives make up the ultimate matrix of human condition. They are basic in the sense in which Danto talks about certain actions as being "basic."[23] A concrete human being is neither reducible to its chemical properties nor to pure consciousness. He is not made up of two irreconcilable substances either. He is only two-dimensional: ontological and semantical. This two-dimensionality is the ultimate feature of human reality, ultimate in the sense that its characteristic is Protean, because as soon as we try to overcome this two dimensionality, we are doomed to find ourselves caught in this two-dimensionality. And this is an undeniable phenomenological fact.

III. ONTO-EPISTEMOLOGICAL CONTROVERSIES

Seen in light of what may be called the onto-semantical matrix of the ontological perspective and the semantical perspective, human reality presents itself in a coherent and comprehensive picture. This onto-semantical matrix describing the human condition also helps us see, in a fresh light, the nature of the perennial controversy over ontology and epistemology and find a solution to it. Let us then consider some of this controversy by focusing on ontology and epistemology.

As Trigg tersely puts it, "what reality is like and how we conceive it are always separate questions."[24] The statement implies that ontology as a branch of philosophy concerned with what there is is logically different from epistemology as another branch dealing with our knowledge. As separate disciplines ontology and epistemology raise seemingly different and independent questions and attempt to give different and separate answers to each other. But, as it will be seen shortly, the two disciplines are in reality not separable and independent from each other, but two different sides of the same discipline. As we have tried to show, although the ontological perspective and the semantical perspective are conceptually different from each other, they are complementary and cannot be made intelligible without the other. In like manner, ontology and epistemology are not possible without implying each other. The perennial character of the ontological and the epistemological controversies can be seen as one of the results of our misunderstanding of the nature of the relationship between ontology and epistemology, the relationship which reflects the relationship between the ontological perspective and the semantical perspective.

The central problem in ontology concerns the two conflicting views about the status of reality. According to one view, reality is independent of our conception of it, and ultimately indescribable in words. Commonsensical views notwithstanding, not only Trigg but also such philosophers as Plotinus, Bergson, Heidegger, and Merleau-Ponty in the West, and Taoists and Buddhists in the East, have been struggling, as it were, to protect the purity of reality from human intrusion, conceptual distortion, and linguistic disruption. According to the other view, what we call reality is nothing more than what we conceive it to be by means of mental categories, conceptual systems, or linguistic nets. In recent years the view has been defended by such diverse and eminently influential philosophers as Quine, Goodman, Kuhn, Rorty, Foucault, and Derrida, to mention only a few.

The former kind of ontology may be characterized as "naturalist," and the latter as "constructionist." The naturalist ontology stresses the distinction between reality and our conception of it, and thus focuses on reality itself. Thus, for instance, Lao Tzu emphasizes the indescribability of reality, and states that "The Tao which is told is not really Tao."[25] We are all familiar with Bergson's denunciation of linguistic distortion of the stream of unbroken reality. In opposition to the ontological naturalist, the ontological constructionist calls attention to the fact that belief in the unconceived and thus unconceptualized reality is a contradiction in terms. Thus Quine concludes that reality is nothing but "posits" or "the value of the linguistic variable"; and Goodman talks about "worldmaking"; Kuhn claims that "paradigm" is constitutive of nature. When Derrida writes that "there is nothing outside texts,"[26] he is advancing an extreme position of the constructionist ontology.

In one sense the ontological naturalist is right. For it is obvious that, as Trigg says, "the way the world is and what we say about it"[27] are distinctive of each other. It seems, then, that the independence of reality from our conception of it is self-evident. However, it is also true that the idea of unconceived, unconceptualized, and thus unknown reality is at best empty if not unintelligible. For, if the naturalist ontology is true, the truth must be known, but knowledge is necessarily our knowledge, which is contingent upon our ability and faculty to know. For this reason in another sense the constructionist ontology is also true. And yet, to conclude with Goodman that the world is made by us with symbols, and with Derrida that there are only texts, is to utter absurdity. As Scheffler pointed out, to follow Goodman is to confuse the "world" in the sense of reality with the "world" in the sense of our conception of the "world" in the first sense.[28] Which ontology is right? Do we have to choose between them? Are they really in conflict?

Upon a little reflection we realize that the conflict between the above two ontologies is not really a conflict between two different views of the nature of reality, but only between two different ways of looking at the same reality. The conflict arises out of the confusion between the ontological perspective and the semantical perspective. The ontological naturalist stands on the ontological perspective whereas the ontological constructionist stands on the semantical perspective. Since there is no conflict between the two perspectives, there is no real conflict between the two ontologies. The trouble and controversies stem from one-sidedness of each of the two ontologies. The confusion of the naturalist ontology derives from the fact that it tries to assimilate the epistemological, that is, the semantical perspective to the reality-perspective, that is, the ontological perspective, whereas the error of the constructionist ontology springs from the fact that it also attempts to reduce the ontological perspective to the epistemological perspective, that is, to the meaning-perspective. However, these two different perspectives *together* constitute the ultimate structure of the human condition, and the ultimate matrix of our thinking, talking, and describing, and thus cannot be reduced to one single perspective. The ontological naturalism is ontologically correct, but semantically false; the ontological constructionism is ontologically false, but semantically correct.

Epistemological controversies can also be analyzed in the same fashion as ontological controversies have been analyzed. These epistemological controversies are just the other side of ontological controversies. On the one hand, the epistemological realist believes that our knowledge is dependent on reality, and, on the other hand, the epistemological nominalist argues that our knowledge is dependent upon our mental structure and conceptual or linguistic system. For the realist truth is independent of our belief "even if we all happened to agree";[29] for the nominalist truth is, in Goodman's word, "habit"[30] and should be determined only with reference to coherence and utility. The realist is right when he says that truth is not merely a matter of shared public belief, but he is in trouble when he has to show how our belief can be checked be some reality that we have presumably not yet known. The realist's notion of the unknown and unconceptualized reality appears empty. The nominalist is right when he insists on the fact that truth refers not to some unknown reality, but to our belief, but he is in pain when he has to admit that truth can be true of nothing.

The fundamental trouble for both the realist and the nominalist stems from the fact that they try to reduce the ontological perspective to the semantical perspective, or the semantical perspective to the ontological perspective. In other words, the realist epistemologist, as it were, "ontologizes" epistemology, whereas the nominalist epistemologist "epistemologizes" ontology. Standing on the ontological perspective, the realist tries to see truth in terms of ontology, whereas standing on the semantical perspective, the nominalist tries to look at truth in terms of epistemology. From the ontological point of view, he is wrong. Analogously, from the epistemological point of view the nominalist is right, but he is wrong from the ontological point of view. Interpreted in this manner, the conflict between realism and nominalism, or between objectivism and relativism in epistemology, is not genuine. It only reflects two different perspectives on the same thing. The apparent conflict has to do with only our conceptual confusion between the ontological perspective and the semantical perspective.

To sum up, not only the nature of the conflict between two opposing theories in ontology and epistemology but also the relationship between ontology and epistemology can be understood in the light of the distinction between the ontological perspective and the semantical perspective, which together constitute the ultimate condition of human reality. Seen in the light of this human condition, namely, the onto-semantical structure of human reality, we can understand better not only the relationship between mind and body, transcendence and immanence, man and nature, but also see ourselves within us and in relation to nature in a truly coherent total picture. Mind and body do not refer to different properties constituting a human being, but to different dimensions of him. As body, a human being is continuous with nature and thus part of it, whereas as mind, he is discontinuous with nature and thus separate from it. If we are two-dimensional and part of nature, the universe as a whole is also two-dimensional. If we are epistemologically doomed to be two-dimensional, the universe at large as well as ourselves is ontologically also fated to be two-dimensional.

Simmons College, Boston, Mass.

NOTES

[1] Hannah Arendt, *The Human Condition* (Chicago: University of Chicago Press, 1958), pp. 10—11.

[2] Anna-Teresa Tymienicka, "Theme," in *Analecta-Husserliana: Phenomenology of Man and Human Condition*, vol. 14 (Dordrecht: D. Reidel, 1982), p. 27.

[3] Jean-Paul Sartre, *Les Carnets de la drole de guerre* (Paris: Gallimard, 1983), p. 34.
[4] Ibid.
[5] See Hilary Putnam, *Reason, Truth and History* (Cambridge: Cambridge University Press, 1981), p. 78.
[6] Richard Rorty, "Mind as Ineffable," in *Mind in Nature*, ed. Richard Q. Elvee (New York: Harper & Row, 1982), p. 76.
[7] Ibid., p. 76.
[8] Ibid., p. 88.
[9] Ibid.
[10] John Searle, "Minds, Brains and Programs," in *The Behavioral & Brain Sciences* 3 (1982): 42.
[11] See P. F. Strawson, *Individuals* (London: Methuen, 1959).
[12] Martin Heidegger, *Time and Being*, trans. Macquarrie and Robinson (New York: Harper & Row, 1962), p. 53.
[13] Maurice Merleau-Ponty, *Sens et nonsens* (Paris: Nagel, 1948), p. 44.
[14] Jean-Paul Sarte, ibid.
[15] See Merleau-Ponty, *The Visible and the Invisible*, trans. A. Lingis (Evanston: Northwestern University Press, 1968).
[16] Merleau-Ponty, *La phenomenologie de la perception* (Paris: Gallimard, 1945); trans. Colin Smith, *Phenomenology of Perception* (New York: Humanities Press, 1963), p. 212.
[17] Ibid., p. xix.
[18] Calvin O. Schrag, *Radical Reflection* (Lafayette: Purdue University Press, 1980), p. 120.
[19] Ibid., p. 11.
[20] Ibid.
[21] Ibid.
[22] Ibid.
[23] See Arthur Danto, *Analytical Philosophy of Action* (Cambridge: Cambridge University Press, 1973).
[24] Roger Trigg, *Reason and Commitment* (Cambridge: Cambridge University Press, 1973), p. 168.
[25] Lao Tzu, *Tao Tech Ching*, trans. James Legge, in *The Texts of Taoism*, vol. I (New York: Dover, 1962).
[26] Jacques Derrida, *De la grammatologie* (Paris: Editions de Minuit, 1967), p. 72.
[27] Trigg, *Reason and Commitment*, p. 1.
[28] Israel Scheffler, "The Wonderful Worlds of Goodman," *Journal of Philosophy*, Nov. 1979.
[29] Ibid., p. 1.
[30] Nelson Goodman, *Ways of Worldmaking* (Cambridge: Hackett, 1978), p. 128.

JOHN FRANCIS BURKE

"THINKING" IN A WORLD OF APPEARANCES*

Hannah Arendt between Karl Jaspers and Martin Heidegger

Thinking does not bring knowledge as do the sciences.
Thinking does not produce usable practical wisdom.
Thinking does not solve the riddles of the universe.
Thinking does not endow us directly with the power to act.

Martin Heidegger[1]

INTRODUCTION

At the conclusion of Hannah Arendt's masterful treatment of the *vita activa, The Human Condition* (1958), she suggests that thinking might very well constitute the most active human activity:

> For if no other test but the experience of being active, no other measure but the extent of sheer activity were to be applied to the various activities within the *vita activa*, it might well be that thinking as such would surpass them all. Whoever has any experience in this matter will know how right Cato was when he said: *Numquam se plus agere quam nihil cum ageret, numquam minus solum esse quam cum solum esset* — "Never is he more active when he does nothing, never is he less alone than when he is by himself."[2]

However, in terms of Arendt's major distinction between the *vita activa* and the *vita contemplativa*, thinking is much more indigenous to the latter. It is not until *The Life of the Mind* (1978) that Arendt unpacks the significance of thinking as suggested in the above quotation.[3]

In the first volume of *The Life of the Mind*, "Thinking," Arendt reconsiders this breakdown between the *vita activa* and the *vita contemplativa*. In particular, she undertakes to establish thinking as an autonomous mental activity apart from its association with contemplation in antiquity, meditation in the medieval period, and science in the contemporary age. Echoing Cato she asks: "What are we 'doing' when we do nothing but think? Where are we when we, normally surrounded by our fellow-men, are together with no one but ourselves?"[4]

A-T. Tymieniecka (ed.), Analecta Husserliana, Vol. XXI, 293—308.

In the intervening years between *The Human Condition* and *The Life of the Mind*, Arendt experiences an event which becomes another motivating factor for her preoccupation with thinking — the trial of Adolf Eichmann. Reporting the trial for the *New Yorker*, Arendt is taken aback by the utterly banal character of this Nazi official who oversaw the executions which ensued in the concentration camps:

> I was struck by a manifest shallowness in the doer that made it impossible to trace the uncontestable evil of his deeds to any deeper level of roots or motives. The deeds were monstrous, but the doer — at least the very effective one on trial — was quite ordinary, commonplace, and neither demonic nor monstrous. There was no sign in him of firm ideological convictions or of specific evil motives, and the only notable characteristic one could detect in his past behavior as well as in his behavior during the trial and throughout the pre-trial examination was something entirely negative: it was not stupidity but thoughtlessness.[5]

Eichmann's frequent use of clichés and stock phrases in response to questions regarding his actions suggests to Arendt that such pat remarks serve as a buffer against reality. By contrast, events and facts prompt us to think.

Arendt concludes that Eichmann's thoughtlessness is a manifestation of the banality of evil. Eichmann was not a stupid person; his trial testimony indicated he understood Kant's categorical imperative. Nevertheless, what he refused to do was think. Arendt's task becomes in "Thinking" to pursue further in what fashion thinking deters evil-doing.[6]

In both the cases of the breakdown of human activities and the question of evil, Arendt is challenging conventional understandings of these issues. She freely admits that she is not a philosopher, but it is precisely her maverick outlook which engenders a fresh appraisal of these issues.

Arendt's reexamination of thinking is reminiscent of the spirit with which Martin Heidegger reconsiders metaphysics. The principal affinity between Arendt and Heidegger is their mutual emphasis on the poetic quality of thinking, in contrast especially to the calculative disposition of modern science and technology. Nevertheless, Arendt maintains a much sharper distinction between the medium in which we think and the world in which we live (or, in essence, the *vita contemplativa* v. the *vita activa*). She attests that this very distinction is the basis of philosophy and metaphysics in the first place. Still, if antiquity emphasizes the attainment of Being through philosophizing, contemporary thought particularly in its positivist mode accents the sensory world. Arendt's endeavor will be to chart a course which avoids these extremes, by acknowledging that thinking and the world of appearances each contain an importance distinct from that of the other.

What is no longer plausible, according to Arendt, is the breakdown of reality in terms of a sensory and a supersensory realm. As Nietzsche explains in "How the True World Finally Became A Fable,"[7] the discrediting of the latter inevitably leads to the discrediting of the former: "We have abolished the true world. What has remained? The apparent one perhaps? Oh no! With the true world we have also abolished the apparent one."[8] Even though traditional metaphysics has come under severe attack in the last century (and by Nietzsche's implication positivism as well), Arendt does not abandon the concerns of metaphysics, but simply their expression in terms of a two-world theory.

If anything, thinking has been liberated through this critical scrutiny of metaphysics. Arendt discloses that thinking has a need to 'transcend' the limitations established by knowledge precisely to pursue many of the issues reviewed in metaphysics. The difference is that thinking's deliberations do not carry cognitive validity, a stance very akin to that of Karl Jaspers' in his *Existenz* philosophy.[9]

In this regard, Arendt and Jaspers are captivated by Kant's sense of critical inquiry. Specifically, Arendt contends that Kant makes a distinction between *Vernunft* and *Verstand*, reason and intellect. Reason through thinking is concerned with meaning; the intellect through cognition is concerned with truth. The essential drawback of metaphysics is that it tries "to interpret meaning on the model of truth."[10]

Granting Arendt her distinction, what does thinking in fact consider meaningful? Arendt makes repeated reference to the traditional issues of God, freedom, and immortality as topics for thinking. But with regard to her intention to draw out the significance of the world of appearances, she includes in the domain of thinking notions which do not appear as phenomena in the world yet are part and parcel of our lived experience. Justice serves as an example of this.

Finally, Arendt cites two positive ramifications for the pursuit of thinking stemming from the disrepute into which metaphysics has fallen. First, the treasures of the past can be rediscovered and reinterpreted apart from petrified traditions. Still, Arendt maintains the deepest respect for the 'invisible concerns' formerly engaged by metaphysics; her invocation of thinking is expressly intended to salvage meaning from oblivion in contemporary times.

Second, she contends that the traditional relegation of thinking to philosophers as opposed to everyone makes no sense anymore. Reiterating her observations on the banality of evil derived from the Eichmann case, Arendt insists that thinking is not merely a matter of intelligence and therefore needs to be done by everyone.[11] Admittedly, thinking is not political in the way she explains action in *The Human Condition*, yet both in terms of her establishment of the autonomy of thinking as a mental activity and this eradication of the few-many distinction, Arendt is dwelling upon the politics of meaning.

The next several pages will build upon Arendt's excavation of thinking from the ruins of metaphysics in two ways. First, it will be shown how Arendt distances thinking from the world of appearances without resorting to a two-world theory. In particular, emphasis will be placed upon Arendt's presentation of the ontological character of Appearance, her mind/soul-body dichotomy, and then the pivotal distinction between thinking and cognition. Second and simultaneous with this descriptive analysis, it will be suggested that although much of the terminology she employs to convey the world of appearances is taken from Heidegger and Maurice Merleau-Ponty, her conception of thinking remains firmly rooted in Jaspers and Kant.

A. THE ONTOLOGICAL CHARACTER OF APPEARANCE

Arendt has insisted in the past that the human condition is one of plurality.[12] Self-disclosure ensues amid the company of others; politics is constituted by the variety of speeches and deeds manifested in the public realm. In "Thinking," plurality becomes the fundamental condition for the disclosure of any phenomenon, for appearingness depends upon observers:

> In this world which we enter, appearing from a nowhere, and from which we disappear into a nowhere, *Being and Appearing coincide.* Dead matter, natural and artificial, changing and unchanging, depends in its being, that is, in its appearingness, on the presence of living creatures. Nothing and nobody exists in this world whose very being does not suppose a spectator. In other words, nothing that is, insofar as it appears, exists in the singular; everything that is is meant to be perceived by somebody. Not man, but men inhabit this planet. Plurality is the law of the earth.[13]

This disclosure of reality through the consciousness-of phenomena, from the standpoint of the spectator, is a standard tenet of phenomenology; the appearing world is constituted through its mutual perception and acknowledgment by spectators.

This world, as Arendt understands it, precedes one's arrival and survives one's departure.[14] Human existence is bounded by one's birth and death; thus, she argues that we are "appearing and disappearing creatures."[15] Arendt's choice of terminology is very Heideggerean, except that she gives it a more literal connotation. Whereas Heidegger maintains that Being is simultaneously revealed and concealed in the manifestation of beings, Arendt contends that Being is disclosed through the appearingness of beings and then is shrouded upon their disappearance.

Arendt's identification of Being with Appearance gives the latter an ontological primacy unbeknown to two-world theories. The question becomes what is the relationship of thinking, which does not appear as an activity, to the world of appearances:

> As Merleau-Ponty once put it, "I can flee being only into being," and since Being and Appearing coincide for men, this means that I can flee appearance only into appearance. And that does not solve the problem, for the problem concerns the fitness of thought to appear at all, and the question is whether thinking and other invisible and soundless mental activities are meant to appear or whether in fact they can never find an adequate home in the world.[16]

Although one always remains rooted in Appearance, Arendt suggests that thinking withdraws from the world to consider questions of meaning. Thinking's deliberation upon 'invisible concerns' does not constitute a separate realm or a fundamental ground though because of the ontological character of Appearance.

Therefore, philosophers and scientists are misguided if they insist upon seeking a fundamental reality behind the world of appearances. Even though a philosopher chooses the life of thinking as his abode, this activity initially departs from and eventually returns to the world. Scientists, as Arendt will subsequently disclose, are even more enmeshed in the world of appearances in the pursuit of their inquiry.

Arendt does not dispute that there are 'causes' beneath appearances, but she is adamant that they do not constitute a higher order. For instance, one's bodily organs which do not appear sustain one's existence, yet this does not lend them a greater ontological significance. In this light, Arendt does speak about the simultaneous revealing and concealing dimensions of Appearance:

... the truth is, not only do appearances never reveal what lies beneath them of their own accord but also, generally speaking, they never just reveal, they also conceal — "No thing, no side of a thing, shows itself except by actively hiding the others." They expose, and they also protect from exposure, and as far as what lies beneath is concerned, this protection may even be their most important function.[17]

The quote within this quotation is made by Merleau-Ponty, but it just as easily could have been stated by Heidegger. Still, Arendt's presentation of the revealing-concealing dimensions of appearances, in terms of the protection argument, constitutes a much more clear-cut understanding of what is to appear and what is not to appear than one would find in either Heidegger or Merleau-Ponty.

At the same time, Arendt recognizes that illusion, deception, and semblance do ensue in the world of appearances. However, the uncovering of such errors leads not to the disclosure of a hidden true realm, but simply to the emergence of a more legitimate appearance. Reality unfolds through appearances supplanting other appearances which prove to be illusions:

For when an illusion dissipates, when an appearance suddenly breaks up, it is always for the profit of a new appearance which takes up again for its own account the ontological function of the first .... The dis-illusion is the loss of one evidence only because it is the acquisition of another evidence ... there is no *Schein* without an *Erscheinung*, every *Schein* is the counterpart of an *Erscheinung*.[18]

These remarks are actually those of Merleau-Ponty, but Arendt's use of them in this instance is perhaps more faithful to Merleau-Ponty's insistence that appearances manifest the multiple possibilities of the world than any other reference she makes to him in her development of the primacy of Appearance.

On the other hand, she is so preoccupied with refuting once and for all the notion that there exists a higher realm beyond the world of appearances, that she makes a distinction between authentic and inauthentic appearances which reinforces the rigidity with which she views what is to appear and what is not to appear. She derives this new distinction from Adolf Portmann who in his biological studies places primary significance on the appearingness of living things. An authentic appearance would be the multi-colored plumage of a particular bird. This appearance displays itself of its own accord and distinguishes this bird from all other phenomena. An inauthentic appearance would be the exposure of bodily organs in surgery. In this case, the organs appear only because of the interference of the surgeon; moreover, they exhibit a uniformity in appearance regardless of whose particular organs they happened to be.

Consonant then with her contention that plurality is the law of the earth, individual distinctiveness is obtained through the urge toward self-display, not in terms of a nebulous 'inner self':

> An 'inside self,' if it exists at all never appears to either the inner or the outward sense, since none of the inner data possess stable, relatively permanent features which, being recognizable and identifiable, characterize individual appearance.[19]

To recapitulate, Arendt is situating the ontological ground of reality in Appearance. In so doing, she rejects all two-world theories which present Being as a higher level of reality than Appearance. She does allow for a distinction between authentic and inauthentic appearances, but this only reinforces her conviction that the significance of reality emerges through the plurality of the world of appearances. Thinking, however, withdraws from the world to conduct its 'invisible' pursuit of meaning. The exact character of this withdrawal is to what we now turn.

B. THE MIND/SOUL-BODY DICHOTOMY

The mind/body dichotomy has been a prevalent part of Western philosophy since Descartes. Arendt makes the consideration of this issue even more complex by bringing a soul as well into the framework of the discussion. Essentially though, for Arendt, the soul in combination with the body are intimately connected with the world of appearances while the mind (or thinking ego in this case) remains at a distance from this world.

The body is basically the receptacle for sensations and impressions. In order for any meaning to be derived from experience, these sensations need to be 'desensed' in the mind where subsequently imagination recreates the impression so that thinking can dwell upon it.

The soul is the arena of inner life where the passions and emotions reside. If man's body organs sustain his existence biologically, the soul serves this end psychologically by keeping emotions in these inner depths. Furthermore, the changing moods of the soul never remain the same long enough to be perceived as phenomena are in the world. The soul can simply express itself by means of a glance, a gesture, or a sound.

The thinking ego, by contrast, never appears in any fashion in the world. As suggested above, the thinking ego is completely removed from sense experience. Because of this gap, metaphors are needed to convey the content of thinking back to the world of appearances.[20]

Unlike the soul, the activity of thinking is characterized by speech or what Arendt terms "the dialogue between me and myself." On the other hand, the thinking ego, as previously disclosed, is not an inner self but, as Kant indicates, simply recognizes that it is: "I am conscious of myself, not as I appear to myself, nor as I am in myself, but only that I am."[21] This disembodied ego lacks qualities as well as a life story; it is both ageless and sexless.

Thinking, then, is characterized by swift, sheer activity, and none of its conclusions remain permanent. Although it is the same subject which participates in both thinking and the world of appearances, Arendt contends that it is not the same person in both arenas, for what one remembers as man is not carried over to the thinking ego. This constitutes the most radical statement Arendt makes regarding the autonomy of the thinking ego. One gets no sense here of any sort of hermeneutical horizon in which personal, cultural, and historical influences would influence the way one thinks.

In view of the autonomous, non-sensational character of Arendt's thinking ego, it should not be surprising that she challenges Merleau-Ponty's notion of a chiasm between mind and body.[22] Certainly, she claims, a case can be made for a chiasm between soul and body, for the moods of the soul never leave the level of sensory expression characteristic of the body; however, the thinking ego is so distanced from the world of appearances, that there is no chance for any "crossings over"[23] between it and the body. Arendt indicates, furthermore, that Merleau-Ponty even acknowledges that "fundamental thought is bottomless."[24]

Merleau-Ponty's intention, though, in depicting thinking as bottomless is precisely to move beyond two-world theories where thinkng is associated with a fundamental ground. But in so doing, he likewise questions a strict objective portrayal of appearances. As a result, he contends the visible and the invisible flow from one into the other: "The invisible is the outline and the depth of the visible. The visible does not admit of pure positivity any more than the invisible does."[25] Similar to her review of the revealing-concealing dimensions of Appearance, Arendt's much more rigid understanding of 'bottomless thinking' prevents her from seriously engaging Merleau-Ponty's chiasm between mind and body.

On the other hand, Kant's notion of a 'thing-in-itself' is strikingly similar to Arendt's thinking ego in that it is, but does not appear. The principal differentiation between them remains that the 'thing-in-itself' causes appearances, thus constituting a variation of the two-world theory. Arendt indicates that traditionally the 'thing-in-itself' has been portrayed in theological terms whereby God is, but does not appear — he is not given to human experience. Admittedly, she adds, God is not for us in his non-appearance, but he is for us insofar as he is an 'idea of reason' engaged by the speculative capacity of thinking which transcends cognition.[26]

Though the activity of the thinking ego attests to the notion of a 'thing-in-itself,' Arendt finds the only parallel to the latter's so-called ability to cause appearances to be in the world of appearances: inauthentic appearances such as the bodily organs, which exist but are not seen, 'cause' or sustain life. Reiterating her critique of two-world theories, Arendt has no problem with this cause-effect sequence, but what she denies is the hierarchical importance given to the cause. Following this hierarchical logic, one would have to say in this instance that the bodily organs are one's "true divinities."[27]

Instead, Arendt traces this hierarchical order back to the experience of the thinking ego which seemingly transcends Appearance in its deliberation upon meaning. The notion that this activity represents a higher realm, she contends, is a semblance of reason. Although this observation certainly discloses the metaphysical fallacy underlying any two-world theory, Arendt's use of semblance in this manner is somewhat troublesome because heretofore she has explained semblance in terms of Appearance. In other words, can an activity which does not appear lead to semblances? One could argue that in fact there is a primordial link which undergirds both the experience of the thinking ego and the experience which is derived from the world of appearances, but Arendt would have a hard time doing so because she does not allow for a chiasm between mind and body. As she distinctly states, all that is left to us then is "the paradoxical condition of a living being that, though itself part of the world of appearances, is in possession of a faculty, the ability to think that permits the mind to withdraw from the world without ever being able to leave or transcend it."[28]

C. THE PIVOTAL DISTINCTION BETWEEN THINKING AND COGNITION

Despite the curious capacity of the thinking ego to withdraw from the world of appearances without in fact leaving it, Arendt reinforces the primacy of Appearance by maintaining that the sheer speculative strength of the thinking ego is incapable of constituting reality. Returning to a theme she raised in her earlier discussion of plurality, she insists that reality instead is constituted by the nexus of subjects and objects: "Objectivity is built into the very subjectivity of consciousness by virtue of intentionality."[29] The "perceptual faith"[30] that the objects we perceive are similarly acknowledged by others enables us to hold steadfast to reality.

By contrast, man's solitude and singularity are accented in thinking. Arendt suggests that the Cartesian *cogito* in its characteristics of bodilessness, senselessness, worldlessness, and self-sufficiency is the example *par excellence* of how the thinking ego is disengaged from reality.[31] On the other hand, Arendt criticizes Descartes for retreating to the *cogito* to establish the certainty he claimed could not be found in the world in view of the unreliability of the senses, for certainty is precisely the one thing thinking can never establish due to its relentless critical capacity:

> What Merleau-Ponty had to say against Descartes is brilliantly right: "To reduce perception to the thought of perceiving . . . is to take out an insurance against doubt whose premiums are more onerous that the loss for which it is to indemnify us: for it is to . . . move to a type of certitude that will never restore to us the 'there is' of the world."[32]

Reality inevitably eludes a consciousness which eschews the world of appearances.

By the same token, Arendt adds, the thinking ego's proclivity toward doubting the world's reality can never lead to an absolute solipsism or nihilism because there remains a prior recognition of realness, an 'I-am,' contained within every 'I-think': "Thinking can seize upon and get hold of everything real — event, object, its own thoughts; their realness is the only property that remains stubbornly beyond its reach."[33] The speculative capacity of man can confirm or reject reality, never create or eradicate it.

The 'perceptual faith' which enables one to grasp realness is reinforced by what Arendt terms the *sensus communis* — a sixth sense which binds the five senses together and confirms "that it is the same object that I see, touch, taste, smell, and hear."[34] This 'sensation of reality' removes the senses from their isolated, private state and makes them communicable in a world shared by all. Though the thinker may have an infinite capacity to dwell among 'invisibles,' he still needs the *sensus communis* to accommodate himself to the world.

Science, in turn, is a refined cognitive mode of the common sense since its fundamental goal is "to see and know the world as it is given to the senses."[35] The crucial criterion for scientific activity is evidence derived from the world of appearances. The speculative capacity of thinking is strictly employed by the scientist to arrive at better methods of implementing experiments. Arendt contends that if a scientist strays too far from a common sense perspective, he will lose sight of the cognitive character of his task.

Cognitively speaking, Arendt insists that science establishes not so much an "abiding truth but a provisional verity that we expect to exchange against other, more accurate verities as knowledge progresses."[36] Traditionally, she maintains, philosophy has distinguished between truths of reasoning and truths of fact. The former have been given a higher ontological dignity because of their self-evident necessity as is the case of logic or mathematics. By contrast, the contingency surrounding factual truths have always made them more suspect than their logical counterparts. However, she claims within the scientific enterprise ultimately there are "no truths beyond and above factual truths";[37] factual occurrences are verifiable precisely because they are situated within the world of appearances. In 'fact,' the brain power which engenders logic and mathematics is also oriented toward dealing with this world; for example, she argues that higher mathematics has played a pivotal role in the scientific exploration of the universe.[38]

What science can never disclose though is the *meaning* of phenomena. Reiterating Kant's distinction between *Verstand* and *Vernunft*, if cognition (*Verstand*) comprehends what is given to the senses, thinking (*Vernunft*) muses over what it means for these phenomena to be. As Arendt exhibits through a poem by Auden:

> We all know:
> Unpredictably, decades ago, You
> arrived among that unending

> cascade of creatures spewed
> from Nature's maw. A random
> event, says Science.
> But that does not prevent us from
> answering with the poet:
> Random my bottom! A true
> miracle, say I for who is not
> certain that he was meant to be?
> But their being "meant to be" is not
> a truth; it is a meaningful
> proposition.[39]

Moreover, she remains convinced that without this essential pursuit of meaning, we would lack even the ability to ask the answerable questions; hence, thinking constitutes the *a priori* condition of cognition.

In view of this close connection between thinking and cognition, Arendt remarks that it is not surprising that philosophers frequently mistake truth as the appropriate criterion for thinking. Cognitive, 'speculative' systems which unwittingly undermine Kant's spirit of critical inquiry, she claims, are particularly constructed by Kant's successors: Fichte, Schelling, and Hegel.

Still, as meaningful an activity as thinking is, the reality of the world of appearances is comprehended by the cognitive outlook of common sense and science. Thinking can produce non-sense and meaninglessness, but never the illusion and deception which is part and parcel of the world of appearances (counter to her earlier discussion of the semblances created by thinking such as the hierarchical order of the two-world theory). In the words of Kant, Arendt concludes that thinking is "occupied with nothing but itself, it can have no other vocation."[40]

D. THE TENSION BETWEEN THINKING AND APPEARANCE

Arendt has established the primacy of Appearance while simultaneously retaining a thinking ego which withdraws from the world to consider questions of meaning in a nonhierarchical fashion. The following chart outlines the various dimensions in which Arendt presents the breakdown between Thinking and Appearance.

speculative 'realm'		world of appearances	
		'inner' dimension	'outer' dimension
mental activities	thinking ego	cognitive intellect	evidence
question of realness	thinking ego	common sense	senses
psychological encounter	thinking ego	psyche/soul	body
biological dimension	thinking ego	bodily organs	skin
ontological consideration	Being as thought-thing[41]	Being as ground of Life[42]	Being as Appearance

The most elusive line on this chart is Arendt's 'ontological consideration.' A little more clarification regarding Arendt's discriminations is gleaned by reviewing Jaspers' distinctions between objective Being, subject Being, and Being-in-itself. Objective Being is disclosed by the multitude of objects in the world of empirical existence or, in part, what Arendt means by Being as Appearance. Subjective Being emphasizes the reality of the self given in 'consciousness at large' though one is inevitably situated in a world; the closest parallel in Arendt would be Being as the ground of Life insofar as Being underlies appearances and yet ultimately cannot be considered apart from them. Finally, Being-in-itself is the boundary concept one anticipates through the realization of the 'I-am' in *Existenz*; in part, for Arendt this would be the dwelling of thinking's 'I-am' on Being as a thought-thing.[43]

For that matter, Jaspers' clear separation of non-objective existential elucidation through possible *Existenz* from the objective consideration of reality by science is simply a forceful vision of Kant's distinction between *Vernunft* and *Verstand* which forms the basis of Arendt's opposition of thinking to cognition. Nevertheless, unlike Kant and Jaspers, Arendt de-emphasizes the transcending character of pure reason and *Existenz* respectively so as to stress the ontological significance of the world of appearances.

In view of her insistence upon the ontological character of Appearance, one could almost argue that whereas Heidegger expresses the ontological difference in terms of Being and beings, Arendt does so in terms of Appearance and appearances. Certainly, her dwelling upon the revealing/concealing dimensions of Appearance indicate her indebtedness both to Heidegger and to Merleau-Ponty. However, Arendt presents a much more literal and restrictive understanding of what is to appear and what is not than one would find in either of these thinkers. Furthermore, her retention of a disembodied thinking ego, from Kant and Jaspers, distances her approach from Merleau-Ponty's notion of a chiasm between mind and body.

Conceivably, Edmund Husserl's phenomenology constitutes a compromise between these two 'camps' which have influenced Arendt's framework. Her observation on the interrelationship of subjectivity and objectivity through intentionality is taken directly from Husserl; at another point, she indicates that his notion of *epoche* is indigenous to the activity of thinking. Still, with Husserl, there remains the transcendental inclination that Arendt is trying to avoid.[44] In fact, one could argue that Husserl's transcendental ego is the last vestige of the 'semblance' of thinking Arendt refers to as the two-world theory.

In conclusion, Arendt intriguingly situates a speculative, disembodied thinking ego within a highly dynamic world of appearances. As a result, a tension ensues between the very disengaged, impersonal temperament of the thinking ego and the exceedingly diverse, pluralistic outlook of the world of appearances.

Southwest Texas State University

NOTES

[1] Hannah Arendt, *Thinking*, Vol. I of *The Life of the Mind*, ed. Mary McCarthy (New York: Harcourt Brace Jovanovich, 1978), p. 1; Martin Heidegger, *What is Called Thinking*?, trans. J. Glenn Gray (New York: Harper & Row, 1968), p. 159. In Gray's translation the third line reads: "Thinking solves no cosmic riddles." Heidegger, p. 159.

[2] Hannah Arendt, *The Human Condition* (Chicago: University of Chicago Press, 1958), p. 325.

[3] Many of Arendt's arguments in *Thinking* were initially formulated in her preceding essay, "Thinking and Moral Considerations," *Social Research* **38** (1971), 417—47.

[4] *Thinking*, p. 8.

[5] *Thinking*, p. 4.

[6] Arendt's concept of the banality of evil constitutes a significant departure from her previous presentation of evil as radical in *The Origin of Totalitarianism* (New York: Harcourt, Brace, and Jovanovich, 1973), p. 459.

[7] Friedrich Nietzsche, "Twilight of the Idols," in *The Portable Nietzsche*, trans. and ed., Walter Kaufmann (New York: Viking Press, 1968), pp. 485—86.

[8] *Thinking*, p. 11; *The Portable Nietzsche*, p. 486.
[9] See Hannah Arendt, "What is Existenz Philosophy?," *Partisan Review* **8** (1946), 51—56; Elisabeth Young-Bruehl, *Freedom and Karl Jaspers' Philosophy* (New Haven: Yale University Press, 1981), pp. 122—29.
[10] *Thinking*, p. 15.
[11] *Thinking*, pp. 12—13.
[12] *The Human Condition*, p. 8.
[13] *Thinking*, p. 19.
[14] *Thinking*, p. 20.
[15] *Thinking*, p. 20.
[16] *Thinking*, p. 23.
[17] *Thinking*, p. 25.
[18] *Thinking*, p. 26.
[19] *Thinking*, p. 39.
[20] *Thinking*, p. 32.
[21] *Thinking*, p. 43.
[22] Maurice Merleau-Ponty, *The Visible and the Invisible*, trans. Alphonso Lingis, ed. Claude Lefort (Evanston, Illinois: Northwestern University Press, 1968), pp. 259—60.
[23] *Thinking*, p. 33.
[24] *Thinking*, p. 33.
[25] Maurice Merleau-Ponty, *Signs*, trans. Richard McCleary (Evanston, Illinois: Northwestern University Press, 1964), pp. 20—21.
[26] *Thinking*, pp. 40—41.
[27] *Thinking*, p. 42.
[28] *Thinking*, p. 45.
[29] *Thinking*, p. 46.
[30] *Thinking*, p. 46. Arendt adopts this term from Merleau-Ponty's discussion of the notion in *The Visible and the Invisible*, pp. 28—49.
[31] *Thinking*, p. 48.
[32] *Thinking*, p. 49.
[33] *Thinking*, p. 49.
[34] *Thinking*, p. 50.
[35] *Thinking*, p. 58.
[36] *Thinking*, p. 61.
[37] *Thinking*, p. 61.
[38] *Thinking*, p. 60.
[39] *Thinking*, pp. 60—61.
[40] *Thinking*, p. 65.
[41] Arendt explains: "But Being, since Parmenides the highest concept of Western Philosophy, is a thought-thing that we do not expect to be perceived by the senses or to cause a sensation, whereas realness is akin to sensation . . . ." *Thinking*, p. 51.
[42] Arendt explains: "In other words, the common philosophical understanding of being as the ground of Appearance is true to the phenomenon of Life, but the same cannot be said of the evaluation of Being *versus* Appearance which is at the bottom of all two-world theories." *Thinking*, p. 42.
[43] Karl Jaspers, *Philosophy*, trans. E. B. Ashton (Chicago: University of Chicago Press, 1969), I, pp. 47—64.

[44] Edmund Husserl, "Consciousness and Natural Reality," in *Philosophy in the Twentieth Century: An Anthology*, ed. William Barrett and Henry Aiken (New York: Random House, 1962), II, pp. 179—205.

HORST WOLFRAM HUBER

THE HERO AS THE SPIRITUAL LEGACY OF HIS CULTURE: WEN T'IEN-HSIANG AND HIS ADMIRERS

The point of departure for the following historical reconsiderations is the central importance of *mimesis* for the constitution of personal identity, in symbolic as well as moral action. For the sake of brevity, and to avoid repetition, I will attempt to present in capsule form some generic considerations that may be kept in mind for my discussion of three documents of 'hero-worship' in medieval China. Interpretive readings of these texts will be sketched in the main part of this study after a brief look at the hero in question and his image in the world of loyalist and official ideology. A minor purpose of these remarks will be to relate the documents chosen for analysis to a group of interpretive concepts recently developed by several authors and thus to indicate a plurality of vantage-points from which our texts may be considered as indicative for the constitution of meaning in history.

At the outset, let me refer to some concepts presented by Jean-Michel Oughourlian and derived mainly from his work with René Girard. Focussed on the role of *mimesis* in a newly conceived "interdividual" psychiatry and psychology, they were formulated for a study of obsessional phenomena, such as sorcery, possession, hypnotism, hysteria and the like.[1] Mimetic interaction is seen as the 'resultant' of two complementary tendencies, that of hetero-suggestion (or suggestion) and that of auto-suggestion (or imitation). *Suggestion* is the mode in which one entity or person acts upon another, which in turn reacts to the first in the mode of *imitation*. In these modes of interaction, mimesis helps both to constitute the net of social interaction as well as to form the individuality of its participants.

The individuals or *holons* (Oughourlian is adapting the term from Arthur Koestler) can be thought of as interacting in the two modes mentioned with the 'culture' to which they belong. At the same time, mimetic interaction between one *holon* and the 'other' occurs within the sphere of 'culture'. Culture is itself constituted by the very mimetic interaction, for which it provides both the background and the 'horizon'.

A-T. Tymieniecka (ed.), Analecta Husserliana, Vol. XXI, 309–336.

In Oughourlian's scheme of obsessional phenomena, the chief entities are thus, in non-technical language, culture, as the manifestation, so to speak, of the "desire" of the social group, and then the *agents*, as well as the *objects*, of possession. We can follow his, and ultimately René Girard's, suggestion that the "désir de la culture" is essentially "de répéter ce qui sauve" (Oughourlian, p. 159). It takes the form of an underlying constant, the perennial demand for re-presentation, for mimetic re-enactment.

Similarly, the structure of the texts analyzed here involves culture, institutionalized in the form of an "espace ou ensemble symbolique" (p. 48), the 'hero-figure' individuated in acts of heroism and sacrifice, and the 'hero-worshipper' individuated in vicarious acts of mimetic re-presentation. Further, Oughourlian suggests, we may discern various levels of intensity in the two-fold process of hetero-suggestion and auto-suggestion; hetero-suggestion proceeds from suggestion through *possession* to incarnation, auto-suggestion likewise from imitation through *identification* to metamorphosis. As our texts indicate, there would be a middle ground of possession and identification where the two vectors interact.

Mimesis includes a temporal dimension; it operates by repetition, through retention and re-presentation. As Husserl saw it, any given phase of experience presents itself as a "Kontinuität von Vergangenheiten";[2] moments are following upon one another in constant modification, shifting, or, if you permit, up-staging the continuities 'of the past' in sequential order. Each and every one of the historical references or 'allusions' in the texts discussed below represents in itself such a continuity of past continuities; the texts themselves are constituted in a sequential (not to be confused with linear) intertextuality.

Apart from Husserl's geological metaphor of the constitution of meaning by "sedimentation" (which he himself attempted to transcend), we might hold on to the notion that in these very "continuities", in their mode of mimetic re-presentation, there occurs the constitution of significant entities, or, in our case, the conceptualisation of past acts, as these entities participate in the "Kontinuität der Vergangenheiten". Thus, in the "espace ou ensemble symbolique", we arrive, to use the wording of Deleuze and Guattari, at the "régime" of *signification*, at a configured universe of meaning. From *signification* emerges the 'post-signficant' "régime" of subjectivation,[3] through and in the acts of heroism or of empathetic mimesis (which is exactly what the statements under study would "lead up to").

This point, I believe, is where Paul Ricoeur's concept of narrative mimesis would seem to enter our deliberations. The step from the 'sphere' of signification to the 'acts' of subjectivation could be seen as mediated by Ricoeur's 'triple mimesis', which I understand to emphasize representation, fictional patterning, and hermeneutic application.[4] Narrative in the strict sense is not the operative concept for this discussion. Rather we deal with the constitution of the moral space, with its configuration in writing and the translation into action as parallel phases.

Quite analogously to Husserl's structure of the internal time-consciousness, the acts and re-enactments, which our texts call for, constitute successive 'continuities' and 'spaces' of signification. In the process, they also enter into the sphere of 'meta-historical' signification. Here, the orders of meaning are not as much organized by, than actively directing the sequence of Husserl's continuities of past moments, which are predicated on the past, present or future 'now'. This 'meta-historical' aspect has always been perceived in Chinese historiography, where hardly a historical act would ever escape judgment, *sub specie*, not of Western 'eternity' but of metahistorical signification and the transcendence of the moral.[5]

WEN T'IEN-HSIANG CANONISED AS AN IMMORTAL OF CONFUCIAN IDEOLOGY

One way of illustrating one of the leading concerns of these deliberations would be to remind the reader of the Hasidic tale with which Joseph Levenson concluded his well-known 'trilogy'[6] and which Pierre Vidal-Naquet used to conclude a lecture some years ago at Harvard.[7]

It deals with ritual performances, with the lighting of a fire, with the prayers it accompanies, the place to which both belong; and it deals with the transmission of the ritual from one teacher to the next, till the last in line has to confess that far from being capable of lighting the fire, of saying the prayers, of locating the place, he could merely tell the story about it. And since the tale is well known, I should not quote it from Gershom Scholem, who himself heard it from S. Y. Agnon,[8] or from Joseph Levenson, or even give a summary of what they quoted. Pierre Vidal-Naquet's use of the tale is most *à propos* here, since his lecture was concerned with the politics behind the supreme self-sacrifice brought by the defenders of Masada. A similar case of heroic and suicidal patriotism, or proto-nationalism, which Sinologists usually refer to as "loyalism", will occupy us here. What we deal with is the spirit of one such individual, a hero for sure, a suicidal martyr in effect, possibly not even in spite of himself. We are primarily concerned with the desire of some of his admirers to keep the spirit, perhaps the soul, of this man alive amongst themselves, and also with their aspiration to live a vicarious existence of and in his spirit. In this way, they also hoped to keep alive the spirit, as they felt it, of the dynastic order that had perished and that their hero had committed himself to defend.

To make a difficult task easier, let me first introduce the *dramatis personae*. Wen T'ien-hsiang (for our present purposes, his life is best sketched in reverse chronological sequence) was executed on what we would count as January 9, 1283 in a market place in Peking, at a location now marked by a commemorative shrine supposedly first established in 1376.[9] In 1983 we might have observed another anniversary, the 700th of Wen's death. Prior to his execution, Wen T'ien-hsiang had spent over three years as a hostage in Peking, more precisely at the capital of the Yüan empire, Ta-tu, then also known by the name of Yen-ching. In 1275, after a decisive final campaign of two years, the Mongol, or Yüan, armies had occupied Lin-an (i.e., Hang-chou), the capital of the Southern Sung dynasty. The Yüan took roughly three years to suppress the last noticeable Sung resistance, including that of Wen T'ien-hsiang, late in 1278 and in early 1279. The end of the Sung in 1275, and the capture of Wen in late 1278 are two of the dates to keep in mind for the following. Earlier, in 1256, Wen had made a brilliant name for himself by passing the '*concours d'état*', the *Chin-shih* and Palace examinations, for the highest degree of the empire as the top-ranked successful candidate, an honor which in itself tied him to the dynasty by a special bond of loyalty. He had led, after only an inchoative career in the civil service, a small semi-irregular military force to the defence of the monarch. Only parts of his army saw action at the unsuccessful attempt to raise the siege of Ch'ang-chou in 1275, and it was dissolved and, after a fashion, reassembled later in 1276. In the interim, Wen T'ien-hsiang had embarked on an unsuccessful and semi-official diplomatic venture to the headquarters of the conquering Yüan armies. The Yüan commander had at first detained him; shortly thereafter Wen was sent north, along with an official delegation from the defeated Chinese court. He had managed to escape *en route*, and then fled adventurously through tightly guarded lines, in the area near the junction of the Grand Canal and the Yangtze River.

From there, he turned south to join the 'remnant court' centered around two under-age imperial princes. Established tenuously first in southern Che-chiang, then in Fu-chien, the last of the Sung later took to the sea, finally seeking refuge in the islands around and to the west of Hong Kong and Canton. As Wen T'ien-hsiang did not get along overly well with the leaders of this court, he had no choice but to move out on his own, forming his second army as mentioned. He began to wage a temporarily promising, but finally disastrous campaign of resistance against the Yüan. A closer look at the record indicates that, apart from its role in the resistance against the invaders, Wen's campaign also served the purpose of suppressing armed uprisings, which sprang up in these parts of southern China as a side-effect of instability and warfare, and which were to keep the Yüan regional government in action long after the Sung resistance itself had been suppressed.

Wen T'ien-hsiang was a highly educated intellectual and literatus, proficient in the classical genres of Chinese literature as they were practiced in his day. His was one of the earliest generations whose intellectual profile took shape under the emerging dominance of the thought of Chu Hsi (1130—1200), which had received its earliest official sanction in 1241, when Wen was five; even though the intellectual filiation of his father and first teacher is not exactly pinned down, it was clearly Neo-Confucian, and his more well-known teacher Ou-yang Shou-tao is known as a second-generation disciple of the Chu school.[10]

Our remaining figures are much less well-known outside of sinological circles. One is Wang Yen-wu, an intellectual from Wen T'ien-hsiang's own home town of Lu-ling, or Chi-chou; Hsieh Ao lived from 1249 to 1295 and Cheng Ssu-hsiao from 1241 to 1318. These two figures have drawn some attention as *i-min*, "survivors" of the Sung, and as "recluses". They were people who kept out of public life and, above all, out of government service — a life-style regarded as sufficiently atypical for an educated person as to require justification. In fact, ever since the net of governmental involvement had begun to tighten around individuals, as well as society as a whole, with the formation of what came to be called the "imperial dynasties" of China, a tradition of such justifications for 'opting-out' had developed. As a sizeable body of literature was generated over time, there was an increasingly large array of topical references to be drawn upon in connection with eremitism. On the other hand, there is not a great deal of truly biographical information regarding the three followers of Wen

T'ien-hsiang discussed here. Much of what we have belongs to the hagiographic traditions of the recluse life or "eremitism"; and the most prominent justification is the motive of dynastic loyalism. The literature of eremitic loyalism tended to be fraught with dialectical, oxymoronic, if not outright paradoxical, interpenetrations among conflicting motives. Eremitism involved seeking a distance from, or even putting up resistance to, state and bureaucracy. Loyalism implied a sometimes extreme identification with the state, specifically a given dynasty, and a sense of belonging to the specific territorial community which the dynasty and its symbols were believed to represent.

Our access to historical understanding is through the very genesis of the elements of such understanding, more precisely the constitution of historical meaning, or signification. The historical documents discussed here happen to give us a privileged opportunity to observe such processes. We can do so by focussing on the transmission — that is, preservation in the course of transfer — of a spiritual entity, intentionally perceived and emotionally experienced. Strictly speaking that spiritual entity can be defined or described in no other way than through the episodic performances of transmittal and preservation. The transmission of the heroic image of Wen T'ien-hsiang will in this sense be our "representative anecdote". In the interest of sharper focus the biographical episodes will not be narrated, but characterized; and to indicate the larger contours of the overall historical process, I will first sketch the transmission of Wen's image in the context of the official ideological establishment.

The line of development that needs to be traced here came to a point of 'cloture', as it were, with the induction of Wen T'ien-hsiang, in 1843—1844, into the official Confucian memorial cult, a shrine, or 'Hall of Fame', if you will, for outstanding Confucian thinkers and moralists.[11] Even though Wen T'ien-hsiang's contribution to philosophy in the technical sense was not of the first order of magnitude, it was the significance of his life for the development of Confucian political morality that brought him into the commemorative shrine. The process of induction involved a presentation from the provincial authorities to the court, which referred the original proposal of induction to the Department of Ritual for a reasoned decision, which was eventually sanctioned by the Emperior in Court. The submission by the province of Chiang-hsi included, apart from the petition itself and a biographical outline, itemised presentations under eight criteria of evaluation. The eight rubrics concerned the contents of T'ien-hsiang's ethical views, the conduct of his life, privately and in office, and importantly, the consistency among his views, as well as that of his conduct, with his basic human nature, or character.[12]

One aspect of political ethics ought to be emphasized here. The authors of the documents involved focussed their attention on an historically decisive point. Wen's life, or more exactly his death, seen against the background of his political and intellectual attitudes, is situated exactly at the juncture of two historical traditions. One concerns the Confucian ethical issue of the relationship of moral principle and personal survival, or more accurately, the principles that should guide the choice between morality and survival, and typically the choice of moral value over survival. The other tradition relates not to an ethical issue, but to one of political, or, if you will, legal custom. The custom as such is not at issue ethically or morally. What can be discussed ethically is whether there are counterindictions against its strict observation, namely values like filial piety or the survival of the state, which might be compelling enough to suspend this otherwise absolute obligation in a given case. I am speaking of the customary rule whereby a servant of a ruler or a dynasty must be prepared to sacrifice his life in situations of extreme danger to the existence of his ruler or dynasty. There is no element of ethical choice in such sacrifice; but what characterises Wen T'ien-hsiang's historical role, as distinct from his 'philosophical' contribution, is that through him (even allowing for a similar role of others as well as concurrent ideological tendencies) the absolutely compelling authority of the decision made ethically is superimposed onto the ingrained compulsory character of duty to the death. His own particular death, in the context of the intellectual and political changes of his time, established in the mind of latter-day loyalists the requirement of total inner assent without reserve — assent even to unjust victimisation in the alleged interest of the state, as we find it in the case of certain political martyrs at the close of the Ming dynasty.[13] Wen's fate came to mark an unprecedented disavowal of individual ethical choice and of the ethical "standing" of the individual as a moral agent in the political arena.

Apart from characterising what the moral 'bottom line' was seen to be, we should also look at the process of 'canonisation'. In order to document Wen T'ien-hsiang's qualification under those eight rubrics, the submitting officials chose a number of passages from Wen's literary writings and from anecdotal reports concerning him and presented these selections out of context and only vaguely identified. On tracing them to the original contexts, one finds several instances where the passages take on meanings that differ from their significance in their original context. This is not to question the methods of the officials, for it is clear that under all their criteria, Wen T'ien-hsiang fulfilled the requirements for induction. The selections were intended merely to fill the slots defined by the eight headings; what emerges is the kind of portrait one could 'draw' by filling in larger or smaller squares on a piece of graph paper. The grid-style likeness rendered the formalized image of Wen T'ien-hsiang's life and mind, as it was perceived and re-presented to be, and it served to guarantee comparability with other figures whose induction into the cult had been granted before or was being contemplated at the time.

Such was the understanding conveyed in the decision rendered by the Department of Cult Affairs. The officials review a whole list of inductions made in the preceding several decades before then taking up the examples of two of the most recent inductions, that of Liu Tsung-chou (1578—1645) and that of Huang Tao-chou (1585—1646), as specific points of comparison. Both men were "loyalists" to the cause of the Ming dynasties, which was overthrown in 1644/45 by the Ch'ing, just as Wen had fought for the Sung at their demise. The same ethical standards to the development of which Wen T'ien-hsiang's life and death had contributed and which had shaped the ethical situation facing these two later loyalists and with it their actions, could then be further interpreted in the way these two men had exemplified them in practice. That interpretation could then be used for an unadmittedly retroactive application to the life and conduct of Wen T'ien-hsiang himself.

The failure to recognize the secondary application of standards as retroactive points to another characteristic of the textual tradition of Confucian history. To human exemplars as the vehicles of transmitting abstract values along with their spirit and their image comes another, more pervasive means of transferral. As befits Confucians, the officials who decided on Wen's case did not create, they transmitted. Apart from the extensive quotations from the works of Wen T'ien-hsiang, their presentation consisted of an array of evocative phrases: building blocks, so to speak, quarried from Wen's own words and from the standard accounts of his life. The resulting mosaic was not, however, the creation of the officials; it too had, with the inevitable variations and deviations, slowly developed over the ages, from the lifetime of our hero down to the nineteenth century. The development took place primarily in a literary medium, appropriately, in the established genre of statements commemorating the establishment, or restoration, of local commemorative shrines. It is instructive to observe how, in their chronological distribution, the memorial writings devoted to Wen tend to parallel the increasing availability of his prose and poetry in the course of the 15th century, while the known collected editions came to be published only in the 16th century. Indeed, the various commemorative statements take their language directly from the original writings, not without occasionally taking note of each other, and following certain trends in their selections. Perused in sequence, these materials witness an evolving intertextuality, which is the prime vehicle of the transmission of the spirit of Wen T'ien-hsiang. For what was decided in 1843 was, after all, whether Wen had indeed fully entered the sphere of the Confucian Sage as that image had traditionally evolved.

There is one further dimension to this intertextuality. Wen T'ien-hsiang, and his admirers, as all educated Chinese writers, were themselves part of a tradition, more precisely of a complex intertextual situation. Wen's writings, and also the accounts of his biographers and admirers, are correspondingly replete with textual references to earlier literary texts, especially those most relevant to the concerns of loyalism. Wen's own expression, his own spirit even, to say nothing of the images of him presented by others, are shaped by the expression and the spirit of others before him. I would like to mention, apart from "the Saint and the Sage", as Wen refers to Confucius and Mencius, authors such as the historian Ssu-ma Ch'ien, the statesman Chu-ko Liang, loyalists like Chang Hsün, Hsü Yüan, Yen Chen-ch'ing and Yen K'o-ch'ing, and very prominently the poet Tu Fu, who shaped much of Wen's own language. In this intertextual tradition emblematic phrases move from context to context. A 'tag', or cue-word from an anecdote concerning Wang Li-hsin, a fellow loyalist, with whom Wen communicated by letter, a cue-word referring to "pure", i.e. still unoccupied land, became part of the 'text', the biographical literature, of Wen T'ien-hsiang in this later tradition, where such shiftings of textual elements are not uncommon.[15]

WANG YEN-WU'S "FUNERARY DEVOTION AT LIFETIME": LITERARY PREFIGURATION AS THE MEDIUM OF MIMETIC RE-PRESENTATION

Now we return to Wang Yen-wu, Wen's contemporary and admirer. The important fact about his text, is its rhetorical situation as a "Funerary Devotion" written for a living person. Bizarre as it may sound to us, to two 14th century writers who prefaced Wang's works it seemed astonishing, but as a document of extreme devotion, not as something inappropriate.[16] Another funerary text "at lifetime" for Wen has been preserved,[17] and Wen himself may have written one for his wife.

The occasion for Wang's votive text was the second captivity of Wen T'ien-hsiang, and his transport *en route* to the Yüan capital in the North, which took him through his home province of Chiang-hsi. Wang Yen-wu addressed Wen in the manner of an open letter, and his message is deceptively simple. Given Wen's previous record of conduct, as well as his own words, given also the tradition in which he had placed himself, it was now, as Wang spells it out, incumbent upon him to commit the honorable suicide of loyalism, most appropriately by self-imposed starvation.

The basic tenor of the argument is developed by placing Wen in the long line of loyalist martyrs of the past, and by thus subsuming his life, in advance, under the appropriate image, namely the stereotype of the devoted servant who gives his life as the ultimate fulfillment of his duty to ruler and state. From this premise followed, quite logically, that Wen had to live the image, or rather, complete it in death. We note a strongly felt need for such sacrifice on the part of the idolized person: his honorable death alone can bring justification to the admirer's own life, and the expiation of guilt (incurred by the mere fact of survival and perhaps by acts of omission or commission). Wang introduces himself into the picture by reporting his own failure to participate more actively in Wen's resistance efforts, a failure motivated by the necessity to take care of his mother in her old age and in failing health. Far from consoling himself with the thought of having met one of two conflicting demands, that of faithful service to parents, while having failed in his duty to the dynasty, he judged himself deficient in filial piety as well. Indeed during the Sung, in a curious reversal, the fulfillment of the duty of filial piety came to presuppose the concurrent fulfillment of the obligation towards the state. The duty to the state was thereby raised to a preemptive position.[18] Having left Wen's campaign to the detriment of Wen's trust and his own self-esteem, Wang felt, in a strange twist of the logic of guilt and expiation, to write the "Funeral Devotion" in order to "accelerate the coming of Wen's death"! And he made every effort to give Wen the benefit of his reasoning by posting copies along the route of Wen's transport.

To understand the situation better, we have to consider that Wen T'ien-hsiang, while in transit through Chiang-hsi, did in fact attempt to starve himself to death in fulfillment of the moral imperatives detailed by Wang. The crucial point of Wen's attempt was that he would meet his death within the borders of his home district of Lu-ling, in accordance, as we would expect, with literary traditions of the starving death of the recluses Po I and Shu Ch'i at the Mountain Shou-yang, and of sayings according to which foxes turn their heads homewards as they were about to die.[19] Again thought and action are prefigured by the literary tradition. As a typical scholar, he understood an episode in the chronicle *Tso Chuan* to mean that seven days of starvation would be sufficient to cause death and timed his fasting accordingly. As he underestimated the travelling speed on the river, the coincidence of time and place was missed. That the suicide attempt was abandoned need not imply a failure of will; the specific circumstances envisaged by the plan, and they alone, could have made this suicide meaningful and appropriate, and they had not been sufficiently met. What counted was the propriety of the gesture and the conformity with ancient precedent and protocol.

The intertextual configuration yields a close match in that Wang Yen-wu's text with its empathetic identification closely parallels the corresponding autobiographical writings of Wen T'ien-hsiang. From the time Wen had been detained on shipboard by order of Chang Hung-fan, the commander of the troops who took him captive, the thought of suicide by starvation on the model of Po I and Shu Ch'i was very much on his mind. We find the elaborate plot of his suicide attempt discussed in several autobiographical statements formulated so as to maintain the link to literary tradition.[20]

Ancient precedent and protocol drawn from the literary tradition characterise, as mentioned above, the structure of Wang's text; apart from the curious argument that Wen had shown sufficient merit as a writer and statesman, as a filial son and patriotic leader, to qualify for heroic suicide as a morally valid choice (even the less drastic decision to live in withdrawal from official life was — in the Confucian sphere — tied to requirements of worthiness), the bulk of his argumentation consists in the casuistic presentation of a very long array of historical precedents that singly as well as cumulatively serve to establish an overwhelming indication in favor of suicide. Still, there is a twist in Wang's use of precedent, which, it seems to me, differentiates this document from many others in Chinese literature where precedent plays a large role. One could see in this exceptionally clear example a model for reading similar highly "allusive" texts, which are most frequently treated as primarily decorative in their stylistic intent.

In Wang's votive text we are dealing not so much with literary references embodying precedents or examples to follow, which would involve a fairly external semantic relationship. Rather, we see a three-sided mimetic adaptation of writer, addressee and historical (or literary) precedent to each other. As prototypical images shape the behavior of the person in action, thus the literary prototypes give verbal form to the image of the hero. They delineate not merely how he should act but define what he should, given the shared cultural heritage, expect himself to be, and as the observer has every right to expect him to be — a model for his own self-cultivation. The sequence of such prototypes spelled out by Wang Yen-wu take us, topically, if not chronologically through the history of loyalism from the brothers Po I and Shu Ch'i (not excluding the traditions concerning steadfast widows), down to the beginning of Wen's own dynasty, the Sung. No reductive method (short of resorting to annotated translation) can do justice to the image, or if you will, wishful image, of Wen T'ien-hsiang, as Wang Yen-wu extrapolates it. Wen is to mould himself on the composite image, by adopting roles outlined for him and to act them through in his own death — that obviously is the only reduction that Wang was presenting as possible.[21]

The curious parallelism concerning Wen's hunger strike is not the only one between Wang's text and Wen's autobiographical writings; another is Wang's highly indirect warning of the murky, intrigue-ridden situation Wen would find himself in, if he did not end his life on his own initiative. By suicide alone could he forestall a large gamut of eventualities by which he might meet an entirely unworthy death. Wang's fearful scenario had the Yüan leaders concerned about insurrection, the deposed Sung ruler still living, while the other Sung leaders, who had capitulated treated each other with mutual hostility and suspicion. Above all there would be the natural antipathy of all of those mentioned, against one who was holding out in defiance. In such a situation it might not be possible for Wen either to die, or survive, without compromise to his integrity. To all appearances, especially considering the turmoil and tension created by the assassination of the powerful, and apparently corrupt, minister Ahmed in 1282, Wen's death actually did ensue from a situation similar to that foreseen in Wang's piece. Four years later, when Wen's death occurred, and Wang heard about it through a mutual friend returning from the capital, he could note, in a "Funerary Devotion written at Long Distance" that Wen's integrity had not been compromised, despite the delay of his death. Though Wang cannot hide some embarrassment about his former unseemly haste, he now feels very much justified.[22] Somehow the image of the loyalist hero had been completed; a piece of the world had been restored to order, albeit tragically. As on several other occasions we encounter here the transfiguration of Wen T'ien-hsiang into a cosmic phenomenon. A change in the color of the light, as it strikes and illuminates the landscape, a particular beam of light in the Big Dipper at night are ascribed to Wen T'ien-hsiang himself and his doing. We would be reckless to dismiss this as merely superstitious; much rather we might interpret this literary expression as the 'objective correlative' to the emotions of the writer and the reader alike. Wang's final image of Wen T'ien-shaing may stand as the culmination of his earlier lengthy elaboration of the 'virtual image' of his hero.

HSIEH AO: MIMETIC RE-ENACTMENT AND THE MEDIUM OF LITERATURE

Wang Yen-wu has given us a retrospective blue-print of the life which was the focus of his own vicarious mode of existence. Through historical prefiguration, he renders a representation of Wen, the hero. With our next author, we find an actual mimetic enactment intended to capture the spirit, indeed the soul, of Wen T'ien-hsiang; we then encounter its re-transferral into literary description, which in turn finds its *décodage* by another author, the commentator, and interpretive representations are given in a long series of explanatory notes by various hands. I do not need to repeat here the anecdotal biographical information we find in the literature about Hsieh Ao (1249—1295), but some selected characterisations might be helpful.[23] Hsieh belongs to the many groups of loyalist 'drop-outs' we find during the decades following the Yüan conquest of southern China; he, too, was frustrated in his endeavors at collaboration with Wen T'ien-hsiang and his resistance campaign. But there is one dominant historical pre-figuration which shaped his life more than has been generally noted. His prototype is the legendary, and historical, figure Ch'ü Yüan, famous for his fate of political frustration. On the other hand, the literary *persona* Ch'ü Yüan, such as he appears in the poem *Li Sao* commonly ascribed to him and in the poetic genre which his poem initiated, is noted particularly for the *motif* of the 'shamanistic quest', which is the quest for the unreachable as such.[24] Ch'ü Yüan's example not only shaped Hsieh Ao in his literary expression, it may have been the inspiration for the restless and rootless life-style he came to adopt particularly after his departure from Wen's campaign. Here we enter into the aspects of Hsieh's life which are described in his "Commemorative Account of the Lamentation at the Western Terrace".[25] This text presents a summation of Hsieh Ao's individual experience with Wen T'ien-shaing, the image — not so much with the living human being and his problems, but rather with the hero in transfiguration. Some time after his separation from Wen, Hsieh set out on the search for a vision that would recreate Wen's image for him, and it was in the landscapes through which his restless wanderings led him that he came to find his vision. Hawkes identified the "goddess", for the shamanistic searchers of the Ch'ü Yüan tradition, as the goal of all wanderings; in Hsieh's case the object is Wen T'ien-hsiang, the hero who had been lost, and whose spirit had to be retrieved. The search is further motivated by the sense of guilt, which Hsieh had in common with Wang Yen-wu. As Hsieh began to seek out his idol Wen T'ien-hsiang in his dreams and in everything he did, places, aspects of landscape, vegetation, seasonal moods came to remind him of the place where he and Wen had parted. Strong impressions and associations with Wen would move Hsieh to profound feelings, even to tears. The 'investment', as it were, of the landscape with emotional content tying Hsieh to his hero is yet one more mode of the representation, or transmission, of the elusive image.

At one time Hsieh came to a place that was strongly associated with his own father as well, and thus a linkage was established between the *motif* of filial piety with that of loyalty, represented by Wen T'ien-hsiang. It was the terrace of the King Fu-ch'ai of Wu, whose famous conflict with his rival, King Kou-ch'ien of Yüeh is one of the most dramatic stories from the past of the particular region. At that place Hsieh broke into protracted lamentations about Wen T'ien-hsiang for the first time. Then, four years later (these chronological indications are as important for Hsieh's text as they have been puzzling), he stages, in a curious symmetry, another lamentation for Wen at the terrace of Yüeh. "Five years later" we come to the main event. In slight deviation from the standard reading I place these events at the turn of 1282/83, 1286/87, 1290/91, to account for the curiously elastic, not exactly inconsistent, chronology. Throughout the text, Hsieh's cryptic chronology (a calculated attempt to avoid reference to the dynastic chronology of the Yüan, whose legitimacy was too repugnant to contemplate) is, above everything else, intended to relate Hsieh's life and its rhythm to the life of Wen T'ien-hsiang. Whereas the standard reading initiates the count from Hsieh's last meeting with Wen, and then starts a new count from Wen's death, I would rather count from the last meeting throughout, incorporating the date of Wen's death as indicated. The structure of the text, regardless of the count adopted, is meant to indicate that the personal relationship to Wen is revived in the rhythm of the chronological sign-posts, which the text, in an intentional cryptogram, has reserved for the participation of "all true sympathizers".[26]

Hsieh's lamentations do not merely serve as expressions of the author's sincerely felt emotions. Such laments were part of the standard funeral performances; and thus, at the latest of the three dates, in the company of three just as cryptically 'identified' friends, Hsieh ascended the "Western Terrace" (possibly the "Western Fishing Place" of the early recluse Yen Kuang, near Yen-chou),[27] and performed, keeping even in details to practices documented for the period of the Sung, the ceremonial known as "Recalling the Soul".[28] To describe it adequately would be a long treatise in itself. The basic notion is that, owing to the fact of Wen's death at a distant place, his soul (more precisely the part of his soul-like attributes known as *hun*) had become separated from his person and needed to be recalled to his proper homeland. The performance is related both to customary religious practices and to the literary tradition dominant in Hsieh's life, that of the land of Ch'u and of Ch'ü Yüan, where it is the topic of two well-known major pieces of literature.[29] He breaks into an improvised song in the Ch'u tradition, beating upon a rock with his bamboo stick, until the song is over and stick and rock are both in pieces — drumming the beat, as if on a shaman's drum. The song itself is hard to render; in essence it points to

Wen T'ien-hsiang's identification with what Edward H. Schafer called the "Vermilion Bird",[30] the asterism associated with the South, with fire, and the Sung Dynasty. As in Wang Yen-wu's text, we find Wen transposed into the heavens, transfigured into a star and a recipient of sacrifice, as are the astral deities. He is, as many authors in the later tradition present him, equated with the Sung, with the order and spiritual meaning they supposedly represented. The transfiguration of the hero implies concomitantly the transfiguration of the dynasty.

In this way we can assume Hsieh had finally found Wen T'ien-hsiang, insofar as he had sought him in "the people and things of nature", which thus, through the cosmic virtue of the "Vermilion Bird", of Wen, were retransfigured into the world of the lost order of the Sung. Involved is a spritualisation of the political which, I dare to suggest, was at a later time paralleled in the secular world, in the abstraction of the ecumenical order from the world of power politics and its transport into the sphere of the cultural and the ideational.[31]

In the mimetic re-enactment the spirit of Wen T'ien-hsiang was to take over, indeed to take possession of his admirers, who in turn sought to be identified with that spirit. In the text, the central passages of which we have discussed, this performance is in turn 'represented', and given the cryptic elements of that text, we need not shy away from the word "encoded". The next stage, then, is the performance of the invocation of Wen's spirit through reading and interpreting. Chang Ting, a practically unknown writer, supplied a commentary to Hsieh's "Record", which, like a similar commentary on another text of Hsieh's, might be dated around 1300—1310.[32] Chang's commentary, as the overwhelming majority of the authors of the appended notes remark,[33] is a veritable *décodage* in the most fundamental sense of the word. It is Chang who actually revealed that the text was concerned with Wen T'ien-hsiang at all (Wen appears camouflaged under the name, more precisely the rank and title, of Yen Chen-ch'ing, a minister and loyalist of the late T'ang). Chang also elucidated the places, names and dates, as well as some deliberately "obscure expressions". To all appearances, Hsieh Ao partly feared political problems if he were to express himself too directly, but possibly he also felt that his most personal experience needed protection from the callous misunderstandings of the indifferent — an additional note by Chang Ting points to such individuals. At such remove, interpretation, exactly because it was difficult, might transmit a truer experience than an 'easy' text, through the reading of the commentary against the original record. Yet one further level in such a preservation of the original spirit, through indirection and interpretation, is given (with express warnings of misunderstanding incurred in literary expression) in the almost two dozen appended notes, extending in time into the early Ming. As with Wang Yen-wu, the technique of symbolic prefiguration is employed. A higher level of indirection is reached when Hsieh Ao, and his relationship to Wen, is described by reference to earlier historical figures, and their relationships. We arrive at an image of Wen T'ien-hsiang, re-reflected in the figures and situations that surround the men to whom Hsieh Ao is now being compared to. Only one of the authors clearly indicates the cathartic character of Hsieh's performance on the Western Terrace;[34] on the other hand, the nature and relative value of vicarious experience come sharply into focus, quite in congruence with the transference that characterises the structural relationship among those appended notes. The affinity as well as distance between vicarious living and genuine, direct involvement is quite clearly represented in the relationship between Hsieh Ao and Wen T'ien-hsiang as in those among the personalities adduced as prototypical. Several of the authors attempt, sometimes forcedly, to present the two modes of experience as equal in value and moral significance, particularly in the guise of the loyalty of political service and that of personal friendship. Occasionally I detect a hint that the latter ranks higher. The cognitive element is greatly emphasized: Friendship is often characterized as a "knowing" (in a phrase used as a standard term for friend). That particular something in the other person that is "known", that can become intelligible or open to experience in interpersonal sensibility, is, I believe, the element of the spirit or mind; and it is in particular the spirit or mind of the admired hero that becomes visible exactly in these exchanges and transferences. In this manner the image of Wen T'ien-hsiang, especially in its identification with the dynasty of the Sung in itself, is represented as the main content and ingredient of the vicarious existence of those "survivors of Sung".

CHENG SSU-HSIAO: FACTS, DAYDREAMS AND TRANSFIGURATION

I have not yet discussed two main terms which are used in our texts to indicate that elusive something, to which I have referred as Wen T'ien-hsiang's spirit, soul or mind. These terms are *hsin*, commonly rendered as "heart", and *i*, commonly translated by "intention", "meaning", "purport". This latter term stands in close relation to the cognitive implications to which I have referred just this moment. *Hsin* is of central importance for the work which remains to be considered here, known as *Hsin Shih*.[35] This title is an index to the importance of the term *hsin* in Confucianism, especially in the tradition of Mencius; and it suggests a link to the school of Hsin Hsüeh as it developed during the Sung, with thinkers like Chen Te-hsiu (1187—1235), an intellectual trend recently studied by Wm. Theodore de Bary.[36] A strong concern of some older schools of thought, where the art of psychological guidance or management was seen as part of the methodological equipment of government, recurs in the thought of the Hsin Hsüeh, which school came to provide much of the rationale of the authoritarian style of government under the Ming. Statements by the Sung statesman Chang Chün (1086—1154), who belongs to Wen T'ien-hsiang's loyalist heritage, manifest such interests. Yet the 'link' of the *Hsin Shih* to this strain of thought may be more indirect, even dialectical. A modern preface actually characterises the *Hsin Shih* as an attempt to transmit *hsin* "outside of history", implying a kind of "Zen" for historiographers. The author sharply distances the *Hsin Shih* from the Hsin Hsüeh, which in his view had become an oppressively dominant school of thought. Its adherents, he quipped, were utterly rejecting 'heart' (*hsin*), while paradoxically claiming to be engaged in *hsüeh* (study, or teaching) while doing so.[37]

Questionable though its authenticity may seem, there may be merit in provisionally accepting the *Hsin Shih* as the work of Cheng Ssu-hsiao (1241—1318),[38] who was also a *i-min*, or "survivor" with a life-style quite as unstable as that of Hsieh Ao. His father had at one time been active, as a teacher of Neo-Confucianism, in places as distant from each other as Shantung and Fu-chien, and Cheng Ssu-hsiao was born while his father held office in Hu-pei. The son tended to link virtually every aspect of life to the principle of *hsiao*, or filial devotion; and the thought attributed to him manifests the fusion of political loyalty and filial devotion that was to encourage, most prominently during the Ming, an intensified attachment to the "ruler-father", signifying a massive trend toward the totalitarian concept of rulership.[39]

At this point I will attempt a brief characterization of the style and method of the *Hsin Shih*, especially as developed in its summary of Wen T'ien-hsiang's life, and also in the sketch of contemporary history which it includes.[40] Both are completely given to recounting discrete details and – contrary to the two authors discussed before – this writer employs strings of seemingly "positivistic" data to convey his message, which becomes at certain points quite bizarre. Most of the details turn out to be truthful in and by themselves, when checked against other accounts. The method employed will at times remind the reader of surrealism, and one is tempted to think of the remark of the Ch'ing author Ch'en Hung-hsü, who pointed out that the currently accepted standard accounts of Wen were totally insufficient to stir the emotions of people; he recommended the alternative accounts from the "Sung survivors" Kung K'ai and Chou Mi instead.[41] In a similar vein, the style of the *Hsin Shih* attempts to attain a heightened emotional charge by means of a barrage of details, which will suddenly explode into wild exaggerations and slide into scenarios more appropriate for daydreams or nightmares. The emotional tension of a straightforward account is felt to be inadequate to convey the fullness of the emotion or the moral spirit. The historiographical practice of the *Hsin Shih* thus appears not categorically different from the official historiography, which resorts to all sorts of rhetorical devices to present its didactic messages, whenever factual narrative is not telling enough. From this point of view, we might be prepared to interpret the *Hsin Shih* 'accounts' of the heated debates with various persons counseling surrender, which Wen allegedly found himself drawn into during his captivity. We might reassess the accusations he is presented as having hurled against Khublai Khan, and also the odd scene, in which Khublai and his close entourage, after Wen's execution, proceed to devour Wen's heart and his lungs. The intent is not so much an imputation of cannibalism, but a characterisation of Wen's charismatic personality. As his indomitable spirit could not be subdued, his captors resorted to 'appropriating' his heroic virtues by 'incorporating' them bodily into their own being. From China, as well as from the West, the notion is familiar that the qualities of the human spirit are embodied in specific organs of the body.[42]

Such excursions into what I would call the daydream world of historiographic surrealism occur quite sporadically in the anonymous *Chao-chung lu*, to which we nevertheless own much solid information on Wen T'ien-hsiang.[43] To place the following example in perspective, two anecdotal details are needed. At one time, Wen (with his consent, according to one version) was to have been released from his captivity in the North in the guise of a Taoist monk, wearing the characteristic "brown cap", but this plan was rejected as a political, if not actually a military, risk. More puzzling is the curious statement attributed to Wen, made as he was being taken from his prison to the execution, saying, indeed, "my work is done", in strange similarity with the story of the death of Jesus. In the *Chao-chung lu* the similarity does not end here. As the soldiers arrive at the prison, they put the aforementioned "brown cap" of the Taoists on Wen's head as a token of humiliation; after that they lock his neck and shoulders into a cangue. Such a load of lumber he then is forced to carry to the place of execution, just as Jesus had been forced to bear his cross. Only after Wen's arrival at the market places does he, in the *Chao-chung lu*, say "my work is done". Seen formally, purely as a dramatic scene, this strikes me as even more similar to the scene of Christ's death.

Although we cannot offer specific evidence of cultural transmission, it is known that Christianity as well as other Near Eastern religions had some circulation in China at that time. The discrete transmission of isolated story elements, would be conceivable, such as the spread of Near Eastern story plots, which were actually presented on the stage in Yüan China.[44] Roughly the reverse phenomenon seems to occur in 17th century Bukhara, where legend speaks of Jesus being born under some kind of bodhi-tree, and of his enemies having sent two assassins to kill him, a plot which is foiled by the divinity; Jesus is taken up into nirvana, we would say, before the assassins can act, but they themselves are changed into the likeness of Jesus to be brutally murdered by the enraged masses.[45] My suggestion is that, as in this tradition Buddhist and other story elements are here attached to the story of Jesus to suit the preconceptions of the audiences, so Christian elements may have been attached to the account of the patriotic martyrdom of Wen T'ien-hsiang.

The *Hsin Shih* contains a truly surprising document in the introduction to a portrait inscription,[46] representing one of the literary genres, in which Wen's image has been transmitted. The introduction centers on the conception of a "numinous" soul, *ling*, which is the human being's share of "heaven" at birth. "Heaven" should be interpreted as a portmanteau term for the various notions of cosmos, nature, or god, according to context, since we are here in the sphere of Confucian-Taoist-Buddhist syncretism, with possibly a smattering of Near Eastern thought added. Our text further states that, so long as the mental and psychic endowment (*hsin*) of the person is developed under the guidance of the precepts of the "Saint and the Sage" (Confucius and Mencius), that is in terms of *li* (intelligible principle), it will be free from privatistic/subjective disfiguration and thus achieve adjustment to larger social concerns. Then, the attitude to life (as the Buddhist ideal of detachment has it) will no more meet new life with joy, nor death with anguish; unworthy people, on the other hand, do not succeed in keeping their share of "heaven" pure and unsullied.

As the animate being (*wu*) undergoes the vicissitudes of life, it may emerge largely unaffected; so long as such is the case, there will be no cause for concern. It is when a 'being' can no longer be ascertained in this process, that we find ourselves in the presence of a truly spiritual being. We are here returning to the question that came to stand, in 1843, before the officials concerned with Wen T'ien-hsiang's induction into the Confucian cult. Since an existence in true conformity to these processes, in the language of the *Hsin Shih*, is accessible only to "Saints and Sages", the question before the officials was exactly whether or not Wen had attained just that level of spirituality. For the *Hsin Shih*, a sure sign for such election was whether the consciousness of a person became completely transparent unto itself. Without this virtue, which is most usually named *ch'eng*, or "sincerity" in Confucian thought, there could not be such a moral attitude as filial devotion and political loyalty, touchstones of morality per se in this sphere of thought. As to the heroic spirit of Wen T'ien-hsiang, there was no question, according to our text, that Wen had attained to this level of "Saints and Sages". He had held out steadfastly through innumerable dangers and threats, had reached back to the limit of his abilities, without however totally exhausting them. He had given the ultimate in loyalty to his ruler and to his parents; he had opened up the 'heavenly element' in the relationship of ruler and servant, parents and sons, and put it into action.

Wen's canonisation in the *Hsin Shih*, anticipating, but going beyond, that by the Ch'ing court, appropriately involves the description of miracles. A special level of consciousness had been attained, where ordinary humans facing their common-life problems, were capable of employing their true potential; they could master all vicissitudes of life as easily as carts could go over land, and ships could cross the waters. True, our text says, people could accomplish this only for moments, not for any length of time; some actually might succeed for some length of time, but not do so from the nature of "heaven" in them (*neng chiu chih, pu neng t'ien chih*). In a phrase that overextends the semantics even of classical Chinese, it says, while normal people were not capable of such a feat, Wen T'ien-hsiang had simply achieved it from the "heaven" in his being (*hsien-sheng t'ien-chih i*). There was in his case, no more question of ability, or of the length of time. Wen had attained the unmoved mind of Mencius;[47] and when he met his death, he died to receive a share in the *t'ien*, "heaven" of the "exalted illumination". The restoration attempt of the Sung, we are then told, had lasted only some days, but Wen's outstanding reputation was unlimited in time, like Heaven, Earth, the Sun and the Moon. It was an empty question, the text concludes, to ask, how an historiographical narrative, with its sequential representation, could encompass Wen's ethical and human qualities in mere words. The ensuing characterisation was written in this spirit, and indeed it need not be discussed at length, since it falls squarely into the tradition of evaluation which we have discussed at the beginning of this study. Nevertheless a word concerning the "heaven" and the "gods" of the "correct exalted illumination" is in order. That cue refers us to an episode represented in the biographical notes on Wen T'ien-hsiang by Teng Kuang-chien, and traceable to the introductory comments to one of two related poems of Wen's own collection.[48] Remarkably, Hsieh Ao, the "survivor of Sung" we have been discussing, included one of these poems in a small selection of poetry transmitted with his own writings.[49] According to these materials, Wen T'ien-hsiang, late in 1279, soon after he had been taken to the capital, met a man of the expressive name "Ling-yang-tzu" (approximately "numinous power of light"), who is also referred to as "Great Old Lü" (Lü ta weng); Lü familiarised Wen with "the correct Dharma of the Exalted Illumination" (*ta kuang-ming cheng fa*). "King of the Exalted Illumination" was one of the names of the Buddha, as the King of the Jambudvipa; the "correct Dharma" is the Dharma which was to exist for five hundred years after the nirvana of the Buddha and was then to be extinguished. This "correct Dharma" is also characterised by the "exalted illumination", which the Buddha, according to tradition, bestows, with his teaching, to all living beings. The "exalted illumination", comes forth as the Buddha himself goes into the final transformation.[50] This is the decisive point. As Wen, who was reasonably well versed in Buddhism suggests, this introduction to the "correct Dharma" was a special preparation for meeting his death. It was not at all a Taoist elixir to make death "disappear", as a later commentator came to emphasize.[51] Then, Wen continues, he "felt as if released, in-between life and death, in a state of spiritual freedom".[52]

Wen's poems indicate that he absorbed the 'Buddhist' content of "Ling-yang-tzu's" message in the (essentially aesthetic) classical way of Chinese literati, which allows divergent philosophical concerns exist independently from each other in their minds. Wen also felt reminded, of his earlier days, when he toyed with the life-style of a cultivated eremitism. While in Wen's mind Confucian, Taoist, and Buddhist elements could exist side by side in this manner, without interference, our *i-min*, "survivor", Cheng was striving to find certainty in a synthetic and somewhat home-spun amalgamation of divergent strains of thought. Cheng's own being and existence was far less assured and established than Wen's own, and so it took Wen's life experience and his death to guide Cheng's own effort at vicarious living. In the end, Wen T'ien-hsiang, in Cheng's meditations, emerges as the figure of a saint, indeed as a savior.

To conclude: We have been moving here in a complex textual configuration, which, as far as the three "survivors" Wang Yen-wu, Hsieh Ao, and Cheng Ssu-hsiao are concerned, is mainly autobiographical. This characterisation applies equally to a large part of the writings of Wen T'ien-hsiang, beginning with his early days, but more outspokenly so since the inception of his career as a resistance leader. The autobiographical mode provides a medium for complex events of transmission, in which the larger cultural tradition, the literary personæ of Wen and his admirers are found in various patterns of interaction. While we may at first hesitate, as I did, to adopt Jean-Michel Oughourlian's terminology of "possession", "identification" etc, I would leave the reader with the thought that the relationships he has outlined in his recent book might be fruitful for a further interpretation of the writings I have presented here. We might take up further interpretive suggestions referred to selectively at the beginning of the present paper, from writings of E. Husserl, Deleuze and Guattari, Paul Ricoeur, and Christian Jambet.

The World Phenomenology Institute

NOTES

[1] Jean-Michel Oughourlian, *Un mime nommé desir: Hysterie, transe, possession,*

adorcisme (Paris: Bernard Grasset, 1982); I will be referring to concepts developed on pp. 13—37, 47—49, 129—132, 159—169, 177—184, 228—232.

2 Edmund Husserl, *Vorlesungen zur Phänomenologie des inneren Zeitbewusstseins*, ed. by Martin Heidegger, 2nd edition (Tübingen: Max Niemeyer, 1980), p. 389.

3 Gilles Deleuze and Felix Guattari, *Capitalisme et Schizophrenie: Mille Plateaux* (Paris: Les Editions de Minuit, 1980), pp. 141—170.

4 Paul Ricoeur, *Temps et récit*, Vol. 1 (Paris: Éditions du Seuil, 1983), pp. 85—129.

5 For the concept of 'meta-history' drawn on here, see specifically Christian Jambet, *La Logique des Orientaux. Henry Corbin et la science des formes* (Paris: Éditions du Seuil, 1983), pp. 288—297. A pertinent discussion of the 'meta-historical' mode in Chinese historiography was presented in David S. Nivison, *The Life and Thought of Chang Hsüeh-ch'eng (1738—1801)* (Stanford: Stanford University Press, 1966), esp. chapters 6 and 8.

6 Joseph R. Levenson, *Confucian China and Its Modern Fate*. Vol. 3: *The Problem of Historical Significance* (Berkeley: University of California Press, 1965), pp. 124—125.

7 Pierre Vidal-Naquet, "Formes d'activité politique dans le monde juif principalement aux environs du Ier siècle de notre ère", *Les Juifs, la mémoire et le présent* (Paris: François Maspero, 1981), p. 41 (English text in *Yale French Studies* **59**, 1980, 105).

8 Gershom Scholem, *Major Trends in Jewish Mysticism* (New York: Schocken Books, 1946), pp. 349—350. It may be worth noting that Scholem, like Levenson and Vidal-Naquet used the story in the function of an epilogue at the end of the works cited; here it would be in the place of a (second) prologue.

9 A summary of the milestones in Wen T'ien-hsiang's career can be found in H. W. Huber, *s.v.*, *Sung Biographies*, ed. by Herbert Franke (Wiesbaden: Franz Steiner Verlag, 1976), Vol. 3, pp. 1187—1201. Cf. Li An, *Wen T'ien-hsiang shih-chi k'ao* (Taipei: Cheng-chung shu-chü, 1972), pp. 187—191 for one of many descriptions of the memorial temple in Peking. The complex bibliography of the *Rezeptionsgeschichte* of Wen has been dealt with thoroughly by William Andreas Brown, *The Biography of Wen T'ien-hsiang in the Sung Shih* (Diss. Harvard University, 1962), of which a printed edition has been announced in Taipei.

10 The historical context of the intellectual currents that swept the thought of Chu Hsi into its later position as an unshakeable orthodoxy has been explicated by John W. Haeger, "The Intellectual Context of Neo-Confucian Syncretism", *The Journal of Asian Studies* **31** (1972), 499—513; James T. C. Liu, "The Road to Neo-Confucian Orthodoxy: An Interpretation", *Philosophy East and West* **23** (1973), 483—505.

11 The following considerations have been given a more restricted treatment summarizing the main documents in Horst Huber, *Wen T'ien-hsiang (1236—1283): Vorstufen zum Verständnis seines Lebens* (Diss. Universität München, 1983), pp. 8—22.

12 In the course of time, the relevant documents came to be included among the prefatory materials of several of the editions of Wen's collected works; more recently the material was anthologized by Hsü Hao-chi, *Wen Wen-shan ch'uan-hsin lu* (Wu-hsing, 1932; repr. Taipei: Chung-hua shu-chü, 1972), ch. 4, pp. 7b—16b; an official summary is reproduced in Yang Te-en, *Wen T'ien-hsiang nien-p'u* (Ch'ang-sha: Shang-wu yin-shu kuan, 1939), pp. 360—362.

13 A more specific, but still preliminary, discussion of Wen T'ien-hsiang's significance for the development of Chinese loyalty concepts was attempted in Horst Huber, *Wen T'ien-hsiang*, pp. 10—13 and 250—259. The victimisation of Confucian 'dissidents' towards the end of the Ming was discussed by Charles O. Hucker, "Confucianism and

the Chinese Censorial System", *Confucianism in Action*, ed. D. S. Nivison and A. F. Wright (Stanford: Stanford University Press, 1960), pp. 208; and id., *The Censorial System of Ming China* (Stanford: Stanford University Press, 1966), pp. 284—285.

[14] The selections are summarized and their contexts interpreted in Huber, *Wen T'ien-hsiang*, pp. 15—17, and identified, pp. 262—243.

[15] This case of intertextual 'grafting' is discussed in Huber, *Wen T'ien-hsiang*, pp. 347—348.

[16] The prefaces by Ou-yang Hsün (1283—1357) and Chieh Hsi-ssu (1274—1344) precede Wang Yen-wu's collection, *Wu-wen kao* (ed. Ssu-pu ts'ung-k'an); Wang's "Devotion", "Sheng chi Wen ch'eng-hsiang wen", ibid., ch. 4, was also printed in Yang Te-en, *Wen T'ien-hsiang nien-p'u*, pp. 278—282.

[17] Wang Yu-sun, "Sheng chi Wen ch'eng-hsiang hsin-kuo-kung wen", found in *Chi-an fu-chih* (1876), ch. 49, p. 60a; a text preserved in Wen's *Chih-nan hou-lu*, in *Wen-shan hsien-sheng ch'üan-chi* (ed. Ssu-pu ts'ung-k'an), ch. 15, p. 5a, is characterised in this way in a biographical text concerning Wen's wife, née Ou-yang, which is part of the addenda to Wen's *Chi-nien lu*, also in *Wen-shan ch'üan-chi*, ch. 17, p. 46a.

[18] This is an historical hypothesis, which will require a separate study to clarify and document; it has been discussed in a preliminary way in Huber, *Wen T'ien-hsiang*, pp. 33, 88 and 281, 320.

[19] The death of Po I and Shu Ch'i, in the 'political' interpretation was in protest against the overthrow of the Shang monarchy by the Chou; but in an anthropological reading they are seen as the representatives of a tribe of people that was extinguished as the result of a campaign of forest clearance, a tribe in whose name the word "fox" has been recognised (Ku-chu); the tradition relating to the fox is also linked to the name of the mountain Shou-yang. Cf. the discussions of Ssu-ma Ch'ien's *Shih-chi*, ch. 61, in Burton Watson, *Ssu-ma Ch'ien: Grand Historian of China* (New York: Columbia University Press, 1958), p. 239, and on Po I and Shu Ch'i generally, Wolfram Eberhard, *Lokalkulturen im alten China*, vol. I: *Die Kulturen des Nordens und Westens* (Leiden: Brill, 1942; Suppl. to *T'oung-pao* 37), pp. 36—37, 41—44, 50. The proverbial statements on the fox occur in *Ch'u-tz'u, Chiu-chang*, v. 32 of "Ai-ying", *Ch'u-Tz'u* (ed. Ssu-pu ts'ung-k'an) ch. 4 p. 18a (D. Hawkes, *Ch'u Tz'u*, p. 67), in *Li Chi, T'an Kung*, part 1 (S. Couvreur, *Mémoires sur les bienséances et les céremonies*, repr. Paris: Cathasia, 1950, Vol. I, p. 131) and other early texts.

[20] The autobiographical statement concerning Wen's attempt to starve himself, in *Chih-nan hou-lu, Wen-shan ch'üan-chi*, ch. 14, pp. 2b—8b; in *Yin-hsiao chi*, ibid., ch. 15, pp. 5a—6b; in *Chi Tu shih*, ibid., ch. 16, pp. 4a—25b; it is further reported in the surviving portions of the Wen-biography by Teng Kuang-chien, which are appended to *Chi-nien lu, Wen-shan ch'üan-chi*, ch. 17, pp. 38a—39b, where Teng also quotes from a letter by Wen to his brother, in which the autobiographical account is repeated. Cf. also note 48 below. The episode in the *Tso Chuan* concerns Shen Pao-hsü, also one of the prototypical heroes in the loyalist tradition, 4th year of Ting Kung, *Ch'un-ch'iu ching-chuan yin-te* (Peking: Harvard-Yenching Institute, 1937), vol. I, p. 466. The two most famous cases of suicide by starvation among the loyalists at the end of the Sung dynasty are Ch'en Wen-lung, *Sung Shih* (Peking: Chung-hua shu-chü, 1977), ch. 451, p. 13279, and Hsieh Fang-te, ibid., ch. 425, p. 12690; Ch'en's death occurred during his transport into captivity, whereas Hsieh starved himself to death, having been taken to the capital in an effort to press him into service; the similarity of these situations with those Wen had to go through is quite striking.

[21] The references and role-models have been identified an interpreted briefly in a tentative *explication du texte* of Wang's "Devotion", in Huber, *Wen T'ien-hsiang*, pp. 24—38, 273—284.

[22] This second "Devotion" by Wang Yen-wu is found in *Wu-wen kao* (ed. Ssu-pu ts'ung-k'an), ch. 4, pp. 10a—11a; also in Yang, *Wen T'ien-hsiang nien-p'u*, pp. 333—334.

[23] Fu Lo-shu, "Teng Mu, A Forgotten Chinese Philosopher", *T'oung Pao* **52** (1965), 46—59, 81—85 gives a characterisation of Hsieh Ao and his loyalism in traditional terms.

[24] David Hawkes, "The Quest of the Goddess", *Studies in Chinese Literary Genres*, ed. by Cyril Birch (Berkeley: University of California Press, 1974), pp. 42—68. The 'epic-elegiac' poem *Li Sao*, in *Ch'u Tz'u* (ed. Ssu-pu ts'ung-k'an), ch. 1, pp. 1a—49a; David Hawkes, tr., *Ch'u Tz'u. The Songs of the South: An Ancient Chinese Anthology* (London: Oxford Unviersity Press, 1959), pp. 21—34.

[25] "Hsi-t'ai t'ung-k'u chi chu", in *Sung i-min lu*, ed. by Ch'eng Min-cheng (ed. Chih-pu-tsu chai ts'ung chu), ch. 3, pp. 1a—4a; a reading of this piece has been attempted in Huber, *Wen T'ien-hsiang*, pp. 43—52, 287—295.

[26] The chronological problems are discussed, ibid., pp. 45—48.

[27] See Fu Lo-shu, "Teng Mu", p. 48, fn. 2.

[28] For this point, see Huber, Wen T'ien-hsiang, pp. 50—52; also J. J. M. de Groot, *The Religious System of China* (repr. ed. Taipei: Ching-wen, 1964), vol. I, pp. 248, 253—254.

[29] "Chao Hun" and "Ta Chao", in *Ch'u Tz'u* (ed. Ssu-pu ts'ung-k'an), ch. 9—10; Hawkes, *Ch'u Tz'u*, pp. 103—114.

[30] Edward H. Schafer, *The Vermilion Bird. T'ang Images of the South* (Berkeley: University of California Press, 1967), pp. 123—125, 259—265. Cf. now David W. Pankenier, "Astronomical Dates in the Shang and Western Chou," *Early China* **7** (1981—1982), pp. 7—9, 13—14.

[31] On this dialectic, cf. Joseph R. Levenson, "*T'ien-hsia* and *kuo*, and the Transvaluation of Values", *The Far Eastern Quarterly* **11** (1952), 447—457, and idem, *Confucian China and Its Modern Fate: The Problem of Intellectual Continuity* (Berkeley: University of California Press, 1958), pp. 95—104.

[32] Chang's notes are printed as an interlinear commentary with Hsieh Ao's text, see above, note 25; the second poem by Hsieh is the "Tung-ch'ing shu yin chu", *Sung i-min lu* (ed. Chih-pu-tsu chai ts'ung-shu), ch. 6, pp. 8a—8b.

[33] The 'hermeneutic' situation of these appended comments, with some of their highlights, is discussed in Huber, *Wen T'ien-hsiang*, pp. 53—60, 295—299; the list of the commentators who emphasise Chang's role, ibid., p. 298, n. 129. Chang also authored one of the appended notes; it follows immediately after Hsieh's text with Chang's commentary, *Sung-i min-lu*, ch. 3, pp. 4a—5a.

[34] This is Liu Yung, ibid., pp. 12b—13a.

[35] (*T'ieh-han*) *Hsin Shih* (repr. Taipei: Shih-chieh shu-chü, 1955 and later).

[36] William Theodore de Bary, *Neo-Confucian Orthodoxy and the Learning of the Mind-and-Heart* (New York: Columbia University Press, 1981). The emphasis on *hsin* is particularly notable in the Mencian strain; cf. Tu Wei-ming, "The Idea of the Human in Mencian Thought. An Approach to Chinese Aesthetics," in *Theories of the Arts in China*, ed. by S. Bush and C. Murck (Princeton: Princeton University Press, 1983), pp.

63—66. For aspects developed in *Hsün-tzu* and the syncretistic work *Kuan-tzu*, cf. Ru Xin, "Mind and Consciousness in Chinese Philosophy: A Historical Survey", *Analecta Husserliana*, Vol. 17 (1984), pp. 77—86.

[37] *Hsin Shih*, preface by Ou-yang chien, p. 2a.

[38] For Cheng Ssu-hsiao's biography in outline, see Chu-tsing Li, s.v. *Sung Biographies*, ed. by Herbert Franke (Wiesbaden L Franz-Steiner-Verlag, 1976), Vol. IV, *Painters*, pp. 15—23; for the attribution problems of the *Hsin Shih*, also John D. Langlois, Jr. "Introduction", *China under Mongol Rule*, ed. by J. D. Langlois, Jr. (Princeton: Princeton University Press, 1981), p. 17, n. 41.

[39] Cf. the references to publications by Charles O. Hucker above, note. 13.

[40] The biographical summary on Wen T'ien-hsiang, *Hsin Shih*, Hsia, pp. 7a—15b is also reprinted in Yang, *Wen T'ien-hsiang nien-p'u*, pp. 363—371. The historical summation, "Ta-i lüeh-hsü", or "Summary outline of the High Principle of Justness", *Hsin Shih*, Hsia, pp. 44a—85a.

[41] Ch'en Hung-hsü, *Han-yeh lu* (ed. Hsüeh-hai lei-pien), Chung, pp. 9—11a.

[42] For the Chinese system of such correspondences, cf. Manfred Porkert, *The Theoretical Foundations of Chinese Medicine: Systems of Correspondence* (Cambridge: The MIT Press, 1974).

[43] *Chao-chung lu* (ed. Ts'ung-shu chi-ch'eng); the work was characterised by Herbert Franke, in *A Sung Bibliography* (*Bibliographie des Sung*), ed. by Y. Hervouet (Hong Kong: The Chinese University Press, 1978), p. 124. The notice on Wen T'ien-hsiang and other resistance leaders, pp. 17—22, was also reprinted in Yang, *Wen T'ien-hsiang nien-p'u*, pp. 375—381.

[44] Cf. Wolfram Eberhard, "Studies of Near Eastern and Chinese Folktales" *Sinologica* **1** (1947), 144—151, and particularly Herbert Franke, "Der gerechte Richter. Eine vorderasiatische Anekdote in chinesischem Gewande," *Asiatische Studien* **4** (1950), 55—59, and "Westöstliche Beziehungen im Zeitalter der Mongolenherrschaft," *Saeculum* **19** (1968), 91—106.

[45] *A General History of the Turks, Moguls, and Tatars*, Vol. II: *An Account of the Present State of Northern Asia* (London: J. and J. Knapton, J. Darby et al., 1729), pp. 478—480.

[46] *Hsin Shih*, Hsia, pp. 40—41a; cf. the somewhat more detailed reading in Huber, *Wen T'ien-hsiang*, pp. 76—81, 310—313.

[47] It is important to note that the reference to the "unmoved mind" (*pu tung chih hsin*), in this context is tantamount to a reference to the "overbounding spirit-substance" (*hao-jan chih ch'i*), Mencius' perfection of all the arts of attaining the unmoved mind; in Neo-Confucianism (Ch'eng I and Chu Hsi), *Hao-jan chih ch'i* came to be identified with the "Spirit-substance of Correctedness" (*Cheng ch'i*). Through Wen T'ien-hsiang's death and his poem *Cheng-ch'i ko, Hao-jan chih ch'i* and *Cheng-ch'i* came to be closely tied to the tradition of 'suicidal' loyalism.

[48] The surviving passages from Teng Kuang-ch'ien's account of Wen T'ien-hsiang are preserved in Wen's 'autobiographical' dossier, *Chi-nien lu*, identified as parts of the appended materials under the years *wu-yin* (1278), *chi-mao* (1279), *jen-wu* (1282). Teng (cf. also Huber, *Wen T'ien-hsiang*, p. 122, 476—477) was commissioned by Wen to write the text of his funeral inscription (*Chi-nien lu*, in *Wen-shan ch'üan chi*, ch. 17, *hsin-ssu* (1281), pp. 40b), and it is gratifying to note that such an inscription has been unearthed in Chi-chou in 1980, see Ch'en Po-ch'üan, "Yüan chih-yüan erh-shih-i nien

[1284] Wen T'ien-hsiang mu-chih ming", *Wen shih* **17** (1983), p. 240. This find corroborates Teng Kuang-chien as an authoritative source on Wen's life, as will be shown in a forthcoming study. The episode we are concerned with is recorded in the *Chi-nien lu* under *chi-mao* (1279), addenda, pp. 38b; the poems and notes of Wen's are in *Yin-hsiao chi, Wen-shan hsien-sheng ch'üan-chi*, ch. 15 pp. 10b—11a; cf. Ch'en Yen-chih, *Wen Wen-shan shih chu* (Ch'ang-sha: Shang-wu yin-shu-kuan, 1939), pp. 307—308.

[49] Hsieh Ao's selection, "T'ien-ti chien chi", in *Hsieh Kao-yü hsi-fa i-chi* (ed. Kuo-ts'ui ts'ung-shu), p. 1b.

[50] Some specific references for the Buddhistic terminology adduced here are given in Huber, *Wen T'ien-hsiang*, p. 311, notes 197, 198.

[51] Ch'en Pang-chan, in *Wen Wen-shan ch'uan-hsin lu*, ch. 7, pp. 18a—18b.

[52] The phrase is from the poem of Wen's quoted above in note 48; it is a further indication that the feeling of existing "in-between" life and death was a fairly frequent component of Wen T'ien-hsiang's life experience; cf. my previous paper included in *Analecta Husserliana*, Vol. 19 (1985), pp. 101—128.

HWA YOL JUNG

THE PIETY OF THINKING: HEIDEGGER'S PATHWAY TO COMPARATIVE PHILOSOPHY

And the fire and the rose are one.
— T. S. Eliot

Uniform ideas originating among entire peoples unknown to each other must have a common ground of truth.
— Giambattista Vico

Only connect!
— E. M. Forster

By "way," or "how," we mean something other than manner or mode. "Way" here means melody, the ring and tone, which is not just a matter of how the saying sounds. The way or how of the saying is the tone from which and to which what is said is attuned.
— Martin Heidegger

I

The influence of Heidegger on twentieth-century thought has been immense, pervasive, and immeasurable. As he sought and brought about a change from theology to literary criticism, he is truly the maker of a philosophical epoch. Once the young German critical theorist Jürgen Habermas, though he was by no means totally or unqualifiedly sympathetic to Heidegger's thought, summarily acknowledged Heidegger as the most influential thinker since Hegel. Heidegger intends to deconstruct or subvert the hegemony of the metaphysical or logocentric tradition of the West from Plato to Nietzsche. By deconstruction, Heidegger means to emphasize a critical procedure in which the accepted concepts are traced back to their sources for the sake of reconstruction. His deconstructive thought poses challenges for us who are living in an epoch which has experienced the globalization or planetarization of Western science and technology, that is, the global domination of what Heidegger himself calls "calculative thinking."

Of course, there has been a long German tradition in the compara-

A-T. Tymieniecka (ed.), Analecta Husserliana, Vol. XXI, 337–368.

tive study of intercultural thematics from Humboldt and Herder to Forke. Since Leibniz, however, there has never, I think, been a mood of Oriental *étrangisme* in philosophy more favorable than the one engendered by Heidegger. According to the legendary reporting by William Barrett of Heidegger's reaction to reading D. T. Suzuki, Heidegger was trying to say exactly what Suzuki said in his writings. It is Japan — Suzuki's birthplace — that has been most receptive to Heidegger in Asia.[1] Most importantly, Heidegger's famous 1953—54 "A Dialogue on Language" was occasioned by the visit of a Japanese scholar that shows the filiation between Heidegger's and Japanese thought toward the affirmation of poetic language as well as his favorable disposition for and familiarity with Japanese indigenous thought. Heidegger expressed his bewilderment that the Japanese forget the beginnings of their own thinking and chase after newest development in European thought.

II

In the wide spectrum of phenomenology as a philosophical movement, Karl Jaspers and Martin Buber wrote extensively on Oriental thought. However, it is most telling for our purpose to elaborate on the openness of Maurice Merleau-Ponty's views regarding Chinese and Indian thought and then to place Heidegger's thought beyond Merleau-Ponty's framework. Of interest in Merleau-Ponty's discussion is not his knowledge of Oriental thought, but the contrast between him and Hegel and Husserl on this subject.

For Merleau-Ponty, Oriental thought is both suggestive and instructive. As he recounts, Hegel viewed Oriental thought in a cavalier fashion. Oriental thought for Hegel is in a perpetual state of "immature" childhood. It is neither philosophy nor religion, since, on the one hand, it is not open to absolute and universal knowledge, its culture being bound by its own assumptions, and on the other, it does not presuppose the principle of freedom and individuality. For Hegel, philosophical truth as absolute and universal knowledge is certified by the Occidental seal of approval alone. The Orient represents the perpetuation of the eternal yesterday; Oriental thought is a philosophical infantilism or paleography in the progression of world history.

Husserl, too, seems to set out on Hegel's path in his judgment of Oriental thought by affirming the "privileged" position of Western philosophy — the entelechy of philosophy and humanity based on Greek *theoria* which is absent in Oriental thought. Husserl traces the origin of Western philosophy in the "theoretical" attitude called *theoria* found in the thought of Plato and Aristotle. Devoid of any practical interest, *theoria* is the fully disinterested way of seeing the world. As he put it in his Vienna lecture of 1935, "Within European civilization, philosophy has constantly to exercise its function as one which is archontic for the civilization as a whole."[2] "Man," he continued,

> becomes gripped by the passion of a world-view and world-knowledge that turns away from all practical interests and, within the closed sphere of its cognitive activity, in the times devoted to it, strives for and achieves nothing but pure *theoria*. In other words, man becomes a non-participating spectator, surveyor of the world; he becomes a philosopher; or rather, from this point on his life becomes receptive to motivations which are possible only in this attitude, motivations for new sorts of goals for thought and methods through which, finally, philosophy comes to be and he becomes a philosopher.[3]

Later, Husserl spoke of the test of European humanity bearing within itself an absolute idea by means of "a universal philosophy which grows through consistent apodictic insight and supplies its own norms through an apodictic method" rather than by merely being the "anthropological types" of China and India. Although the common features between Oriental thought and Greek philosophy are evident, for Husserl the essential distinction lies in the fact that the purely theoretical attitude has its historical origin in Greece, and the perfecting of that attitude has become the unending task of the successive generations of Western inquirers.

For Merleau-Ponty, however, there is a fundamental difference between Hegel and Husserl in their respective views of Oriental thought. Even if Husserl, like Hegel, retains the privileged position of Western philosophy, "he does so not by virtue of its right to it — as if its possession of the principles of all possible cultures were absolutely evident — but in the name of a fact, and in order to assign a task to it."[4] That is to say, for Husserl the privileged position of Western philosophy is not simply proclaimed but must be proven and witnessed.

For Merleau-Ponty, moreover, all thought is part of the life-world (*Lebenswelt*) as a sociocultural world; all philosophies are anthropological types and none has any special right to the monopoly of truth. Husserl admitted the value of "primitive cultures" for an understanding of our own type of life-world and the meaning of the life-world as the invariant form of generality. "If Western thought is what it claims to be," Merleau-Ponty contends, "it must prove it by understanding all 'life-worlds.'"

For Merleau-Ponty, the arrogant path of Hegel that excludes Oriental thought from absolute and universal knowledge and draws "a geographical frontier between philosophy and non-philosophy" also excludes a good part of the Western past. Philosophy as a perpetual beginning is destined to examine its own idea of truth again and again because truth is "a treasure scattered about in human life prior to all philosophy and not divided among doctrines." Thus the life-world — and its different versions both Occidental and Oriental — is the source from which truth emerges. If so, Western philosophy is destined to reexamine not only its own idea of truth but also related matters and institutions such as science, economy, and politics. Merleau-Ponty declares:

> From this angle, civilizations lacking our philosophical or economic equipment take on an instructive value. It is not a matter of going in search of truth or salvation in what falls short of science or philosophical awareness, or of dragging chunks of mythology as such into our philosophy, but of acquiring — in the presence of these variants of humanity that we are so far from — a sense of the theoretical and practical problems our institutions are faced with, and of rediscovering the existential field that they were born in and that their long success has led us to forget. The Orient's "childishness" has something to teach us, if it were nothing more than the narrowness of our adult ideas. The relationship between Orient and Occident, like that between child and adult, is not that of ignorance to knowledge or non-philosophy to philosophy; it is much more subtle, making room on the part of the Orient for all anticipations and "prematurations." Simply rallying and subordinating "non-philosophy" to true philosophy will not create the unity of the human spirit. It already exists in each culture's lateral relationships to the others, in the echoes one awakes in the other.

In this search of truth, nothing should be taken for granted or prejudged. It is just here that Merleau-Ponty makes a decisive break with Hegel. As childhood and adulthood are one inseparable ontological order of man, so Oriental and Occidental cultures are one integral order of humanity which together point to philosophical universality. For we learn as much from primitive cultures as from modern cultures regarding the human condition. Merleau-Ponty declares that "There is not *a* philosophy which contains all philosophies; philosophy as a whole is at certain moments in each philosophy. *To take up the celebrated phrase again philosophy's center is everywhere and its circumference nowhere*" (my italics). In the final analysis, for Merleau-

Ponty the Orient must also have a place in the museum of philosophies to celebrate its hitherto "secret, muted contribution to philosophy." He writes, "Indian and Chinese philosophies have tried not so much to dominate existence as to be the echo or the sounding board of our relationship to being. Western philosophy can learn from them to rediscover the relationship to being and initial option which gave it birth, and to estimate the possibilities we have shut ourselves off from in becoming 'Westerners' and perhaps reopen them." In this spirit and ambiance, we shall judge the contribution of Heidegger's thought to comparative philosophy.

III

There are formidable, if not insurmountable, difficulties in comparing Heidegger's with Asian thought in a short article. First, it is obvious that Heidegger himself is a prolific and difficult writer. Second, there is the famous "turn" or "divide" (*Kehre*) between the early and the later Heidegger or, to use William Richardson's expression, "through phenomenology to thought." There is the "left-handed" (i.e., "sinister") compliment — certainly, in favor of our present study — that Heidegger's Being is really the Chinese *Tao* which has no comparable term in Western thought. Even concerning Heidegger's *Being and Time*, there are two diametrically opposed or conflicting interpretations: some view it as "historicist" because it fails to advance the timeless or eternal truth beyond all ages and cultures, whereas others conclude that it is "ahistoricist" because in it the structures of human reality are meant to be true of all ages and cultures. Third, there is a multitude of philosophical thought, methods, and approaches in Asia. Several writers have proposed the compatibility of Heidegger's thought with various aspects of Eastern thought. There have been important studies that compare Heidegger with individual Oriental thinkers: the most favorite comparison, for good reason, is between Heidegger and Kitaro Nishida. The broadest thematic category of East Asian thought comparable to Heidegger's thought would probably be "Sinism" — the title of the early and pioneering work of Herrlee Creel. "Sinism" was meant to accentuate what is peculiarly or uniquely Chinese in Confucianism, Taoism, and Ch'an or Zen Buddhism and was formulated to encompass the important aspects of philosophical, religious, and ethical thought of both Japan and Korea as well as the Chinese language or ideography.

From the perspective of Heidegger's early thought, the only legitimate way to comparative philosophy is the identification of the philosophical universe as "pluralistic." Heidegger indirectly intimated this when he expressed bewilderment that the Japanese chased after everything new in European thought. Why such a rush for (Western) truth unless they believe that it is superior to their own? Nishida — the most important twentieth-century Japanese philosopher who did more than any other thinker in integrating Western and Oriental thought — also believes in cultural pluralism in that the unity of a world culture would be based on the specificity of an individual culture. Parenthetically, it must be pointed out from the very outset that from the standpoint of Heidegger's thought there can be no separation between "truth" and "method" as there is equally no separation between language and thought on the one hand and between language and reality on the other. After all, Heidegger himself advances the notion in *Being and Time* that ontology is possible only as phenomenology. The separation between truth and method has been consummated with the rise of positivism or scientism; while the unitary philosophical claim that the methods of the natural sciences, especially physics, are paradigmatic to all truth-claims has been contested by Hans-Georg Gadamer in *Truth and Method*, which makes an effective argument within the context of Heidegger's philosophical hermeneutics.

Let us transpose the notion of what the one-time student of Heidegger, Hannah Arendt, calls "human plurality" (in her seminal work *The Human Condition*) onto cultural pluralism. According to Arendt, human plurality is the basic condition of both speech and action, and it has the twofold character of equality and distinction. The first refers to the existence of commonality among human beings and human cultures without which we would not be able to understand one another whether they be "primitive" and "modern," or "Western" and "Eastern." On the other hand, without distinction we would need neither speech nor action to make ourselves understood. This is in essence, I propose, the *principium* of "ontological difference" in Heidegger's thought. The true merit of this philosophical pluralism — the Heideggerian principle of "ontological difference" and the Chinese idea of the complementarity of the two opposites (*yin* and *yang*) — lies in a logic of correlation.[5] Following cues from Hayden White, I have also called this logic of correlation "diatactics." White wishes it to replace — appropriately, I think — the term "dialectics" in order to avoid the ideological overtone of Marx and the transcendental overtone of Hegel's thought: diatactics is neither "paratactical" (conceptually underdetermined) nor "hypotactical" (conceptually overdetermined). As a nonreductive method of correlation, diatactics confirms the complementarity of any two given disparate phenomena. It is also tactile to the extent to which it arouses the intimate sense of touch and accentuates the dialogic, pragmatic side of language-in-use. In this regard, the principle of diatactics, as we will later see, is fundamentally Heideggerian as well when Heidegger defines thinking as a "handicraft." Ultimately, in the logic of correlation — unlike the dialectics of Hegel and Marx — the positive (*yang*) and the negative (*yin*) are not resolved categorially or otherwise in terms of a higher synthesis but preserve the unending flow of their opposition as complementary. In short, the conceptual advantage that diatactics proffers is the following: it rejects the facile monism and dualism as well as the reductionism of all kinds which are likely to fall into the dichotomy of superiority and inferiority in grafting knowledge and action, mind and body, subjectivity and objectivity, language and reality, speech and writing, man and nature, and so on.

IV

Questioning, Heidegger declares, is the "piety of thinking" ("Frömmigkeit des Denkens"). Indeed, questioning has been the quintessence of phenomenology as a philosophical movement. Phenomenology is neither a school, a set of dogmas, a method, nor a technique in the same way that mathematics is a technique for the physical sciences today. The founder of phenomenology, Edmund Husserl — who is also Heidegger's mentor — himself conceives of philosophy as a perpetual beginning, as Socratic ignorance. For the "true thinker," philosophy never ceases to be a riddle and wonder to itself. As a vigilance, philosophy seeks the source of knowledge in the "things themselves." Husserl often used the term "introduction" to characterize a new beginning of phenomenology in his work. Merleau-Ponty was fond of using the term "interrogation" for his philosophical research to insist that the questioner himself is "always already" (*immer schon*) — to use Heidegger's expression — or implicated in the question he raises. To paraphrase Merleau-Ponty slightly in order to accent this genealogical spirit in phenomenological thought, the end of phenomenological philosophy is the account of its beginning. In essence, the true phenomenologist is necessarily a perpetual beginner. In this selfsame spirit, Heidegger too defines phenomenology as the possibility of thinking.

The "piety of thinking" is the thematic *abecedarium* of comparing Heidegger with Eastern thought in this paper. As the *principium* of this paper, it heralds the voice of poetic language in the conversation of humankind and ultimately acknowledges the "thinking" poet as the legislator of the world. By delimiting our discussion to the single theme, "the piety of thinking," we do not betray Heidegger's spirit but instead uphold it when he himself affirms that, "To think is to confine yourself to a single thought that one day stands still like a star in the world's sky."[6]

In the crisis of man and (his) humanity is foremost the crisis of thinking. We prefer the word "thinking" to "thought" for the simple reason that while the latter is a passive end-product, the former denotes the active process of unending activity: it is indeed "wayward," "on the way." It is, in other words, the *Tao* — the unending, often meandering "road." In *What Is Called Thinking*? Heidegger himself explains the twofold meaning of participles which "participate" or "take part" in two meanings in which the one refers to the other as a pair. In this case, thinking — like "blossoming" — is something thinking and the act of thinking. Be that as it may, what is most needed in our time is a way to think appropriately. There is interest in thinking today as it is engendered by thought-provoking events. Paradoxically, however, Heidegger rightly observes that in our age the most thought-provoking event is that we are still not thinking, a dangerous condition which leads itself readily to "the banality of evil" (to borrow a phrase from Hannah Arendt's controversial reporting of Adolf Eichmann after the model of Heidegger's anonymous "they" [*das Man*]). In such an impoverished state of thinking, we must first clarify the conditions of our thinking; this is the primary function of philosophy. The opposite of thinking or "thoughtfulness" is, of course, thoughtlessness. The "banality of evil" refers to the unprecedented phenomenon of evil deeds committed by Eichmann on a gigantic, hideous scale which cannot be traced to any wickedness, pathology, or ideological conviction but to the extraordinary shallowness which is neither monstrous or demonic but rather the "authentic inability to think." What Arendt is interested in pointing out here is not our ability to theorize abstractly or think in abstraction but our ability to make moral judgments based on the *sensus communis* (as in Socratic wisdom, Aristotelian *phronesis*, or Kantian moral judgment) as the abode of man's humanity. It is thinking as a natural necessity or duty of "examined" human life which is a faculty of every man who belongs to the species called human. To talk about thinking as the pillar of morality is not to neutralize it but to dig its *arche* itself: thoughtlessness is indeed a moral latitudinarianism. On the other hand, to moralize is to trivialize the essence of human thinking.

According to Heidegger, thinking as autonomous activity is distinguished from both "knowing" (in an epistemological sense) and "acting" (in a pragmatic sense). For it neither produces solely epistemological truths nor is merely instrumental to action whose manifest aim is measured by its pragmatic cash value. Rather, "knowing" and "acting" presuppose "thinking." Thinking as *aletheia* is presupposed in both knowledge and action. As such it is neither scientific, technical, nor utilitarian. In short, genuine thinking is not "calculative." The alternative to calculative thinking (*rechnendes Denken*) is for Heidegger meditative thinking (*besinnliches Denken*). In the following pages, I will try to show the "meaningful" nexus of thinking as *poiesis, praxis*, and *techne* in the original Greek sense of the terms.

In Heidegger's ontology, thinking and language are inseparable: language is the house of Being whose guardian man is. We "always already" find ourselves in the midst of language: we are attuned to language as reality rather than as a surrogate of reality. Let us call man's attunement to language "linguisticality" (*Sprachlichkeit*) — that which makes thinking possible. By linguisticality we denote the following four inseparably related properties: (1) language is intrinsic to human specificity as social; (2) it is inseparable from our conception of reality, that is, it is not a mirror to reflect, copy, or represent it but is reality itself; (3) it is the primary medium of human communication; and (4) it is an embodied phenomenon. In this connection, the most revealing challenge Heidegger poses for comparative philosophy in relation to thinking and language concerns whether Western languages are ineluctably tied to Western metaphysical thinking or instead offer other possibilities of thinking. In Heidegger's thought one would suspect that they are more or less two sides of the same coin.

For Heidegger, our attunement to language and thus to the world is not simply "rationalistic" in the traditional sense of Western metaphysics. Instead, language is an embodied phenomenon, for which reason, as I have already intimated, I use the term "diatactics." In the West, it was Vico's linguistic *autopsia* (or anatomy) — in the sense of seeking the evidence of the senses — which discovered language originally as embodied speech (*Sprachleiblichkeit*). For Vico, words are carried over from human bodies to signify nature where the "body language" animates inanimate things and from the properties of human bodies to signify the institutions of the mind, including, of course, what we call the "rhetoric" and "grammar" of language. For Heidegger, too, the hand is more than an organ, altogether it is peculiarly human, that is, the quiddity of being human. In Heidegger's *What Is Called Thinking*? occurs a marvelous passage concerning the many ways of the hand that culminate in thinking as a "handicraft" (handy craft) in the same way that the use of language — both speaking and writing — is the *techne* (art) of the body. It defies paraphrasing and is worth quoting in full:

> We are trying to learn thinking. Perhaps thinking, too, is just something like building a cabinet. At any rate, it is a craft, a "handicraft." "Craft" literally means the strength and skill in our hands. The hand is a peculiar thing. In the common view, the hand is part of our bodily organism. But the hand's essence can never be determined, or explained, by its being an organ which can grasp. Apes, too, have organs that can grasp, but they do not have hands. The hand is infinitely different from all grasping organs — paws, claws, or fangs — different by an abyss of essence. Only a being who can speak, that is, think, can have hands and can be handy in achieving works of handicraft. But the craft of the hand is richer than we commonly imagine. The hand does not only grasp and catch, or push and pull. The hand reaches and extends, receives, and welcomes — and not just things: the hand extends itself, and receives its own welcome in the hands of others. The hand holds. The hand carries. The hand designs and signs, presumably because man is a sign. Two hands fold into one, a gesture meant to carry man into the great oneness. The hand is all this, and this is the handicraft. Everything is rooted here that is commonly known as handicraft, and commonly we go no further. But the hand's gestures run everywhere through language, in their most perfect purity precisely when man speaks by being silent. And only when man speaks, does he think — not the other way around, as metaphysics still believes. Every motion of the hand in every one of its works carries itself through the element of thinking, every bearing of the hand bears itself in that element. All the work of the hand is rooted in thinking. Therefore, thinking itself is man's simplest, and for that reason hardest, handiwork, if it would be accomplished at its proper time.[7]

In Heidegger, therefore, thinking, speaking, and the hand form a filial unity. He unifies the sayings of Anaxagoras and Democritus which may appear contradictory at first glance: Anaxagoras, who said that man is intelligent because he has a mouth rather than hands, and Democritus, who said that human progress depends on the working of the hand rather than the mind. Heidegger's thought is indeed the diatactics of Anaxagoras and Democritus. As the hand is tactile, so are thinking and speaking. If thought is a handy craft, its opposite, thoughtlessness – which is pervasive in the modern world, as evidenced in the disembodied (alienated) language and behavior of Eichmann — is an infliction of cutaneous alagia, the condition of feeling no pain in the skin. The "thinking hand" or thinking as a "handy work," of which Heidegger speaks, confers upon us the work of the hand as embodied conduct. As such the hand is not just an "extension" of the body but is the body incorporated. The hand *is* the lived body; it is an organized "corporation." As embodied conduct, the hand is synesthetic. It activates the workings of the other senses, for example, hearing, seeing, saying, and singing. Indeed, it embodies the sociability of the senses.

The import of Heidegger's diatactics of thinking and language – especially speaking — is, I submit, enormous for comparative philosophy. As Heidegger himself is an etymological master, the etymological and paleographical study of Chinese ideography, known as "etymosinology," which began with Ernest Fenollosa and influenced later the poetry of Ezra Pound (e.g., the *Cantos*), should not be left unnoticed. Doubly important, as we will later see in detail, is Fenollosa's view of the Chinese ideogram as an authentic medium of poetry or the poeticized language par excellence. If the notion of "eco/nomy" betokens the genius and gift of Chinese ideograms, Chinese ideography is the exemplar of poetic language. Etymosinology is a dazzling display of the ideogrammic abracadabra by means of which Fenollosa tried to celebrate the meeting of the *yin* of the "markedly feminine" East and the *yang* of the "markedly masculine" West. Although the feminine and masculine allusions may not be totally appropriate, the correlative logic of the *yin* and the *yang* as one unity is brought close to Heidegger's "ontological difference."[8]

Chinese ideography — calligraphy in particular — is a kinetic art: it is the human body in graceful motion. And the Chinese revere the art of calligraphy as much as painting: calligraphy is the painting of ideograms or — to intimate the spiritual and personal dimension of artists themselves — "characters." In very significant measure, Chinese ideography is a choreography of human gestures and, as a family of signifiers, "a conversation of gestures" which, because of the presence of meaning, is not the reduction of ideograms to human physiology.

Picasso's *Swimmer* (1929) and *Acrobat* (1930) are two choreographs of the human body in fluent and rhythmic motion which are approaching ideography or calligraphy. They are, in short, dancing or frolicking anthropograms. In *The Principles of Art*, R. G. Collingwood observes that every language is a specialized form of bodily gesture and, as such, that dance is the mother of all languages. With the Vichian Samuel Beckett we can say that in language as gestures the spoken and the written are identical. Since the dance is the motion of the body in time and space, it may legitimately be claimed as the "birth-place" of all visual and auditory arts (e.g., painting and music). Moreover, Marshall McLuhan, who had the romantic vision of writing his antitypographic treatise, *The Gutenberg Galaxy*, in the ideogrammic medium, thought of the Chinese ideogram as a vortex of corporate energy — "corporate" in the twofold sense of being bodily and collective. Several years ago, the Japanese journalist Takao Tokuoka reported with humor in the editorial column of the *New York Times* the inventive word *towelket* — a shop sign written in Japanese phonetic characters (*kana*) — which is a combination of the two English words "towel" and "blanket" (towelket = towel + (blan)ket). What born Joyceans the Japanese are! Of course, what I have in mind is Joyce's talent or inventiveness for composing chordal vocabularies in *Ulysses* and *Finnegans Wake*. The Joycean "decomposition" or musicalization of the "abcedminded" world or the "popeyed world" of the "scribblative" has its structural parallel in Fenollosa's deconstruction or demystification of Chinese ideograms as simple or complex "corporations." Joyceanism or Joyce's unique stylistics stands out for its tremendous proclivities for making visible words chordal and for thinking in chords after the acoustic model of music, that is, for the production of chordal meanings by obliterating (auditory) content and (visual) form: thornghts (thorn + thoughts), rhythmatick (rhythm + mathematics), paupulation (paucity + population), evoluation (evolution + evaluation), cerebration (cerebrum + celebration), etc. If Fenollosa's etymological and paleographic deconstruction of Chinese ideograms (*kanji*) is correct, the composition of "towelket" is a result of the natural *habitus* of the Japanese mind — in fact, of the Oriental mind that is accustomed to the vorticism of Chinese ideograms. I find the etymolinguistics of Chinese ideography both enlightening and fascinating. Consider the following examples of "hypograms" as the corporate insemination of other ideograms provided by Fenollosa himself: "East" is an entangling of "sun" with

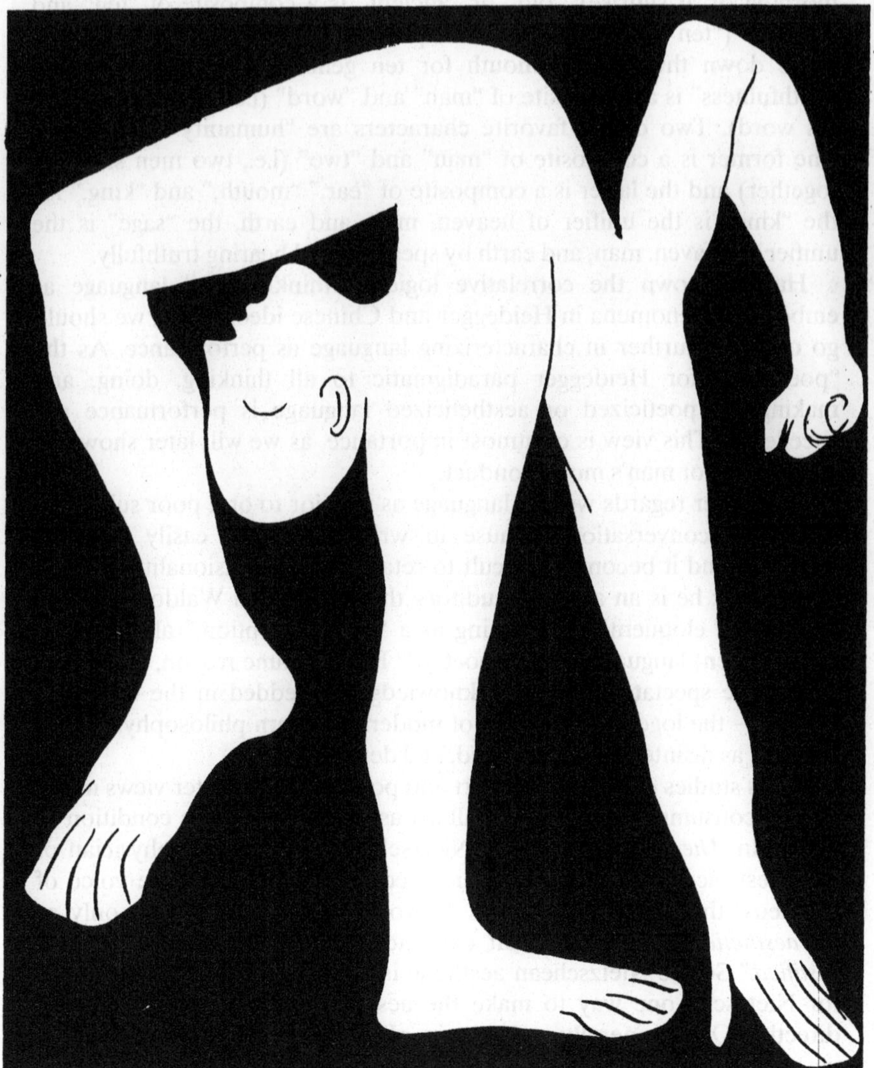

Picasso, *Acrobat* (Upside Down: A "Break Dancer"). © S.P.A.D.E.M., Paris/V.A.G.A., New York 1986.

"tree" (i.e., the sun entangled in the branches of a tree in the early morning or at sunrise); "old" or "ancient" is a composite of "ten" and "mouth" ("ten" *over* "mouth," i.e., presumably referring to what has come down through the mouth for ten generations); and "truth" or "faithfulness" is a composite of "man" and "word" (i.e., man standing *by* his word). Two of my favorite characters are "humanity" and "sage." The former is a composite of "man" and "two" (i.e., two men standing together) and the latter is a composite of "ear," "mouth," and "king." As the "king" is the unifier of heaven, man, and earth, the "sage" is the unifier of heaven, man, and earth by speaking and hearing truthfully.

Having shown the correlative logic of thinking and language as embodied phenomena in Heidegger and Chinese ideography, we should go one step further in characterizing language as performance. As the "poetic" is for Heidegger paradigmatic to all thinking, doing, and making, so poeticized or aestheticized language is performance par excellence. This view is of utmost importance, as we will later show, for accounting for man's moral conduct.

Heidegger regards written language as inferior to or a poor substitute for direct conversation because in writing thinking easily loses its flexibility and it becomes difficult to retain multidimensionality peculiar to thinking: he is an aural or auditory thinker.[9] Ralph Waldo Emerson, who spoke eloquently of thinking as a "pious reception," also thought that (written) language is fossil poetry.[10] For the same reason, Heidegger rejects the spectatorial view of knowledge embedded in the Cartesian *Cogito* — the logocentric kernel of modern Western philosophy, that is, thinking as disinterested, distanced, and detached.

In his studies of Renaissance art and poetry, Walter Pater views music as the "consummate art" in that all art aspires to attain the condition of music. In *The Birth of Tragedy*, Nietzsche — the last metaphysician of the West, according to Heidegger — echoes and rejoices the voice of Orpheus, the singer who moves "a world in darkness": "it is only as an *aesthetic phenomenon* that existence and the world are eternally *justified*." So the Nietzschean aesthetic is epitomized by music: music is for Nietzsche one way to make the aesthetic intelligible and grasp it directly: "Quite generally, only music, placed beside the world, can give us an idea of what is meant by the justification of the world as an aesthetic phenomenon."[11] The prevailing use of auditory metaphors in Heidegger's thought is congruent with his critique of the modern age as the age of "world picture" (*Weltbild*), his general conception of language, thought, and the world in relatin to Being, for example, "mood" or "dis/position" and "attunement" (*Bestimmtsein*). There is indeed the semantic kinship or filiation of the German words "to hear" (*hören*), "to hearken" (*horchen*), "to belong" (*gehören*), and "to obey" (*gehorchen*). Moreover, "hearing" and "obeying" have the same etymological root: the Latin *obaudire*, to listen from below, is to obey. In *Being and Time*, Heidegger conceives of mood (*Stimmung*) as a primordial, existential character of man. As a *factum brutum*, mood has an ontological bearing because it is a basic mode of "There-being" (*Dasein*). As *Befindlichkeit*, it is the basic way of "finding" ourselves in the world. Through it, we attune ourselves (i.e., "obey") to the world. As disposition, mood is not just the sensitive and emotional side of human consciousness, but the essential nature (*animus*) of man. "Mood," Heidegger writes, "is never merely a way of being determined in our inner being for ourselves. It is above all a way of being attuned, and letting ourselves be attuned, in this or that way in mood. Mood is precisely the basic way in which we are *outside* ourselves. But that is the way we are essentially and constantly."[12] No wonder, being *outside* ourselves, mood — both sorrowful and joyful alike — is highly contagious in a small circle of spouses and close friends and relatives. In short, mood is the *tonality* of *Dasein* as "Being-in-the-world" (*in-der-Welt-sein*).

The structure of oral poetry is indeed paradigmatic to language as performance. Emerson captures the performative spirit of language when he describes words as actions and conversely actions as a kind of words. In oral poetry, *poiesis, praxis*, and *techne* are rolled into one. In the genealogy of language (as communication), oral poetry is the "first language," and the oral poet is the "first born." The persona and (oral) poetry of Homer signify the dawn of Western thinking itself. Like Hermes, Orpheus, and even the Muses (the daughters of Mnemosyne or Memory), Homer was a singer or bard. The name of the god Hermes, in whose name hermeneutics celebrates the sacred and magical canon of his spoken words, stems from *herma*, the mute stone monument which also corresponds phonetically to the Latin *sermo* ("speech") — as we know, muteness or silence is the "mother tongue." Hermes is a *hermeneus* — a "linguistic mediator." According to legend, Hermes was the inventor of language and the "magician" of spoken words who was "spellbinding." According to the Homeric "Hymn to Hermes," the Hermetic legend is synonymous with the discovery of the universe (uni-verse) as the sounding orbit. Hermes was a herald (*keryx*): he was a virtuoso of sound-making whose *kerygma* (message) was aired by and encoded in the sound of spoken words. The term "herald" is related to the Latin *carmen* ("song") and the Sanskrit *karuh* ("to sing") and *karus* ("bard"), whose sole virtue scales the pitch of the "signifying excellence of voice." As the master of speech, Hermes was the messenger of Zeus; in his capacity as a herald he gave Pandora her voice. No wonder, then, this genuine Olympian was "the friendliest of the gods to men."

The tintinabulation of oral poetry is consonant with the poetic voice of Heidegger which is resonant with the poet Rilke: *Gesang ist Dasein* where *poiesis, praxis*, and *techne* — the sung word, the deed, and the art of word-making — come together as unity in the organization of existence or Being.[13] First, oral poetry is oral and auditory (i.e., acoustic), that is, in it the mouth and the ear are the main organs of communication. In Homeric oral poetry as shown in the works of Milman Parry, Albert B. Lord, Marshall McLuhan, Eric A. Havelock, Walter J. Ong, and others, speech and music are inseparable simply because it is sung. Second, in oral poetry there is no separation between composition and performance: composition is spontaneous performance. The oral poem — like jazz or any improvised music — is composed not for but in performance. Third, in the Homeric culture, the oral *techne* of poetry was the instrument of preserving the dynamic flow of a collective life. *Poiesis* was *techne*, and *mimesis* was the poetic technique of communication to preserve the acoustic effects (*pragmata*) of an oral culture. Poetry is an oral performance in the service of the acoustic effects of culture. The dynamic flow of this oral culture was preserved in the cornucopia of poetry, that is, in its repetition, redundancy, and verboseness. The Homeric epic was metaphorically called a "river of song." In essence, oral poetry is speech sung as performative utterances. When a person is making a performative utterance, he is doing something rather than merely saying something that may be true or false, that is, describing or reporting the state of affairs. In language as speech acts, "performance" and "promise" are no longer antithetical lexicons. The spoken word as performative utterance becomes an index of moral value once it is extended to the realm of human conduct: we are, morally or otherwise, what we say (as doing).

The Chinese syntax itself, as etymosinologists have tried to show, *is* action or performance. The Chinese cardinal moral virtue called "sincerity" is spelled syntactically "word-performed" (for which the Japanese "tragic heroes" from ancient to modern times sacrificed their lives). As it means what we say or perform what we promise in words, it personifies the Chinese conception of the circular unity of knowledge and action in the sense that knowledge is the beginning of action and action the completion of knowledge. There is, of course, a union or family of moral notions connected with sincerity — *jen*, piety, faithfulness, rite, etc. Faithfulness or fidelity literally means "man-standing-by-his-word," and it refers to the responsibility of the speaker to his word as ethical performance. The Confucian formulation of the "rectification of names" (*cheng ming*) — calling things by their right (rite) names — exemplifies the ethics of language-in-performance. In the *Analects*, Confucius said: "without knowing the (performative) power of words, it is impossible to know man." This affirms the following: the centrality of language to human conduct, the ethics of language embodying the humanity of man, and the spoken word as pre-scriptive. The following passages from the *Analects* express the same idea: (1) when the superior man "is heard to speak, his language is firm and decided"; (2) "The wise err neither in regard to their man nor to their words"; (3) "The virtuous will be sure to speak *correctly*, but those whose speech is good may not always be virtuous"; (4) "The superior man is modest in his speech, but exceeds in his actions"; and (5) friendship with "the glib-tongued" is injurious.

It should not be left unnoticed that in recent years Western grammatologists in literary theory have become fascinated with Chinese ideography because it is perceived mistakenly to have the character of "pure writing" (*écriture*) — the writing that exorcises the phonetic or is devoid of phonetic genealogy. They echo the voice of Oswald Spengler (the author of *The Decline of the West*). Spengler acknowledges writing as an entirely new kind of language that brings about a complete change in the conscious relationships and associations of men and women. The cardinal virtue of writing, as he sees it, is the liberation of human consciousness from the bondage and tyranny of the present embedded in speech. Writing resembles architecture, monument, pyramid, tapestry, cathedral, or the Acropolis where the immortal god of abstraction resides. It is the idea of writing as immortal that Spengler wishes to propagate. Timeless truths are embodied not in speech but only in script. He emphasizes that writing is "the grand symbol of the Far" — the Far East, perhaps — and the will to eternity. In search of the *karma* of the ideogram, the thirteenth-century Chinese scholar Tai T'ung wrote *Six Scripts* after, by his own admission, thirty years of arduous thinking and strenuous work. It is the exegesis of the six different etymological and paleographic roots of Chinese ideography whose tradition has been continued and preserved in the works of Bernard Karlgren. For our analysis here, we ought to mention only the "phonetic" principle or rootage of Chinese writing which traces the "origin" of written figures to spoken sounds. Tai is critical of those etymologists who knew nothing of the principle of phonetic composition. According to him, the written is preceded by the spoken in Chinese: writing makes speech visible. The invisible vapor of speech is distilled, as it were, in visible writing. Chinese ideography is a silent speech as performative. This does not mean, of course, that Chinese writing is reduced to the physiology of phonation. Rather, it only means that Chinese writing is, in important measure, the "abstract" form of the "concrete" speech. It is no accident that the Chinese expressions that denote linguistic activity contain the radical (spoken) "word." Correlatively, however, without the written, the spoken could not be represented to the eye. The circle of language, whose epicenter is sociality, is made up of the spoken *yin* and the written *yang* as complementary.

V

What is piety? How can piety be related to thinking in this comparative study? To answer these questions, we must return to the idea that the opposite of "thoughtlessness" is "thoughtfulness." Discussing Eichmann's real thoughtlessness which was led to the "banality of evil," Arendt contends that it is his inability to think from the standpoint of somebody else. That is to say, Eichmann was not thoughtful. He lacked receptive reciprocity for the thought of the other which is tantamount to the absence of genuine communication. "No communication," Arendt writes, "was possible with him [Eichmann], not because he lied but because he was surrounded by the most reliable of all safeguards against the words and the presence of others, and hence against reality as such."[14]

The definition of piety as receptive reciprocity is like the *yin* principle as contrasted with the *yang* principle of the *I Ching*. In his illuminating discussion of Greek and Roman religion, C. Kerenyi relates the Latin *pietas* to *religio.* In fact, some writers (e.g., Cicero) used the former synonymously with the latter. For the "ritually" performed act (rite) belongs to the sphere of *pietas.* By examining the legend of the temple which was built in Rome for the goddess Pietas, Kerenyi observes that

> On the site of this temple, so it was related, a mother had once been imprisoned and had been kept alive by the milk of her own daughter's breast. The story may have been adapted from a Greek original, though this is by no means certain. But it would have been pointless, had it not represented *pietas* in the ideal form in which it appeared to the Romans. The special thing which here stands out is something bodily and spiritual at the same time. *Pietas* here shows itself as a form of absolute reciprocity in nature, a completely closed circle of giving and receiving. In some variants of the story of mother's place is taken by the father. But the example thus revered is always this same natural circle of reciprocity. While *aidos* presupposes that one can also stand outside it, *pietas* as a matter of course unites those who give nourishment with those who in uninterrupted thankfulness return it, unites the source of life which its creatures from which its sources receive life. The *pii* and *piae* are completely enclosed in this circle.[15]

It is no accident that economists discuss "exchanges" as the reciprocal relations of economic life both primitive and modern, and modern sociologists and anthropologists recognize the "gift" as an anthropological specimen for obligatory reciprocity in human relations. Marcel Mauss's classic study of the "gift" of archaic societies such as Polynesia, Melanesia, and Northwest America is relevant to the understanding of reciprocity in its primordial form. As a form of obligation, the gift is the "necessary form of exchange" and "one of the bases of social life" as carried out by the people on the communal basis in the form of courtesies, entertainment, ritual, dances, feasts, military assistance, marriages, succession to wealth, etc.

For Heidegger, too, thinking as pious is receptive and reverential. It is a receptive response to the "call" or "voice" of Being: authentic thinking "hears" the voice of Being and "belongs" to it. What must be made absolutely clear from the outset concerning piety as reciprocity is this: like "mood" as the "attunement" or "being attuned" (*Bestimmtsein*), it is not egocentric (autocentric) but heterocentric (allocentric). For Heidegger as for Kerenyi, piety as absolute reciprocity approximates — as milking nourishes and is being nourished — the "touching" sense of intimacy and thus stands for the incorporation of the spiritual in the corporeal. For, after all, thinking is for Heidegger a "handicraft." There is in Heidegger as in Kerenyi — each in his own different way

— a filiality between the "piety of thinking" and "thanking." For Heidegger, the old German word *thanc* is memory, and thinking that recalls is thanking. Moreover, Cicero, who associates *pietas* with *religio*, describes the ears, not the eyes, as religious (i.e., the *aures religiosae* of the Attic orators). Although the scriptural words are the best guide for the understanding of God's divine purpose in the universe, St. Augustine, who knew Cicero's writings well and from whom we inherited some of Cicero's lost writings, thought that art such as music and painting contribute to it. In *De Musica*, however, he contended that music (invisible) is a better guide than painting (visible) for the attainment of man's religious spirituality and the understanding of the divine order. Thus, the former is a higher art than the latter. Piety as associated with "hearing" signifies "nearness," whereas *theoria* as associated with vision implies distance. It is no accident, therefore, that Heidegger, who regards "speaking" as the primary essence of language and resorts frequently to aural or auditory metaphors, relates reciprocity to the sense of "hearing," "hearkening," "belonging," and "obeying." Hearing is by nature socializing or reciprocating, whereas vision is inherently narcissistic and isolating.[16]

Piety as reciprocity is also genuinely Sinistic — Confucian, Taoist, and Zenish. Indeed, it is the thread that weaves all the Sinistic fabric of such important moral notions as humaneness, filial piety, propriety, and rite as "neighboring" concepts. Piety as reciprocity signifies the sacrament of coexistence or, as I call it elsewhere, "social principle."[17] The "social principle" has two distinguishable but inseparable elemental components: (1) homopiety and (2) geopiety. Homopiety refers to the convivial relationships between (wo)man and (wo)man, which may also be called "sociopiety," whereas geopiety is that part of the "social principle" which is concerned with the connatural reciprocity between (wo)man and nature or the earth.[18] Homopiety is dominantly a Confucian principle, whereas geopiety is a Taoist and Zen principle. One may hastily conclude, therefore, that the early "humanistic" Heidegger would be closely Confucian, whereas the later "ontological" Heidegger would be closely Taoist and Zenish. Besides the *Tao Te Ching*, to which Chung-yuan Chang has made numerous allusions to Heidegger's "new way of thinking," there are the sayings of Chuang Tzu: man is truly a "cosmion" — "Heaven and earth were born at the same time I was, and ten thousand things [i.e., Nature as material world] are one with me." However, Confucianism does not exclude geopiety, nor does Taoism and Ch'an or Zen Buddhism exclude homopiety. In *Li Chi*, Confucius writes: "To fell a single tree, or kill a single animal, not at the proper season, is contrary to filial piety." In this way, the moral objective of filial piety is not confined to the effect of one man's conduct on another but is extended to the effect of man's conduct on other nonhuman beings and things. Conversely, there is no shortage, in Taoism and Ch'an or Zen Buddhism, of allusions to human reciprocity. By the same token, it is wrong to associate the early Heidegger with "homocentricity" or the later Heidegger with the "end" or renunciation of man as such. After all, the "fundamental ontology" of *Being and Time* is the affirmation of man as an earth dweller: *homo* is rooted in *humus*.

Harmony, which is the pitch of musicality and the paradigm certainly of all "performing arts" and perhaps of the aesthetic itself, underscores the piety of thinking and thus the performances of both homopiety and geopiety. It is a "gathering" of many as an ordered whole: it is, musically speaking, a chorus, polyphonic chord, or orchestration of the differentiated many or, as the Chinese would say, "ten thousand things" — including the making of our thoughts. To gather our thoughts is to be thoughtful. Gathering denotes "nearness" which approximates auditory rather than visual space.[19] There is an episodic account by J. Glenn Gray of the German word Heidegger uses: *versammeln* which was translated, after considerable deliberation and with Heidegger's own approval, as "to gather." It is rooted in the old German *gattern* ("to couple," "to espouse," or "to join in marriage") which in turn was related to the Greek *to agathon* ("the good").[20] The Greek *logos* originates from *lego* ("to speak") and the root *leg-* is "to gather." *Logos*, like the Hebrew *dabhar*, originally meant "to gather," "to speak," and "to think." The Neapolitan philosopher Vico, who heralded poetry as the "first language" and poetic wisdom as the fountainhead of philosophical thought and to whose rhetorical tradition Heidegger himself belongs, notes the rustic origins of the Latin *lex* as "gathering" or "collecting" and then later "law," the "assemblage" of people, and "reading" (*legere* or "collecting letters").[21]

In Sinism there is a filiality between the aesthetic and the ethical: the beautiful and the good are synonymous. In characterizing Asiatic thought simply as aesthetic in *The Meeting of East and West*, F. S. C. Northrop was not far from describing its essence. Since the seventeenth century many European minds often created the image of the Orient as the aesthetic "paradise on earth." As the aesthetic is the connatural harmony of man and the wilderness of nature (i.e., geopiety), so is the good the convivial relationship between (wo)man and (wo)man (i.e., homopiety). Harmony is the essence not only of the aesthetic but also of the social, that is to say, it is both geopious and homopious. It cannot be overemphasized that "serenity," "releasement," or "composure" (*Gelassenheit*) is the keyboard on which the idea of harmony — both connaturality and conviviality — are being played out. For it "poeticizes" or "aestheticizes" man and the earth. From *Li Chi*, we also find the ancient Chinese sense of things as an ordered whole in which music plays an integrating role:

> There are heaven above and earth below, and between them are distributed all the (various) beings with their different (natures and qualities): — in accordance with this proceeded the framing of ceremonies. (The influence of) heaven and earth flow forth and never cease; and by their united action (the phenomena of) production and change ensure: — in accordance with this music arose. The process of growth in spring, and of maturing in summer (suggest the idea of) benevolence; those of in-gathering in autumn and of storing in winter, suggest righteousness. Benevolence is akin to music, and righteousness to ceremonies.[22]

Meditative thinking is characterized by serenity — the receptive reverence of things as they are in themselves, while calculative thinking implies dominance, manipulation, and utility. Serenity is nonwilled, nonforced, and nonconcerned activity, that is, an active responsiveness in man to the "natural light" of a thing. It is then "a will-less letting in of everything," the spontaneity that sets a thing free to be nothing but itself. "A higher acting," Heidegger writes, "is concealed in releasement than is found in all the actions within the world and in the machinations of all mankind."[23] Here Heidegger's notion of serenity or releasement parallels the Taoist idea of *wu wei* and the Zen way of thinking and doing — the way of refraining from thinking and doing contrary to the natural and spontaneous way of things. Just as Heidegger's serenity is "a higher acting," so is *wu wei* both nonanthropocentric in man's relationships with nature and nonindividualistic in our multiple relationships with one another (i.e., it is both geopious and homopious). It is like the flow of a river that meanders unconcernedly along the contours of the landscape or the splitting of a bamboo with its natural grain rather than against it. The Zen way of perfecting archery, as shown in Eugen Herrigel's classic study, lies in unconcernedness or "purposeless detachment." It is the "artless," effortless, and self-forgetting immersion of the archer in his performance without any ulterior, utilitarian ends in view.

The notion of Nature — *tzu-jan* in Chinese, *ji-nen* in Japanese, and *ja-yun* in Korean — signifies a "serene" or "reverential" composure for the "natural spontaneity" of all things — the "thisness" or "thatness" (i.e., "thusness") of everything living or nonliving. As it is af-filiated with serenity, the poetic discloses or discovers the topology of Being — the topology that tells the "whereabouts" of Being's presence. The poet is destined to dis-cover Being — the ultimacy of thinking — and its true nature which has too long been hidden in the metaphysical closet of Western philosophy. He is the thinker personified who "advocates" the fourfold unity (*Einheit*) of sky, earth, gods, and mortals which allows their free interplay. In the *Tao Te Ching*, the same unity is expressed in the circular chain of man that follows the earth; earth that follows heaven; heaven that follows the Tao; and, finally, Tao that follows Nature or the natural order of things (*tzu-jan*).[24] This splendid, simple gathering of the elements in "serenity" is expressed in a haikuesque quality of meaning, style, and subjects in the last lines found in Heidegger's discussion on "The Thinker as Poet":

Forests spread
Brooks plunge
Rocks persist
Mist diffuses

Meadows wait
Springs well
Winds dwell
Blessing muses[25]

Compare Heidegger's lines with the famous haiku of Basho, "Furu ikeya / Kawazu tobikomu / Mizu no oto," which should be read and heard as "the sound of a frog jumping in the water of the old pond," as a concordant continuum of the natural elements in harmony. So the simplicity and wilderness of *oto* ("sound") is the elemental, all-embracing *principium* of the haiku. The Zenish splendor of the simple and wild in this Basho's haiku airs and echoes the sonorous mood that is "being attuned" to the seasonable serenade of Being or Nature. The harmony of the elements is the great, topological chain of Being where the reverberation of the water's sound is perceived — better, "conceived" — by the poet in the little creature's consonance with nature or the whole uni-verse as the sounding background of tranquillity or beatific repose. The poetic, which is attuned to the topology of Being, is the acme of thinking. It does not surmount the earth in order to exploit and conquer it, but rather it brings man onto, and makes him belong to, the earth: it brings him dwelling on earth. The poetic embodies the quintessence of geopiety.

The aesthetic or poetic may now be contrasted with the technological: it is a stark contrast between *Gelassenheit* and *Gestell* in Heidegger's thought. One is meditative (i.e., spontaneous and natural), whereas the other calculative (i.e., exploitative and utilitarian). Very few, by now, argue against the conclusion that technology is the consummation of modern civilization, East or West, South or North. It destroys both homopiety and geopiety. Technology is, as it were, the "ontological fix" of modern man everywhere. Thus Heidegger writes:

> What now *is*, is marked by the dominance of the active nature of modern technology. This dominance is already representing itself in all areas of life, by various identifiable traits such as functionalization, systematic improvement, automation, bureaucratization, communications. Just as we call the idea of living things biology, just so the presentation and full articulation of all beings, dominated as they now are everywhere by the nature of the technical, may be called technology. The expression may serve as a term for the metaphysics of the atomic age. Viewed from the present and drawn from our insight into the present, the step back out of metaphysics into the essential nature of metaphysics is the step out of technology and technological description and interpretation of the age, into the *essence* of modern technology which is still to be thought.[26]

Heidegger contends that we have not yet come fully to grips with the nature of modern technocentric or technomorphic culture in which man himself is the "functionary" of technology. To say that the essence of technology is not technological is to deny the idea that technology is merely instrumental. The lethal network of globalized technology or, as Heidegger calls it, "technological frenzy" is a historical nightmare from which we wish to awaken or narcolepsy from which we would never awaken. Its unthinkable potential lies in bringing the entire earth into a necropolis. For Heidegger, it is infinitely inadequate to view modern technology merely as *instrumentum* or instrumental facilitation. For it has now acquired the ontological priority over every form of thinking, making and doing — including scientific activity. The worst consequence of misunderstanding the true nature (i.e., essence) of technology merely as instrumental rationality is bound to neutralize it as amoral. Heidegger proposes the idea of "autonomous technology" in place of "instrumental technology" to understand the essence of technology itself. Technology as autonomous is no longer simply a means to human activity. For it is not merely the application of mathematical physical science to *praxis*, but is an autonomous *praxis* itself. The traditional rationale of technology as *instrumentum* that serves the *telos* of man is obsolete and anachronistic. Means has indeed become end. This, in fact, is the "categorical imperative" of technology as autonomous.

"Representation" is the epistemological keyword, after the Cartesian legacy of the unblinking gaze of the mind, in the metaphysics of technology as autonomous. This "representational" metaphysics of technology may be best shown in the television technology that "enframes" modern society. It epitomizes man's "channeled existence" and his visual narcissism in the modern age as the age of "world picture" (*Weltbild*). Representational metaphysics is a continuation or an extension, therefore, of the Cartesian *Cogito* in which the subjective ("I") and the visual ("eye") viewpoint are brought together in the absolute affirmation and certainty of cognition as "clear and distinct ideas" (i.e., the Cartesian legacy of visual and narcissistic metaphysics).[27] The world viewed as picture in television is an aspect (idea) of *Gestell* ("enframing"). The essence of technology is *Gestell* which, as "to set in place" (*stellen*), is eminently spatial and visual. Television both spotlights and highlights this enframing in our age as the age of world picture. As the essence of all technology as enframing is representing, television too represents as picture or structured image (*Gebild*) the original presence of the world as it is in itself. To be sure, there is the fundamental distinction between "seeing" and "seeing as," that is, seeing a real object and seeing it as a picture, an image, or a mirror. One is fully visual, fully thought, whereas the other only half visual, half thought. We must understand here that to capture the essence of reality by means of representational metaphysics, that is, an anthology of images, is a forever Sisyphean task, for image has only a semblance of knowledge, subtraction of reality, and an appearance of participation.[28]

The advent of technology, as we have stressed, begins to destroy both geopiety and homopiety. For technology "scatters" rather than "gathers" things and our thoughts. The mathematical is basic to the grammar of modern technology as autonomous *praxis*. However, the language of modern mathematics or of the mathematical sciences has little in common with the original expression of the mathematical by Pythagoras or the ancient Pythagoreans — "All is Number" or "Number is all things." With them, it was meant to express the perfect, ontological ordering of invisible (temporal) *harmonia* but not visible (spatial) equations, that is, universal *harmonia* embodied in the mathematical. In the Renaissance, too, the Vitruvian figure inscribed in a square and a circle was a symbol of "sympathy" (*symphonia*) between microcosm and macrocosm, and Leonardo da Vinci obtained a harmonious smile in *La Gioconda* by means of the geometry of a circle. So with Pythagoras the *ratio* or harmonic proportionality of the octave (1 : 2), of the fourth (4 : 3) and of the fifth (3 : 2) was meant to perfect the ordering of the mathematical. The sum of the numbers used in these ratios — 1, 2, 3, and 4 — is 10, which is for Pythagoras the perfect number. To Pythagoras, this mathematical order came as a revelation of the *harmonia* of the natural (heavenly) and the moral world — each individually and both together. The celestial *harmonia* is the "sound" or "music" of the heavenly bodies. However, we do not hear this harmonious sound of music only because from our birth we hear it constantly. What happens in the modern mathematical sciences with and since Galileo and Descartes, however, is another story. With them the Pythagorean mathematical became converted into a technical method rather than an ontological ordering of the universe. For Husserl, modern mathematics or the mathematization of nature by Galileo into the geometric boxes of triangles, circles, and squares inscribed in the "Book of Nature" is "a garb of ideas" (*Ideenkleid*) in which "true being" (or reality) has been taken for a method. It is the mathematical sublimation of nature which "mirrors" and "represents," that is, falsifies reality, nature, and the world — it falsifies because what is "mirrored" or "represented" is only half true. In modern mathematics and the mathematical sciences, the ontological generality of the mathematical has been replaced by the generality of method: the birth of methodolatry or the mathematical truth sublimated in method. For this reason, it is often said that the invention of method but not the use of mathematics is the hallmark of modern science.

VI

In 1969 an important conference on Heidegger and Eastern thought was held at the University of Hawaii to honor Heidegger on the occasion of his eightieth birthday; the invited speakers included the late

J. Glenn Gray, Calvin O. Schrag, and J. L. Mehta.[29] Following Heidegger's own hint that a dialogue between him and Eastern thought was inevitable, the conferees expressed a strong sentiment that not only might Heidegger serve to build a bridge between East and West but also his thought might, more importantly, serve as the basis of uniting world philosophies. Today, after fifteen years, a conclusive study on this projected task remains unavailable, and many hard years of work on it are already cut out for students of comparative philosophy and culture. The manifest kinship between Heidegger's thought and Sinism can, of course, be stretched to include not only the philosophical landscape of China as a region but also the other regions of East Asia where Chinese ideography — the indispensable signpost of Chinese influence — has long been accepted as their regular linguistic diet.

Heidegger is truly a philosophical deconstructionist — in exactly the same sense as he himself defines the term — who reenvisions and subverts the long cherished metaphysical tradition of the West — logocentrism that coincides often with ethnocentrism or Occidentalism of one kind or another (e.g., Hegelianism). It is that tradition of inherited thought which is at the brink of unleashing awesome, destructive forces ranging from thermonuclear power to cybernetics — that almighty tradition of calculative power which may summarily be called "politology." An alternative to illusionary and fateful politology is the "piety of thinking" as an *ultima philosophia* which may serve as the yarn to weave the woof of Heidegger's own thought and the warf of Eastern thought — Confucianism, Taoism, Ch'an or Zen Buddhism, and others.

Thought-provoking events should, according to Heidegger, insure the provocation of eventful thought. The eventful decades in which we live would certainly provoke serious thinking. As such Heidegger's conception of the "piety of thinking" is a seminal thought that teaches us both what and how to think, where truth and method are integrated into one unity. It is the pathway to gather or orchestrate our thoughts — both Eastern and Western. Pious thinking is all poetic thinking that leads both to homopiety and geopiety. This poetic thinking may be against "humanism" or the homocentric view of the world but is never against man as such, that is, for the abolition of man. Following the poet Hölderlin, Heidegger is convinced that what saves grows out of danger, impending danger or what the Chinese call "crisis" — that idea which combines "danger" and "opportunity" or proposes an opportunity in the presence of danger. This saving power does not come about haphazardly or by "makeshift." Rather, it emanates from a radical turn of the mind by "being attuned" reverentially to Being or what the Orient calls Nature. This saving power is predicated upon the ready acknowledgment and pious reception of the poet as the supreme and consummate legislator of the world. By speaking the voice of poetry in the conversation of humankind, the thinker as artist is able to subvert the monopoly of established reality by virtue of the reversibility of Heidegger's thought and the thought of East or Orient which, as Heidegger calls it, is the "land of dawn."

Moravian College

NOTES

[1] In a recent collection of essays that explore the status of phenomenology in Japan, it is claimed that in Japan phenomenology became "a field of scholarship they [the Japanese] could explore, evaluate and appropriate in their own terms: ultimately, as it seems, it became a genuine mode of Japanese philosophizing. The result is a *Japanese* phenomenology, that is to say: a reflection which is unmistakenly heir to Husserl, but reflects as much the Japanese intellectual legacy and the philosophical quest of contemporary Japan, from the Meiji era to World War II and the present" (ed. Yoshihiro Nitta and Hirotaka Tatematsu, *Japanese Phenomenology*, Analecta Husserliana, vol. 8, Dordrecht: Reidel, 1979), p. x.

[2] Edmund Husserl, *The Crisis of European Sciences and Transcendental Phenomenology*, trans. David Carr (Evanston: Northwestern University Press, 1970), p. 289.

[3] Ibid., p. 285.

[4] The passages of Maurice Merleau-Ponty cited below are from *Signs*, trans. Richard C. McCleary (Evanston: Northwestern University Press, 1964), chap. 5, "Everywhere and Nowhere," pp. 126—58.

[5] The twentieth-century Chinese philosopher Tung-sun Chang propounds the logic of correlation as uniquely Chinese. See "A Chinese Philosopher's Theory of Knowledge," in *Our Language and Our World*, ed. S. I. Hayakawa (New York: Harper, 1959), pp. 299—324.

[6] Martin Heidegger, *Poetry, Language, Thought*, trans. Albert Hofstadter (New York: Harper & Row, 1971), p. 4.

[7] Martin Heidegger, *What Is Called Thinking*? trans. Fred D. Wieck and J. Glenn Gray (New York: Harper & Row, 1968), pp. 16—17..

[8] See my "Misreading the Ideogram: From Fenollosa to Derrida and McLuhan," *Paideuma* (forthcoming). For a comparative analysis of Heidegger's thought and the Chinese language, see Johannes Lohmann, "M. Heidegger's 'Ontological Difference' and Language," in *On Heidegger and Language*, trans. and ed. Joseph J. Kockelmans (Evanston: Northwestern University Press, 1972), pp. 303—63. Lohmann writes: "A dominating trait manifests itself here in the relationship between Chinese thought and

Indo-European, or Greek and Western, thought, a trait which is manifested equally in the relationship between Chinese and European cultures. As Humboldt has pointed out for languages, this trait is unique in the relationships of languages and of cultures to one another; it can be briefly characterized as a total difference within the context of an equally total comparability. One can observe this trait in various aspects of a culture — from such formalities as the mourning color (white or black) or the place of honor (left or right) to the overall *habitus* of philosophical thought and of the conception of life and world view. There is obviously not only a question of 'two extreme realizations of the possibilities of man's language which are related to one another in polar opposition,' but a question of two realizations of human possibilities, each of which in its own way is perfect. From this the incomparable, paradigmatic value which the Chinese language as well as the Chinese culture and its history have for us becomes evident" (pp. 337—38). For the most comprehensive study of Oriental thought in this connection, see Hajime Nakamura, *Ways of Thinking of Eastern Peoples*, ed. Philip P. Wiener (Honolulu: East-West Center Press, 1964).

9 In this Heideggerian tradition, Albert Hofstadter writes that "the statement, as a form of oral language, articulates in sound. Not only does it articulate being, both objective and subjective in their unity, but it makes it audible. *Written language is a score for oral performance*, except in extreme cases of artificial symbolic language-construction. Human being becomes audible in the articulate form of the declarative sentence. The serial ordering of different phonemes, making use of their sound qualities, rhythms, and other characteristics, constitutes an utterance which is heard as the uttering of the self-world, subject-object complex just described, and which therefore is the means by which that complex is itself heard as just that complex. . . . In the end, every feature of language, from vocabulary, case, tense, and mood to synonymy-antonymy, logic, rhyme, rhythm, and rhetorical order, is intelligible as a functioning constituent of a medium that articulates human existence. Language is the act by which man brings himself out as man" (*Truth and Art* [New York: Columbia University Press, 1965], pp. 82—83; my italics).

10 The following passage from "The Poet" is one of the most eloquent passages in the entire corpus of Emerson's writings: "The poets made all the words, and therefore, language is the archives of history, and, if we must say it, a sort of tomb of the muses. For, though the origin of most of our words is forgotten, each word was at first a stroke of genius, and obtained currency, because for the moment it symbolized the world to the first speaker and to the hearer. The etymologist finds the deadest word to have been once a brilliant picture. *Language is fossil poetry.* As the limestone of the continent consists of infinite masses of the shells of animalcules, so language is made up of images, or tropes, which now, in their secondary use, have long ceased to remind us of their poetic origin. But the poet names the thing because he sees it, or comes one step nearer to it than any other" (*Essays: Second Series* [New York: Lovell, Coryell, n.d.], p. 21; my italics).

11 Friedrich Nietzsche, *The Birth of Tragedy*, trans. Walter Kaufmann (New York: Random House, 1967), pp. 52, 141.

12 Martin Heidegger, *Nietzsche*, vol. 1, *The Will to Power as Art*, trans. David Farrell Krell (New York: Harper & Row, 1979), p. 99. The acoustic side of Zen or Zen's "record of things heard" is brought forth by David Applebaum in "On Turning a Zen Ear," *Philosophy East and West* **33** (April 1983): 115—22.

[13] See my "Martin Heidegger and the Homecoming of Oral Poetry," *Philosophy Today* **26** (1982): 148—70.

[14] Hannah Arendt, *Eichmann in Jerusalem: A Report on the Banality of Evil*, rev. ed. (New York: Penguin Books, 1977), p. 49.

[15] C. Kerenyi, *The Religion of the Greeks and Romans*, trans. Christopher Holme (New York: Dutton, 1962), p. 119. For my treatment of reciprocity or sociality as the absolute ground of Sinism in general and Confucianism in particular, see "*Jen*: An Existential and Phenomenological Problem of Intersubjectivity," *Philosophy East and West* **16** (July—October 1966):169—88 and "Confucianism and Existentialism: Intersubjectivity as the Way of Man," *Philosophy and Phenomenological Research* **30** (1969): 186—202.

[16] We have already noted etymological and semantic filiation of the German words *hören* ("to hear"), *horchen* ("to hearken"), *gehören* ("to belong"), and *gehorchen* ("to obey") in addition to the Latin *obaudire* that has the double meaning of hearing and obeying. Hans-Georg Gadamer points further to the familial intimacy of these auditory words when he notes: "It is the Greek word *oikeion*, i.e., that which pertains to the household, to the *oikos*. It is an ordinary expression for relatives and house friends, i.e., for all who belong to the household. Oikos, household, thus has the broad sense of an economic unit such as the Greek household characteristically was. But oikeion is just as much an expression for that place where one feels at home, where one belongs and where everything is familiar. We too have usages similar to the usage of the Greek oikeion which display this double aspect in the conceptual field of household. In German, *hoi oikeioi* is rendered as *die Angehörigen*, and in abstraction from this normal usage we have come to speak of *das Angehörige*, meaning everything which pertains to the household and not only those people who belong to it" (*Dialogue and Dialectic*, trans. P. Christopher Smith [New Haven: Yale University Press, 1980], p. 18). For a discussion of ecology based on the auditory model of sound as performance, see my "The Orphic Voice and Ecology," *Environmental Ethics* **3** (1981):329—40.

[17] See Hwa Yol Jung and Petee Jung, "Toward a New Humanism: The Politics of Civility in a 'No-Growth' Society," *Man and World* **9** (1976):283—306.

[18] I am greatly indebted to the geographer Yi-Fu Tuan for my understanding of geopiety, which is distinctly Heideggerian. From reading his works, I learned that geography is not confined to the reading maps but is a philosophical and humanistic discipline. Among his writings, see especially "Geopiety: A Theme in Man's Attachment to Nature and to Place," in *Geographies of the Mind: Essays in Historical Geosophy in Honor of John Kirtland Wright*, ed. David Lowenthal and Martyn J. Bowden (New York: Oxford University Press, 1976), pp. 11—39. For expositions on geopiety in Heidegger's thought, see J. Glenn Gray, "Heidegger's Course: From Human Existence to Nature," *Journal of Philosophy* **54** (1957):197—207; Vincent Vycinas, *Earth and Gods* (The Hague: Martinus Nijhoff, 1961); and Michael E. Zimmerman, "Toward a Heideggerean *Ethos* for Radical Environmentalism," *Environmental Ethics* **5** (1983):99—131.

[19] Color and sound are two radically different ways of organizing the human sensorium and the world. There is a qualitative difference in human experience between the visual and the acoustic. As music is the organized movement of sound, the spatiality of sound is most fully actualized in the tones of music. Color does not separate itself from the

object, whereas sound separates itself from its source (e.g., voice or the sound of a musical instrument). In other words, color is a dependent attribute of an object, whereas sound is not. While the color we see is the property of a thing itself and we confront color in space, the tone we hear is not the property of anything and we encounter it out of or from space. Color is locatable and localizable in one single position with the object, whereas sound, once separated from its source, has no definite topological property or determination although its source is locatable. Sound travels in no one direction, it travels in all directions. Musical tones have no locatable places: they are neither "here" nor "there" but everywhere (i.e., placeless or ubiquitous). In *Poetic Thinking: An Approach to Heidegger* (Chicago: University of Chicago Press, 1981), David Halliburton writes that "In the performance of a symphony, . . . responsibility may be seen in the interconnecting indebtedness of each constituent: the musicians, as users of equipment (instruments, chairs, music stands, and the like), together with their skills; the artisans responsible for the preparation of the equipment; the members of the audience, together with their capacity to hear and to sustain attention; the score, a being with a thingly character that allies it with equipment even as it carries an already constituted inclination (the totality of the composer's notations); the composer, as one who brings forth within the same order as the artisan; that artisan who is the printer of the score; the manner (in the sense of melody, timbre, tone) of the score as performed; the space of time in which that manner emerges through the concerted composure of performance; the space of time of the tradition without which the music could not move into its own articulation — without which, as the temporal structure that preserves the reciprocal responsibility of all the constituents, it would not be music; and finally, the space of time which is the world play's manner of moving, through all that is thus indebted, to its own disclosure" (p. 217).

[20] J. Glenn Gray, "Heidegger in Remembering and Remembering Heidegger," *Man and World* **10** (1977):62–63.

[21] See my "Vico's Rhetoric: A Note on Verene's *Vico's Science of Imagination*," *Philosophy and Rhetoric* **15** (1982):187–202.

[22] *I Chi: Book of Rites*, 2 vols., trans. James Legge (New Hyde Park: University Books, 1967), 2:102–3. Speaking of the locus of the personal as moral performance embodied in the Confucian thought of rite, Herbert Fingarette writes: "We would do well to take music, of which Confucius was a devotee, as our model here. We distinguish sensitive and intelligent musical performances from dull and unperceptive ones; and we detect in the performance confidence and integrity, or perhaps hesitation, conflict, 'faking,' 'sentimentalizing.' We detect all this *in* the performance; we do not have to look into the psyche or personality of the performer. It is all 'there,' public. Although it is there *in* the performance, it is apparent to us when we consider the performance not as 'the Beethoven Opus 3' (that is, from the composer perspective), nor as a 'public concert' (the *li* perspective), nor as a 'post-Mozartian opus' (the style perspective), but primarily as this particular person's performance (the personal perspective)" (*Confucius — The Secular as Sacred* [New York: Harper & Row, 1972], p. 53).

[23] Heidegger, *Poetry, Language, Thought*, p. 61.

[24] David Farrell Krell sketches the fourfold as envisioned by Heidegger in the following pictogram in which the crossing of Being is not a crossing *out* (*Durchstrechung*) but a crossing *through* (*Durchkreuzen*):

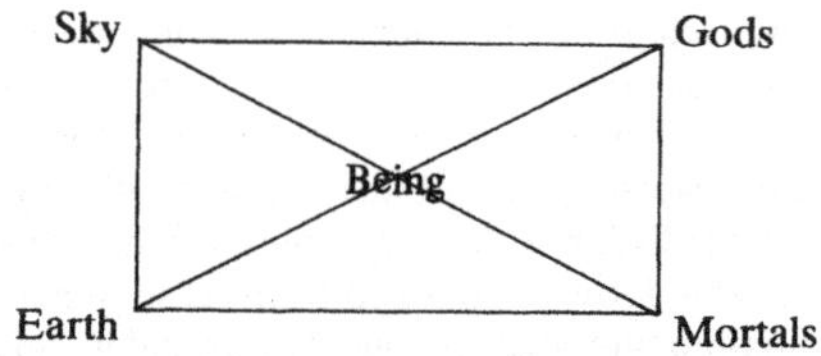

See "Analysis," in Martin Heidegger, *Nietzsche*, vol. 4, *Nihilism*, trans. Frank A. Capuzzi (New York: Harper & Row, 1982), p. 289. For a fine discussion of Taoism in this connection, see Kuang-ming Wu, *Chuang Tzu: World Philosopher at Play* (New York: Crossroad, 1982).

25 Heidegger, *Poetry, Language, Thought*, p. 14.

26 Martin Heidegger, *Identity and Difference*, trans. Joan Stambaugh (New York: Harper & Row, 1969), pp. 51–52.

27 In *The World Viewed*, rev. ed. (Cambridge: Harvard University Press, 1979), Stanley Cavell uses the term "automatism." The significance of automatism for our discussion here is twofold. First, as mechanical reproduction or representation, photography, for example, is of reality but not reality itself, photographic automatism removes the human agent from the task of reproduction or representation. Second, photography is not hand-made but mechanically reproduced — and what is mechanically reproduced is an image of original reality, that is, the mechanical fact of photography points to the absence of the human hand in forming its objects. To put it in the Heideggerian terminology, in photography thinking is no longer a "handicraft."

28 See Susan Sontag, *On Photography* (New York: Dell, 1977).

29 See *Philosophy East and West* **20** (July 1970).

HIROSHI KOJIMA

THE MODERN AGE AS A TRANSITIONAL PERIOD

An Essay on Metaphenomenology

It is a common practice to suppose that the modern ages, together with antiquity and the Middle Ages constitute, three periods of human history. Since this is a roughest demarcation, the actual contents and characterization of these periods may vary depending on the country and people. Differences between the East and the West is also significant. However, in the following discussion I shall confine myself to describing these three periods as they appeared in Europe because my interest in this paper is in the modern period in Western Europe, which had a profound influence on the entire world, engulfing various civilizations including the writer own.

To be more exact, my purpose is to cast doubt on the status of the modern age as one of the three periods of human history. Could the modern age be a historical period in the same sense in which antiquity and the Middle Ages are? There are two reasons for this doubt.

First, there is the issue of length. Identifying the beginning of antiquity is difficult even if we narrow our discussion to Europe. For instance, the Archaic Period, when Homer is supposed to have appeared, began around 800 B.C., while Cretan-Mycenaean civilization dates back to 1900 B.C. In any case, we may say that European antiquity extends over a thousand years if we assume as is commonly thought that it ended in the fourth century A.D. The Middle Ages also lasted about a thousand years, from the fifth century (when Christian dominance was established) until the peak of the Renaissance at the end of the fifteenth century.

By contrast, no more than 500 years have passed since the beginning of the modern age, and we already have various symptoms and arguments that we will soon reach a deadlock. The modern age seems to be too much shorter than antiquity and the Middle Ages.

A second and more important reason for doubting that the modern age is a true historical period. This relates to a fundamental conception governing a historical period. Whether or not there is conception by which we may characterize European antiquity throughout, and if any, what it is like, are difficult problems. We lack sufficient knowledge for

A-T. Tymieniecka (ed.), Analecta Husserliana, Vol. XXI, 369—382.

detailed discussion. For the present purpose it suffices to follow a commonly held hypothesis. The "form-matter" conception established by Aristotle in the fourth century B.C. represents the whole of antiquity in its essence. This conception was comprised of various precedent thoughts and marked a culmination in the Golden Age of antiquity. According to Martin Buber, in *What is Man?*, "The tendency of the Greeks to understand the world as a self-contained space, in which man too has his fixed place, was perfected in Aristotle's geocentric spherical system. . . . Even Plato's world of ideas is a visual world, a world of forms that are seen. But it is not before Aristotle that the visual image of the universe is realized in unsurpassable clarity as a universe of *things*, and now man is a thing among these things of the universe. . . ." Indeed, it still surprises us how skillfully Aristotle explained all strata of existence from minerals, plants, and animals up to human beings with only one "form-matter" conception. A notable feature of this conception is that it does not constitute a dualism because "form," which always holds priority as *energeia*, regulates "matter," i.e., *dynamis*, and therefore "form" as *telos* leads "matter." We might say that ancient European culture flourished when these two causes were balanced in a delicate way, just like Greek sculptures, to form a unity.

When this ancient conception lost its delicate balance, the transition from antiquity to the Middle Ages began. Thus form and matter changed into a dualism. Matter stopped being subject to form as it used to be and began to challenging the dominance of form on an equal footing. This kind of radical dualism which characterized the transitional period between antiquity and the Middle Ages is generally called "Gnosticism." Buber writes,

> The solitude out of which he [Augustine] asked the [anthropological] question can only be understood when one realizes that that round and unified world of Aristotle had long since collapsed. It collapsed because the soul of man, divided against itself, could no longer grasp as truth anything but a world which was divided against itself. In place of the sphere which had collapsed there now arose two autonomous and mutually hostile kingdoms, the kingdom of light and the kingdom of darkness. We meet them again in almost every system of that widespread and manifold spiritual movement of gnosis, which at that time seized the embarrassed heirs of the great oriental and antique cultures, split the godhead and emptied value from creation. . . . (*What Is Man?*)

The gnostic dualism distinguishes the God of salvation from the God of creation, and regards as good pneumatic or spiritual beings which belong to the former, while regarding as evil material or fleshy beings which belong to the latter. Thus the Gnostics tried hard to extricate themselves from the divided state, to which the earthly world and man are doomed, through "mystic knowledge" (gnosis) of the salvational God. Valentinus, Bardesanes, and Manichaeus are known to us as representative Gnostic thinkers of the second and third centuries A.D. It is also well-known that before his conversion to Christianity, St. Augustine had subscribed to Gnostic beliefs (first enunciated by Manichaeus).

It was only Christianity that succeeded in reuniting the divided Gods of salvation and creation, or form and matter. Christianity achieved the final victory over the Gnostic dualism after a fierce battle. In particular, as is commonly known, the Augustinian idea of "creation of the world from nothing" provided a new direction for the reunification. Toward the end of his *Confessions*, Augustine quotes the first passage of Genesis, "In the beginning God created the heaven and the earth," and offers an interpretation that "heaven" in this sentence means "pure form," i.e., "angel," while the "earth" means "primary matter." Thus he maintained that God created both form and matter from "nothing" in its literal sense, and that during the next five days God molded all the creatures ranging from minerals, plants, and animals up to human beings by combining form and matter. In this way, Augustine could deny that matter (material things) is evil. For matter was also created by the one and only God which is both a creator and a savior. This grandiose idea found later a more articulate philosophical expression in Thomas Aquinas's conception of "essence-existence." According to Aquinas, "creation from nothing" means that God's love and free will endow existence to the essence (form and matter) of creatures. The essence was already in God. Thus he thought only God possesses existence in its essence (essentially). (God is the unity of essence and existence.) In this way the conception of "essence-existence," which incorporates the ancient conception of "form-matter" in a new theory of creation, was established as the fundamental frame of thought in the Middle Ages.

We saw that the "form-matter" conception which represented ancient thought changed into the Gnostic dualism in its last stage, the Hellenistic age, and that a fierce battle was fought for centuries between Gnosticism and Christianity. We now see what happened to the "essence-existence" conception in the last stage of the Middle Ages. Buber writes,

Aristotle's cosmological image of the universe breaks up from within, through the soul's experience of the problem of evil in its depth, and through its feeling of being surrounded by a divided universe; Aquinas's theological image of the universe breaks up from without, through the universe manifesting itself as unlimited. What causes the crisis is on the one occasion a myth, the dualistic myth of gnosis, on the other occasion it is the cosmos of science itself, no longer clothed with any myth. (*Ibid.*)

It is true that the collapse of the medieval world had much to do with the scientific discovery of the infinite universe. However, we should notice that its deeper meaning lies in a change in the theory of creation. For not only the self-contained finite world became exposed to the infinite vacancy, but more significantly, man, who had been one kind of creature among others, now emerged as a creator. The discovery of the infinite universe is important only because it occasioned this profound change.

Unlike God, man cannot create from "nothing," but in the infinite space he decomposes everything into "material elements" and undertakes the "creation from elements." In this sense, the discovery of the infinite universe was an indispensable precondition to the emergence of man as "creator from elements."

Prior to this, Nicolaus Cusanus had expressed the idea of infinite space in the early fifteenth century. However, the idea was not given a definite form until the work of Giordano Bruno and Galileo Galilei at the end of the sixteenth century.

Just as the growth of the "matter" moment in the Gnostic dualism had destroyed the ancient unity based on the Aristotelian conception of "form-matter," this time the discovery of the "infinite space," a precondition to a new theory of creation, caused the rapid disintegration of the medieval theory of creation. For in the "infinite space" the medieval conception of "essence-existence" ceased to work. Instead, newly obtained scientific laws based on what might be called the conception of "accident-extension" dominates throughout. The universe regarded as a huge "clockwork" leaves no room for God's intervention.

However, the idea of "man as creator" was not formulated overnight. Prior to this formulation occurred a few centuries during which man was a "decomposer into elements." This shift period is a strong indication of the transitional character of the modern age. In other words, the modern age experienced a "pseudodualism" before the appearance of a fully mature, genuine modern dualism. By "pseudodualism" I mean the epistemological dualism that was given an articulate form by Descartes.

It is worthwhile to ask why Descartes persistently adhered to the absolute distinction of mind and body, as the full title of his book *Meditations* suggests. I believe it is evidence of the great impact that the novel concept of the universe as "infinite space" had during that period. Blaise Pascal's well-known passage, "The eternal silence of these infinite spaces terrifies me," is another indication. We know that Pascal held that "Man is a thinking reed" in order to overcome this terror. He says, "if the universe were to crush him, man would still be nobler than his destroyer, because he knows that he dies, and also the advantage that the universe has over him; but the universe knows nothing of this." Apparently both Descartes and Pascal were impelled by the anthropological motivation of protecting man's autonomy and dignity against the "clockwork" of the universe and against the human body which, as a thing, is subject to the clockwork. Thus it is obvious that the modern notion of *cogito*, which tries to detach itself strictly from any material things including the human body, is motivated by the dread and detestation of the unrestricted "materialization" that the emergence of the infinite universe had made possible.

A "pseudodualism" appears when this "modern (self-)consciousness" attains the status of the "epistemological subject" in natural science.

Modern self-consciousness is a "pure consciousness" deprived of all corporeal or spatial aspects because it is a result of an escape or evacuation from the infinite universe. Obviously this is the origin of "modern idealism." However, whether or not this "idealistic subject" is also an epistemological subject in natural science is not so obvious. In other words, a purified subject without objective corporeity faces an aporia of how it relates itself to the objective world. The difficulty is essentially the same as that which Descartes encountered in the mind-body problem. Just as a mind strictly distinguished from its body could not reunite itself with the body, it seems that pure consciousness which has once rejected spatial properties cannot have any relations with extended things again. Therefore, it is also impossible for pure consciousness to perceive external things in natural science.

It is, accordingly, not surprising that the mainstream idealists of modern philosophy since Descartes could neither solve the mind-body problem nor complete their epistemology. Underneath both problems lies the pseudodualism between "pure consciousness" and "extended things (or a body)."

Let us now see how representative modern idealists trapped in this fictitious pseudodualism sought in vain for an epistemology where there were no ways out.

Extension for Descartes is primarily an "idea" that exists within consciousness as an object of pure mathematics; whether there "really exists" an extended thing in space which corresponds to the "clear and distinct" idea is by no means obvious. For we just "seem to see clearly that the idea of it comes from external things, which it perfectly represents" (*Principles of Philosophy*). That this "seeming" is not an illusion is assured only by "veracity of God" for which a medieval type of proof was necessary. In short, the only ground of the argument was that "God cannot be a deceiver."

Even after the existence of corporeal things was assured by this roundabout argument, Descartes had a deep distrust of "sensation," because he thought that "the perceptions of our senses pertain only to this union of a human body with a mind" and that they are confused accidental cognitions by nature. Hence his conclusion that we should reject prejudices arising from sensations and should depend solely on understanding even for knowing the nature of external things. Indeed, Descartes declares, "I do not accept or desire in physics any other principles than in Geometry or abstract Mathematics." He also states, "nothing concerning these [divisions, shapes, and movements of material substance] can be accepted as true unless it is deduced from common notions, whose truth we cannot doubt, with such certainty that it must be considered as a Mathematical demonstration." As is now well known, this underestimation of sensation, or adherence to deductionsm, is responsible for many of dogmatic misconceptions Descartes held in physics. (See, for instance, Yôitsu Kondo's *Descartes's View of Nature*.)

Descartes, who lived in the seventeenth century, could still resort to "God" for the mediation between "pure consciousness" and "extended things," but a century later Kant could no longer use the notion of "God" and was forced to rely on "transcendental apperception" instead. Kant successfully showed that this reflexive self, which is the center of consciousness in general and which includes in it extensional space (and time) as a priori forms of intuition, can bridge the epistemological gap between particular empirical consciousness and extended substances. However, this success cost him decomposition of corporeal things into "unknowable things-in-themselves" and "sensuous manifold without unity."

Thus we can see Kant's so-called "Copernican turn" was nothing but an attempt to overcome the pseudodualism, i.e., the antithesis between "pure consciousness" and "extended things." Kant deprives the latter of extensionality to change it into sensuous manifold (and things-in-themselves), and then restores the extensionality to it through a priori forms of intuition and categorial synthesis. In other words, extended things forsake their extensionality only to attain it (or something similar) again by a surprisingly circuitous route.

The residuum of this process, things-in-themselves, can be regarded as an evidence that the mediating transcendental apperception is finite human understanding which is naturally short of the infinite consciousness of the Creator, and further that (as Kant himself admits) the space confined to finite consciousness no longer serves as a form of reality, which must be infinite.

Kant's remarkable feat, however, lost most of its epistemological significance when the laws of physical science turned out not to be results of a priori synthetic judgments as Kant thought they were, but only a kind of probable relation. In order to know nature, scientists today no longer seek for a circuitous route that the transcendental apperception was expected to provide, because there is no need for incorporation into consciousness of space as a form of intuition.

Husserl provides another example of the desperate efforts. In his theory, the mediator between pure consciousness and extended things is "intentionality," which means the act of consciousness to transcend itself. The notion is manifested in his basic thesis that "Consciousness is consciousness of something." The operation called "phenomenological reduction" reveals how the mediation takes place. In his *Ideas — General Introduction to Pure Phenomenology*, Husserl defined this operation as a "bracketing" of beliefs in general held in the natural attitude, and thought that he could obtain "transcendental pure consciousness" as a phenomenological residuum of this reduction. This consciousness differs from Kant's consciousness in general because it is not trans-individual nor does it include space in it. However, they are similar in that both are supposed to "constitute" the world as an intentional object. By the phenomenological reduction a corporeal object is first decomposed into an "object X" and its continuous appearance, that is, a current of hyle (which exists within consciousness). The hyle is activated by the sense-giving intentional act (*Noesis*) of pure consciousness, and is formed into an intentional unity. Thus it provides those features that are to fill the immanently transcendental objective meaning (*Noema*) attached to *X*. In this way a corporeal object is constituted as a noematic object which is a correlative to pure consciousness. Here Husserl seems to take extensionality of an object as one of the noematic features. (I think he could have understood extensionality of a corporeal object as attained by an intentional act independent of the sense-giving one. Why he disregarded this way is unclear.) For the constitution of the objective three-dimensional space, it would be necessary for Husserl to take a further detour of intersubjectivity based on empathy. Thus, in the same manner as Kant, Husserl deprives an extended thing of extensionality by decomposing it into an "object *X*," which has no other features than self-identity, and a hyle in a current of consciousness. Then, after a long detour through the workings of pure consciousness, he restores extensionality by constituting a noematic extended thing.

To make matters worse, Husserl's explanations are inadequate on such crucial issues as "how the transcendental object *X* can be clearly given as the center of identity beyond the continuous current of consciousness" and "whether the noematic objective sense to be attached to the *X* by pure consciousness can be obtained prior to, and independently of, the filling of the hyle." These problems have much to do with a later change in his theory. Thus in *The First Philosophy* he called his earlier theory of reduction developed in *Ideas* a "Cartesian way to reduction" and replaced it by a modified version called a "new way."

In the foregoing we have examined the development of transcendental philosophy, or transcendental idealism, during the three hundred years from Descartes to Husserl. This brief history should have revealed that transcendental philosophy was a result of the pseudodualism or the fiction of pure consciousness, which had to be provided as an escape from the idea of infinite space. It should also be clear that transcendental idealism failed to provide an epistemology capable of explaining cognition in natural science, because purified consciousness lost real contact with extended things.

I will now explain why the antithesis between pure consciousness and extended things is a pseudodualism. There are two reasons. First, these two correlatives do not constitute a genuine antithesis; second, a genuine antithesis which characterizes the modern age consists of a different pair. I shall first elaborate on the former reason.

Our perceptive consciousness is not pure consciousness which, having no extensionality, is opposed to extended things. Such pure *cogito* does not exist. The truth is that perceptive consciousness is always a kind of "bodily consciousness" united with an extended body. More precisely, it is united with a schematic apperceptive body which formulates particular extended body-images. This concept of schematic apperceptive body resists our understanding because it is an unperceivable body which never renders itself objective in its action. Nevertheless, it is certain that when we perceive a certain thing before us, our perceptive consciousness grasps this object through the workings of our apperceptive body. In a sense one perceives things by analogy with the body-image of oneself. In other words, one regards an object as a branch, or a transformation, of one's three-dimensional body-image. It is for this reason that an object of perception is given as a whole solid with its unobserved reverse side, and further that we can connect an object to our proper body-image to make the object our tool. For example, a ballpoint pen serves its purpose when it connects with our hand and becomes its extension. Also, to be a good driver, we have to make the body of the car an extended body-image of ourselves. Thus both our extended body and things surrounding us are subject to the same bodily apperception. It is very reasonable that in European languages the same words such as *corpus* or *Körper* refer to both of them.

Without this apperceptive body there is no perceptive consciousness at all and the extensionality formulated by the apperceptive body is neither an idea nor a form of intuitions. Therefore, the seeming dualism between pure consciousness and extended things does not really exist. Since perceptive consciousness is a subject inseparable from the apperceptive body, it is also objective and has extensionality in it. Thus, when we acquire a motion pattern by repeated practice, the pattern is "planted" into our apperceptive body as a memory. Later, when we plunge into water, for instance, we can display a series of extensions of the body by recalling various motion patterns from the stock of planted memories.

However, as the example shows, in an action the bodily extension (image) produced by the apperceptive body does not have perceptual objectivity. It is considered an object only when perceptive consciousness takes a reflexive stance. If a perceptual consciousness in this reflexive stance keeps an infinite distance from its body-image and regards both its own image and others' in the same way, this perceptual consciousness becomes an intersubjective consciousness with an "infinite space" in its perspective.

Therefore, the emergence of the "infinite space" which terrified Descartes and Pascal at the beginning of the modern age was nothing but a transition of human perceptual consciousness from its prereflexive stage to the infinite reflexive stage. People in those days, not knowing the crucial role of the apperceptive body to mediate reflexive consciousness and objective body-images even in infinite space, isolated the former as a transcendental subject in order to attain a complete separation. However, the truth is that reflexive consciousness is united with body-images in space by the mediation of the apperceptive body, and the infinite space is one of the perspectives of our reflexive consciousness. This infinite space, considered as a perspective, is the "universe," where the intersubjective reflexive consciousness works as a cognitive subject in natural sciences. Thus the universe of cognition (consciousness) and the universe of existence (space) are one and the same. A change in the way of cognition means a change in the mode of existence. When we depart from sensuous cognition to attain more abstract knowledge, the mode of existence of objects changes accordingly.

The discovery of the apperceptive body thus abolished the dualism between pure consciousness and extended things, or subject and object, which had predominated in modern thinking for a long time. It turned out to be a pseudodualism based on a fictitious notion of consciousness.

Let us now turn to the new stage of modern thought. The apperceptive body tends to shrink into a "point" when it takes the perspective of the infinite reflexive consciousness and engages itself in rational abstraction. Accordingly things also shrink into points or sets of points. To these points are assigned homogeneous properties such as mass, electric charge, and velocity. Being homogeneous, they allow measurement or numerical accounts in terms of standardized units. The cognitive universe of natural science is the universe of things rendered to "points" with measurable properties. This is exactly what I earlier called the world's "decomposition into elements." Creation by man is not "creation from nothing" but "creation from elements" in the sense that we create a new set of points on the basis of laws that are found to hold among these abstract points with numerical properties. For instance, when we launch an earth satellite, we first reduce the movements of heavenly bodies into those of points, discover the laws of celestial mechanics, and then combine the elements by consulting the laws. In like manner we analyze a given material into molecules and atoms, reveal their structural correlations in numerical terms, and then create a new kind of compound by synthesis. Thus these technological creations are possible only when we have the "cognitive universe of sciences," which is a numerical abstraction of the infinite space. However, it is obvious that this is not the whole story and a fuller explanation of our age is to be given.

We see that genuine modern dualism has finally reached maturity in our age when technological creation by man has become a central issue of philosophy. This dualism consists of two moments of human creation that has replaced God's creation. They are the moment of "apperceptive body united with infinite reflexive consciousness" and the moment of "imaginative will to join the world via images." In the history of philosophy, the former has had a close connection with idealism since Descartes, while the latter includes all the positive results obtained in materialism, philosophy of life, and existentialism.

The latter material moment of human creation originates in "matter." However, unlike that in materialism, matter is here a pure "source of force," that is, "nature as productive." It does not have a form but produces various forms (images). The will of life stands on this material ground. Life is a force immanently united with will, a force with certain direction and preference. By nature it aims at retention and expansion of the force. Furthermore, in humanity it is immanently united with "imagination." The workings of the will of life aiming at retention and expansion of the force are not based on mere impulses or conditional reflexions as in animals. They are realized via production of "images." Thus the will of life in humanity formulates its desires as a "projection" toward the future. Projection of the purpose as a definite image enables us to find appropriate combinations of "means" for its attainment.

Thus the matter-moment of human technological creation is "projection" based on matter and life, and hence is "imagination of means" made possible by the projection. Projection aiming at retention and expansion of the force is in essence an attempt by the human being to return to his "source of force." Therefore, it is in a sense an attempt at a "reunification with the whole material world." It is suggestive that Karl Marx wrote, "Animals produce only themselves, but man reproduces the whole nature" (*Economic and Philosophic Manuscripts*). We can say that human technological creation has its ultimate motivation in natural history.

Despite this motivation, means for realization were not available before the industrial revolution. The motivation could express itself only imaginatively in myths and fairy tales. It is the "decomposition into material elements" in natural science since the beginning of the modern age that provided the long-awaited means for its realization. In this sense, the decomposition of the world made possible by the infinite reflexive consciousness and by the workings of the apperceptive body can be called the "form-moment" of human technological creation.

Owing to the discovery of the causal laws governing things regarded as points with physical properties, we can now predict from certain causes certain probable effects as well as mutual relations in numerical terms, that is, in terms of differential equations. Imagination, which plays a leading part of the matter-moment in technological creation, extrapolates these numerically expressed causal relations from the existing objective space to an imaginary future space. As a result, cause-effect relations turn into means-end relations. For instance, a causal relation between the "rapid evaporation of a certain liquid" and the "absorption of heat" turns into a relation between the "mechanism of an air-conditioner" and the "cooling-down of the air." As this example indicates, a machine is generally a package containing a bundle of causal phenomena arranged toward one direction for a certain purpose.

In this way human technological imagination reorganizes the world decomposed by natural science into a bundle of causal relations and builds up a "linkage of tools" for certain purposes. Among them those to be used repeatedly are molded into packages, that is, machines. Thus, as a result of this effort over two centuries, we are now in the latter half of the twentieth century when the world is filled with a variety of machines and linkages of tools. This is exactly what I meant by human "creation from elements."

The description might seem to lead to the view that the form-moment of technological creation: the apperceptive body and the matter-moment: the imaginative will are complementary with no dualistic antithesis between them. However, as was already said, the imaginative will originates in the sheer material basis and therefore is half-blind. It always insists on its principle, that is, retention and expansion of the force, whatever contents an individual person's imaginative will might have. In history, it is typically seen in the development of capitalism. However, in this respect, the situation is essentially the same in socialistic economies. For although a socialistic economy places restrictions on private projections, the state as a whole is engaged in a projection on the basis of the same principle, "retention and expansion of production." The history of the modern age shows that these half-blind monsters of projection have tried hard for their expansion, competing with one another and sometimes even resorting to war to defeat others. In the meantime, the form-moment seems to have been subject to the matter-moment, serving as a "means" for the latter. However, its principle, "decomposition into elements," showed its nature for the first time when "nuclear weapons" appeared (a development which took place regardless of individual scientists' intention). That was in a sense an appearance of "nothingness" that supports natural science from behind. Nuclear weapons, use of which would decompose all the will of life into homogeneous elements, threatens the very existence of the matter-moment. The imaginative will is now confronting the possibility of destruction by its own technological creation. The radical and genuine dualism of the modern age has now revealed itself. Form and matter, that is, ever-decomposing natural science and half-blind material production, are no longer complementary, but are now in principle opposed to each other.

Since nuclear weapons are a culmination of human technological creation and a manifestation of the antithesis between natural science and material production — which have hitherto been considered complementary — it is obvious that efforts to "abolish" this specific weaponry, if dissociated from the wider context, would be futile. We have to understand the situation as the beginning of the struggle between the two principles, both of which are rooted in the modern age. In other words, this is the inaugural appearance of the dualism between the form-moment of "decomposition into elements" and the matter-moment of "retention and expansion of the force," a dualism as radical as that of Gnosticism at the end of antiquity. This modern dualism will soon be manifest in other fields such as the life sciences. We can foresee a new conflict between science trying to carry through the decomposition of life into "homogeneous elements" and unknown phenomena of life which by nature refuse the decomposition.

This dualism, though latent, has always existed behind the pseudo-dualism from the beginning of the modern age, as we have seen in the foregoing. It is not a recent production. It has surfaced after gradual maturation during the first few centuries of the modern age.

It should also be noted that the two moments of this dualism correspond to the two structural moments of the human body, the "apperceptive body" and the "imaginative will of life." Thus as the antithesis deepens, limitless development of natural science and expansion of material production in society tend to divide the human personality in two directions. As a result, it has become difficult to find a "great personality" around us. This is interesting, especially when we recall that the Gnostic dualism in like manner was also apt to deeply divide humanity into a spirit and a body.

Our problem in this paper was whether the modern age could be a historical period as are antiquity and the Middle Ages. It seems that man is now in the stage where pseudodualism is being replaced by the hitherto latent genuine modern dualism as human technological creation preceeds. This suggests that the modern age from the beginning lacked a principle of unity similar to that of antiquity and the Middle Ages. It seems that modern men have mistaken for "progress" the unstable rapid changes of the modern age caused by the lack of a principle.

Thus I conclude that the modern age should be regarded as a transitional period infected throughout with a dualism which should be resolved in principle.

Niigata University

(Translated by T. Shôgenji)

CLOSURE

ANNA-TERESA TYMIENIECKA

THE INTERDISCIPLINARY PHENOMENOLOGY OF MAN AND OF THE HUMAN CONDITION

*A Dialogue between Chinese and Occidental Philosophy in Meeting the Challenge of Our Times**

INTRODUCTION: TOWARD THE REVIVAL OF PHILOSOPHY'S ROLE

Among the variety of contemporary philosophies the Chinese and Occidental in particular stand out in respect of the wealth of their historical inheritance. In spite of their respective differentiations into trends, schools, and systems of ideas they oppose each other because of their striking originalities. This contrast, which is attributed to their so-called different "cultural frameworks," is ultimately the fruit of the difference in man's attitude to the cosmic forces, Nature, his fellowman, and his own personal destiny; this attitude culminates in man's estimation of his role with respect to his Human Condition. These contrasting differences, which are rooted in man's vital responses to the circumambient factors in his life-struggle and which are experiencing deep transformatory processes as well as mutual influences, prevailed until our times.

This contrast acquires a special significance when we become aware that both of these great philosophies find themselves — albeit in a seemingly radically divergent fashion — at the turning point of their natural historical unfolding. Let us leave aside the much discussed issue of the continuity/discontinuity of culture and focus upon the crucial issue:[1] the role of contemporary philosophy and its capacity to fulfill its appointed mission.

The question of rationality, which has occupied us for some time and which has now achieved prominence in general discussion,[2] has in recent decades acquired a specific form and has reached a point of unprecedented acuteness; this situation calls upon philosophy to resolve it. It is not only the need to interpret and assimilate the continuing flow of scientific "rationalities," on the one hand, and, on the other hand, the corresponding technological attempts to transform the rationale of human life. It is the source, nature, and significance of reason itself that is at stake.

A-T. Tymieniecka (ed.), Analecta Husserliana, Vol. XXI, 385—396.

Both Occidental and Chinese philosophy have not only to confront the drastic transformations in the ensuing human approach toward Nature, cosmos, life, and human relations; foremostly, they must deal with the enormous outpouring of new knowledge with its "multiple rationalities," which shake and put in question the most basic traditional preconceptions and stands upon reason and rationality as such. The role of philosophy is put to the task.

Yet, because of the different development of the two cultures, Chinese and Occidental, within which their respective roles are to be performed, each of these two philosophies is affected in a contrasting way by the radical challenge mentioned above.

In what follows I will attempt, first, to outline the modalities of the challenge to philosophy of our times; second, by stressing the contrasting status quo of Chinese and Occidental philosophy at the present time, I propose to indicate a point for a stimulating and constructive dialogue between them. Last, I will sketch an outline of a meeting ground for such a dialogue.

Indeed, the object of this contribution is not to dwell upon the issue of rationality, which is being discussed in various circles at random, but to analyze the *specific* form of the challenge which it presents to the role of both Chinese and Occidental philosophy, and to present the new philosophical approach which may alone meet it, namely, the INTERDISCIPLINARY PHENOMENOLOGY OF MAN AND OF THE HUMAN CONDITION.[3] This philosophy, now in its second decade as a "third phase" of classic phenomenology (founded by Edmund Husserl), is already in its essence conceived as an "interdisciplinary" philosophy. This fact speaks for itself. Philosophy can reach to its source in the Human Condition only by taking into account from the start the enormous number of new insights, approaches, and results that contemporary science/scholarship continues to bring forth; it has to pursue its investigation in an interchange with scientific research.[4] Such an interdisciplinary phenomenology lays bare the primeval intuitions of mankind as the ground for the meeting of Chinese and Occidental thought in a way that is fruitful for both. This meeting has already been initiated by the World Institute for Advanced Phenomenological Research and Learning.[5]

I. THE CHALLENGE OF OUR TIME

A. *The Role of Philosophy*

Let us first reflect upon the task of philosophy as such. From time immemorial the role of philosophy was, in fact, twofold. Leaving aside the "marveling" reflection, originating in the human being who is overtaken by awe before the vast universe facing him with its unfathomable mysteries and wonders, philosophy performed in the first place, a crucial role with respect to man's quest after understanding/knowledge; in the second place, it served as a guide for man's conduct in life.

It could have performed this role because of its unique nature. Although philosophy participates in man's quest for understanding/knowledge, it pursues it in a unique fashion. In contrast to any other pursuit it does not stop at fragmentary grasp of things or facts, at none of their many different levels; furthermore, philosophy will not be satisfied with "appearances." It has to reach intuitively the "heart of things" in their originary causes, principles, and reasons. Consequently, while all other intellectual pursuits remain satisfied with answers to a narrow range of questions, philosophy encompasses all and reaches to the very "heart of things;" it thus seeks to supply the human being with answers to the most puzzling questions concerning the ultimate status of life and the meaning of existence.

Last, it has been the aspiration of every great philosophy to be not only the crowning point for all human knowledge but also its foundational ground. Although differing among themselves, the various branches of knowledge (science, scholarship) have sought a common ground in philosophy.

However, since the first decades of this century philosophy has been overwhelmed by the enormous development of science and has pleaded to be relieved of this perennial aspiration under the pretext that "who could penetrate and encompass the variety and richness of the new ideas, approaches, and methods of science? What philosophy could match the precision of the scientific observation of facts, the verifiability of scientific interpretation, and the practical application of its insight?"

But is this tendency to give up its millennial vocation toward the human being justified? If philosophy were to abandon this threefold aspiration, would it not leave the human being like a ship without a helm on an unchartered ocean? And yet this millennial vocation of philosophy has to be assumed in its entirety or to be abandoned. No piecemeal or roundabout approaches to the universe, nature, and human life will do.

I submit that in taking up again this great role both Chinese and Occidental philosophy may meet and interchange their respective sociocultural experiences in the task of rekindling the great intuitions at the "heart of things."

B. *Philosophy from an Occidental Perspective*

Contemporary Occidental philosophy has, as is well known, attempted to go along with scientific developments. Whether outright or indirectly it has imitated — and still imitates — the criteria governing scientific/ scholarly appreciation of the validity of their observations and findings, as well as the methods of their presentation.

While some "schools" of philosophy (e.g., logical positivism) have renounced a universal role and defined their limited object as the "clarification of concepts," Occidental philosophy at large has diversified its field, following the scientific proliferation of seemingly independent *rationales*; it has fragmentized its hitherto holistic — unified — word-view and discourse into separate areas. Such philosophical branches as logic, ethics, aesthetics, psychology, cosmology, epistemology, methodology, etc., have become independent fields of research. Moreover, the scientific assumption of a great wealth of knowledge has led philosophy to increasing specialization; scholars have begun to devote themselves to narrowed fields of inquiry without paying attention to interlinkings with other issues (e.g., philosophical anthropology with philosophy of life or aesthetics, etc.).

And yet Edmund Husserl made a monumental effort to revamp the entire philosophical field. He proposed laying foundations for a new "holistic" philosophy, which under the name of "phenomenology" would lay bare the ground of the primeval philosophical intuitions of mankind — intuitions either lost or obscured by the developments of the human intellect — in a radical way, so that it could again assume its leading role with respect to human life and the universe of knowledge.

Phenomenology had been conceived by Husserl as the *mathesis universalis*, that is, a foundational clarification of all knowledge. Such a gigantic project could not have been accomplished by the master alone;

and yet, except for a handful of associates of Husserl (e.g., A. Pfaender, M. Scheler, H. Conrad-Martius), Husserl's successors have differentiated his enterprise into a spectrum of specialized fields (ontologies, anthropologies, aesthetics, etc.). Although all of them maintain some links with the basic phenomenological assumption, they have worked on their own in an original fashion and loosened their ties. It may be said that ultimately they have freed themselves from the originally envisaged total phenomenological man-and-world view, the focus upon which has since been lost from sight.

However, what is most striking for my present argument is that in the further development of the philosophies deriving from Husserlian phenomenology and crucial assumption of Husserl, namely, recourse to direct intuition or intuitive penetration "in the heart of things," or as he called it, "things themselves," has been forgotten. With this loss, the return to the "primeval intuitions of mankind," as I call it, has been "deferred". Instead, intermediary steps in the exfoliating and formulating of the life-world and man's role in it has been brought into focus. The direct intuitive penetration into the nature of things, conceived by Husserl as capable of reaching primeval insights through the sedimentations of meanings in which cultures have clad them, has been deviated into lingusitic, structuralistic, and poststructuralistic, hermeneutic, and semiotic perspectives. This recent prevailing tendency to shift the focus from the proper objectives of philosophy to their means of mediation supersedes the phenomenological analysis of immediate intuitions, meant to give its "objects" in their "bodily selfhood;" the phenomenological approach dissolves into an extremely ingenious, mind-seducing, and philosophically empty verbalism.

These philosophies may expect that the point of the unity of things can ultimately be reached, yet this point remains distant. Most importantly, however, I challenge any such expectations. In agreement with the millennial philosophies, I claim that in order to arrive at a holistic grasp of all things, we have to begin our philosophical itinerary at an ultimate, Archimedean point and not, in reverse, expect to find it at the end of the road. It must be sought in the primeval intuitions at the heart of things, but contemporary Occidental thought has lost sight of these intuitions.

The contemporary challenge to Occidental philosophy consists, I maintain, first, in *reestablishing the communicative links among the diversified fields of knowledge/scholarship*; second, *in regaining the life-significance of all knowledge*; *this life-significance, hitherto absent from scientific pursuits as well as from technological praxis, concerns the Human Condition.* It is crucial for Occidental man to regain its lost bearings within Nature, the world, and societal existence: the conditions and meaning of life.[5] The foremost task of *Occidental philosophy thus lies in returning to the primeval philosophical intuitions that have been concealed or distorted by the twin artifices of culture and intellect.*

Only in rediscovering these direct intuitions at the heart of things, or of "philosophical facts" — which means at the heart of man's commerce with natural forces and of the societal conditions of life — may philosophy again resume its proper role. These postulates point to phenomenology in its "third phase" of development, namely, to the "phenomenology of man and of the Human Condition," as particularly apt to fulfill this task. They also point to Chinese philosophy as a source of inspiration for rekindling "philosophical faith."

C. *Present-day Chinese Philosophy*

While Occidental philosophy has obscured, if not altogether lost its unity, Chinese philosophy, in striking contrast, has through many centuries preserved it intact. Approaches, ideas, and systems of thought succeed each other in their momentous appeal to man caught in one or other type of vital situation of the moment, but Chinese philosophy has maintained its holistic nature until now.

In another striking contrast to Occidental philosophy, Chinese culture has not undergone the profoundly disorienting impact of scientific and technological change. Only now does it stand at the threshold of a radical overturning of the relationship between nature and the social world, and life, and thus face a transformation of ways of life, work, aspirations, etc., as a result of scientific and technological innovation. By the end of this century Chinese philosophy — despite its effort to preserve the thread of continuity between past and present phases of culture — will confront the need to discriminate, digest, and process into proper formulations the enormous reservoir of knowledge associated with the new technologies. The scholar, the man in the street, and the philosopher will have to meet the radical challenge of "multiple rationalities" and thus have to deal with their effect on traditional rationality.

In spite of this contrast, the task which lies ahead of Chinese philosophy is in the main the same as that facing Occidental philosophy. Along with new means to conduct the business of life and the new knowledge and technologies that will lead to the transformation of the entire sociomaterial existence, Chinese thought will have to confront new desires, new tendencies, and new ways of evaluation. It means that these new rationalities erupting within traditional sets of values will have to be endowed with new "life-significance." These will have to be brought to an Archimedean point of primeval intuitions. However, even if the intuitive source of philosophy has not been obscured, in the midst of a new inventory of knowledge to be interpreted and assimilated on the one hand, and directed toward inventing new ways to deal with life, on the other, its intuitive insights will have to undergo an appropriate reworking. To be capable of endowing the entire life activities of man with an appropriate meaning of life, new conceptual formulations will have to be forged.

This response to the radical challenge to human reason must come solely from interior Chinese culture. Scientific results and their technological implementation alone enjoy a cultural neutrality (tractors, computers, etc., function equally well in every culture). In sharp opposition, philosophy as the "wisdom of a culture" is the thread spun from all the old and new material which is being absorbed by the personal and societal life of human beings and which functions as the innermost conductor of cultural continuity — continuity of the meaning of life for people in transition from one form of life to another. The Human Condition and the intuitive insights into the heart of things, that is, into the primeval links of the living human being with the natural and societal forces out of which he delineates his existence, is always the same as long as the human being persists in being "human;" but the forms which these intuitions have to acquire maturate in the transformation processes of culture. It is by forging new responses to the demands of time that philosophical intuitions acquire meaningfulness for human life.

Both Chinese and Occidental philosophy have to forge these new forms; both have to find and/or reformulate the Archimedean point of unity of the entire human discourse as the key to the primeval unity of life, Nature, and the human existence.

II. PHENOMENOLOGY OF THE CREATIVE ACT OF MAN AS THE KEY TO THE UNITY OF REASON

A. *The Creative Context for New "Investigation of Things"*

As mentioned above a dialogue has been conducted, on a modest scale, within the framework of the research projects of the World Institute for Advanced Phenomenological Research and Learning in Belmont, MA.[7] The main task of the Institute's research is the "reconstruction" of a philosophy for our time. First, in a new phenomenological approach it takes into account the great inheritance of the phenomenological investigations devising an all-encompassing framework in which the fragmentized researches of the great phenomenologists of the past find their proper place. Assuming fully the Husserlian radicalism, a new starting point for this investigation is established: the one-sided intellectual cogito of Husserl is here replaced by the *Creative act* as the *prototype* of all cognition and action.[8] The absolute allegiance to the Husserlian faith in the immediate intuitions and "originary giveness," which it yields is thereby reinforced. Man's creative act is operative in the emergence of the "source-experience," which establishes the living being as human. Thus it offers us the key to the primeval intuitions which stem from the junctions of the conditions of life and of the specifically Human Condition *before* the cultural differentiations may occur; intuitions shared by *all mankind*.[9]

Within the *creative context* of investigation, thus circumscribed by the spread of the creative act, we find the great intuitions of Chinese philosophy: spontaneity, naturalness, creativity, and moral sense. In their pristine, culturally unbiased nature, they offer themselves for a new investigation of things.

Most importantly, however, the *creative act offers us the key to the unity of human knowledge to be reinterpreted.*

B. *The Retrieval of the Archimedean Point of the Unity of Reason within its Life-spread*

With all this in mind, let us return to the challenge of the "multiple rationalities" of the scientific/scholarly outburst. I have proposed that in order to answer the challenge facing traditional conceptions of rationality reestablishing the lost links among various types of *rationale*, we must find access to the springs of all meaningfulness, to the *reason of life itself*. We must disentangle the knot which life in its constructive games ties among its innumerable factors. What is the key to disentangle it?[10]

As the answer to these questions, I have proposed above, first, to pursue the constructive progress of life through *man's self-individualization*; second, I have emphasized that it is only through the key of man's creative act that we may enter the self-individualization progress of the living being, where all the ties of life are joined.[11]

It may well be true that all knowledge is *human knowledge*, and that the "reasons," "rationale," or "order of life" and of the world manifest themselves only within human consciousness as it emerges in the evolutive constructivism of life's progress. Nevertheless, when we start from man's creative act and follow downward this constructivism of life, we will distinguish several types of *rationale* according to which it progresses. The German biologist Uexkul pointed out at the end of the nineteenth century that the living being articulates his circumambient conditions through his life-progress. I can corroborate his intuition by conceiving of this articulation as the effect of the *self-individualization* of the living being in which his life-progress consists. To promote his self-individualizing unfolding from within the circumambient conditions, the living being proceeds by an incessant series of responses to them: reactions, solicitations, discriminations and selections, assimilations and rejections. It is through these responses that he differentiates the *elemental life forces* and *circumambient elements*. With respect to the specific types of relevances to his vital needs that they offer, the living being endows them with types of significance for his life-progress. Thus his constructive life-progress establishes simultaneously among them an *order of life-significance*, and a *vital rationality* thereby emerges.[12] It differentiates with respect to the divergent modalities of the relevance that is projected by the progressing complexity of the individualization process. That is, within the vital rationality there are to be differentiated — as scientific research progressively brings forth — various modes of *ratio, reasons*, and *rationale*. Seemingly independent, they are in fact all intertwined within the constructive progress of the individualization of life; as its various *rationale* within the same constructive design, they codetermine each other. Lastly, they share a common origin in the primeval schema of life itself.

However — and here we strike the pulsating chord of the heart of things or of the philosophical facts — we would not have been able to reach the *constructive outline* of life's progress had it not been by following the threads of the *creative act* of man. As I have maintained elsewhere, with the unfolding of the constructive complexity of life's individualization — or "man's self-interpretation-in-existence — the phase at which the invention/creation function emerges within the functional system is reached. There it is that carrying the individualization process of life onward, the specifically "human" constructivism of life begins. In the creative act of man we touch the springs of its complete synthesis: the generative synthesis of all rationalities.

More specifically, the creative act orchestrates all the functions operative in the constructive progress of self-individualization into a new synthesis of the "source-experience," within which are unfolding the specifically human faculties: imaginatio creatrix, aesthetic sensibilities, moral deliberation, and the objectifying intellect. With the aid of this apparatus, the creative act promotes an infinitely varied proliferation of man-invented avenues of life; each of them introduces its rationale on the basis of its intrinsic significance for human life.

As the new forge of life's modalities, the creative act synthesizes the entire horizon of life-significant relevances, reasons, ratios, etc., and projects new rational links among the elements of life's constructivism. Simultaneously, it merges the invented rationalities with the vitally established ones. They all intertwine and intergenerate within the synthesis of the creative act of man.

C. *Interdisciplinary Phenomenology of Man and of the Human Condition as the Ground for the Reconstruction of Philosophy for Our Time*

As I have been pointing out in my writings during the past two decades, it is neither any form of cognition — whether empirical or transcendental — nor any type of human action — whether ethical, artistic, practical — that is capable of offering us access to the center of man's involvement with life, Nature, and his fellowman that may reveal to us the Archimedean point of life's order.[13] Yet it is this involvement that establishes this "order" projecting network of rationalities within which the living human being evolves. From the creative act of man, as from a vortex of the gigantic system of life's individualization-in-progress which synthesizes the entire reach of life's rationality, we are at last in position to uncover and circumscribe a field of an unprejudiced phenomenological investigation of all the insights which science and scholarship bring. Simultaneously, we approach them from the perspective of the originary synthesis of life's self-interpretative reasons, and throw our philosophical hawser upon the source of the primeval intuitions of mankind.

Extended to the entire field of man's creative genius, and plunging into the recesses of the Human Condition, interdisciplinary philosophy becomes the ground of encounter as well as of the reconstruction of philosophy for our time.

With these principles, the dialogue between Chinese and Occidental philosophy in their respective tasks of revival has been opened.

The World Phenomenology Institute

NOTES

* This paper has been read at the "Chinese Philosophy Round Table" at the World Congress of Philosophy, Montreal, August 1983.

[1] *The Crisis of Culture, Analecta Husserliana*, vol. V, D. Reidel, Dordrecht, 1976.

[2] Anna-Teresa Tymieniecka, "The Initial Spontaneity," ibid.

[3] Anna-Teresa Tymieniecka, "Phenomenology of Man and of the Human Condition," in *Phenomenology of Man and of the Human Condition; Plotting the Territory for Interdisciplinary Communication, Analecta Husserliana*, vol. XIV, 1983.

[4] Anna-Teresa Tymieniecka, "The Phenomenology of Man and of the Human Condition in Communication with the Human Sciences," ibid.

[5] *Phenomenology of Life in a Dialogue between Chinese and Occidental Philosophy, Analecta Husserliana*, vol. XVII, 1984.

[6] Anna-Teresa Tymieniecka, "Phenomenology of Man and of the Human Condition in Communication with the Human Sciences"; also cf. Angela Ales Bello, "Phenomenology of Man and of the Human Condition", a review-article in *Phenomenology Information Bulletin, a Review of Philosophical Ideas and Trends*, issue 7, 1983, World Phenomenology Institute, 348 Payson Road, Belmont, MA.

[7] *Phenomenology of Life in a Dialogue between Chinese and Occidental Philosophy.*

[8] Anna-Teresa Tymieniecka, "Beyond Ingarden's Idealism/Realism Controversy with Husserl; the Third Phase of Phenomenology," a monograph in *Ingardeniana, Analecta Husserliana*, vol. IV, 1976.

[9] Anna-Teresa Tymieniecka, "Natural Spontaneity in the Translacing Continuity of Beingness," in *Analecta Husserliana*, vol. XIV, 1983.

[10] Ibid.

[11] Anna-Teresa Tymieniecka, "Harmony in Becoming and Self-individualization of Life," *Phenomenology of Life in a Dialogue between Chinese and Occidental Philosophy, Analecta Husserliana*, Vol. XVII, 1983.

[12] Anna-Teresa Tymieniecka, "The Moral Sense in the Foundations of the Social World," a monograph, *Analecta Husserliana*, vol. XV, 1984.
[13] Anna-Teresa Tymieniecka, "The Prototype of Action: Ethical or Creative," *Analecta Husserliana*, vol. VII, 1978.

INDEX OF NAMES

ANALECTA HUSSERLIANA

The Yearbook of Phenomenological Research

I. *Analecta Husserliana.* 1971.
II. *The Later Husserl and the Idea of Phenomenology. Idealism-Realism, Historicity and Nature.* 1972.
III. *The Phenomenological Realism of the Possible Worlds. The 'A Priori', Activity and Passivity of Consciousness, Phenomenology and Nature.* 1974.
IV. *Ingardeniana. A Spectrum of Specialised Studies Establishing the Field of Research.* 1976.
V. *The Crisis of Culture. Steps to Re-Open the Phenomenological Investigation of Man.* 1976.
VI. *The Self and the Other. The Irreducible Element in Man, Part I: The 'Crisis of Man'.* 1977.
VII. *The Human Being in Action. The Irreducible Element in Man, Part II: Investigations at the Intersection of Philosophy and Psychiatry.* 1978.
VIII. Yoshihiro Nitta and Hirotaka Tatematsu (eds.), *Japanese Phenomenology. Phenomenology as the Trans-Cultural Philosophical Approach.* 1979.
IX. *The Teleologies in Husserlian Phenomenology. The Irreducible Element in Man, Part III: 'Telos' as the Pivotal Factor of Contextual Phenomenology.* 1979.
X. Karol Wojtyła, *The Acting Person.* 1979.
XI. Angela Ales Bello (ed.), *The Great Chain of Being* and *Italian Phenomenology.* 1981.
XII. *The Philosophical Reflection of Man in Literature.* 1982.
XIII. Eugene F. Kaelin, *The Unhappy Consciousness. The Poetic Plight of Samuel Beckett, An Inquiry at the Intersection of Phenomenology and Literature.* 1981.
XIV. *The Phenomenology of Man and of the Human Condition. Individualisation of Nature and the Human Being. I. Plotting the Territory for Interdisciplinary Communication.* 1983.
XV. *Foundations of Morality, Human Rights, and the Human Sciences. Phenomenology in a Foundational Dialogue with the Human Sciences.* 1983.
XVI. *Soul and Body in Husserlian Phenomenology. Man and Nature.* 1983.
XVII. *Phenomenology of Life in a Dialogue Between Chinese and Occidental Philosophy.* 1984.
XVIII. *The Existential Coordinates of the Human Condition: Poetic – Epic – Tragic. The Literary Genre.* 1984.
XIX. *Poetics of the Elements in the Human Condition: The Sea.* 1985.
XX. *The Moral Sense in the Communal Significance of Life. Investigations in Phenomenological Praxeology: Psychiatric Therapeutics, Medical Ethics and Social Praxis within the Life- and Communal World.* 1986.
XXI. *The Phenomenology of Man and of the Human Condition. II. The Meeting Point Between Occidental and Oriental Philosophies.* 1986.

ANALECTA HUSSERLIANA

THE YEARBOOK OF PHENOMENOLOGICAL RESEARCH

Editor:

ANNA-TERESA TYMIENIECKA

The World Institute for Advanced Phenomenological Research and Learning, Belmont, Massachusetts

I. *Analecta Husserliana.* 1971.

II. *The Later Husserl and the Idea of Phenomenology. Idealism–Realism, Historicity and Nature.* 1972.

III. *The Phenomenological Realism of the Possible Worlds. The 'A Priori', Activity and Passivity of Consciousness, Phenomenology and Nature.* 1974.

IV. *Ingardeniana. A Spectrum of Specialised Studies Establishing the Field of Research.* 1976.

V. *The Crisis of Culture. Steps to Re-Open the Phenomenological Investigation of Man.* 1976.

VI. *The Self and the Other. The Irreducible Element in Man. Part I. The 'Crisis of Man'.* 1977.

VII. *The Human Being in Action. The Irreducible Element in Man. Part II. Investigations at the Intersection of Philosophy and Psychiatry.* 1978.

VIII. Yoshihiro Nitta and Hirotaka Tatematsu (eds.), *Japanese Phenomenology. Phenomenology as the Trans-Cultural Philosophical Approach.* 1979.

IX. *The Teleologies in Husserlian Phenomenology. The Irreducible Element in Man. Part III. 'Telos' as the Pivotal Factor of Contextual Phenomenology.* 1979.

X. Karol Wojtyla, *The Acting Person.* 1979.

XI. Angela Ales Bello (ed.), *The Great Chain of Being and Italian Phenomenology.* 1984.

XII. *The Philosophical Reflection of Man in Literature.* 1982.

XIII. Eugene F. Kaelin, *The Unhappy Consciousness. The Poetic Plight of Samuel Beckett. An Inquiry at the Intersection of Phenomenology and Literature.* 1981.

XIV. *The Phenomenology of Man and of the Human Condition. Individualisation of Nature and the Human Being. I. Plotting the Territory for Interdisciplinary Communication.* 1983.

XV. *Foundations of Morality, Human Rights, and the Human Sciences. Phenomenology in a Foundational Dialogue with the Human Sciences.* 1983.

XVI. *Soul and Body in Husserlian Phenomenology. Man and Nature.* 1983.

XVII. *Phenomenology of Life in a Dialogue Between Chinese and Occidental Philosophy.* 1984.

XVIII. *The Existential Coordinates of the Human Condition. Poetic – Epic – Tragic. The Literary Genre.* 1984.

XIX. *Poetics of the Elements in the Human Condition. The Sea.* 1985.

XX. *The Moral Sense in the Communal Significance of Life. Investigations in Phenomenological Praxeology: Psychiatric Therapeutics, Medical Ethics and Social Praxis within the Life- and Communal World.* 1986.

XXI. *The Phenomenology of Man and of the Human Condition. II. The Meeting Point Between Occidental and Oriental Philosophies.* 1986.

The manufacturer's authorised representative in the EU is Springer Nature Customer Service Centre GmbH, Europaplatz 3, 69115 Heidelberg, Germany. If you have any concerns regarding our products, please contact ProductSafety@springernature.com

Printed and bound by CPI Group (UK) Ltd, Croydon, CR0 4YY

30/09/2026

02225510-0015